Probability and Social Science

METHODOS SERIES

VOLUME 10

Editor

DANIEL COURGEAU, *Institut National d'Études Démographiques*
ROBERT FRANCK, *Université Catholique de Louvain*

Editorial Advisory Board

PETER ABELL, *London School of Economics*
PATRICK DOREIAN, *University of Pittsburgh*
SANDER GREENLAND, *UCLA School of Public Health*
RAY PAWSON, *Leeds University*
CEES VAN DER EIJK, *University of Amsterdam*
BERNARD WALLISER, *Ecole Nationale des Ponts et Chaussées, Paris*
BJÖRN WITTROCK, *Uppsala University*
GUILLAUME WUNSCH, *Université Catholique de Louvain*

This Book Series is devoted to examining and solving the major methodological problems social sciences are facing. Take for example the gap between empirical and theoretical research, the explanatory power of models, the relevance of multilevel analysis, the weakness of cumulative knowledge, the role of ordinary knowledge in the research process, or the place which should be reserved to "time, change and history" when explaining social facts. These problems are well known and yet they are seldom treated in depth in scientific literature because of their general nature.

So that these problems may be examined and solutions found, the series prompts and fosters the setting-up of international multidisciplinary research teams, and it is work by these teams that appears in the Book Series. The series can also host books produced by a single author which follow the same objectives. Proposals for manuscripts and plans for collective books will be carefully examined.

The epistemological scope of these methodological problems is obvious and resorting to Philosophy of Science becomes a necessity. The main objective of the Series remains however the methodological solutions that can be applied to the problems in hand. Therefore the books of the Series are closely connected to the research practices.

For further volumes:
http://www.springer.com/series/6279

Daniel Courgeau

Probability and Social Science

Methodological Relationships between the two Approaches

Daniel Courgeau
Institut National d'Etudes
Démographiques (INED)
Boulevard Davout 133
75980 Paris Cedex 20
France

ISBN 978-94-007-2878-3   e-ISBN 978-94-007-2879-0
DOI 10.1007/978-94-007-2879-0
Springer Dordrecht Heidelberg London New York

Library of Congress Control Number: 2012932214

© Springer Science+Business Media B.V. 2012
No part of this work may be reproduced, stored in a retrieval system, or transmitted in any form or by any means, electronic, mechanical, photocopying, microfilming, recording or otherwise, without written permission from the Publisher, with the exception of any material supplied specifically for the purpose of being entered and executed on a computer system, for exclusive use by the purchaser of the work.

Printed on acid-free paper

Springer is part of Springer Science+Business Media (www.springer.com)

To my beloved wife, Hella

Acknowledgements

I would like to thank all the persons with whom I exchanged ideas and discussed different problems raised in this volume during the last 10 years: from the point of view of probability and statistics (first Marc BARBUT, unfortunately deceased just before the publication of this book, who opened the way to this volume, as editor of *Mathematics and Social Sciences -Mathématiques et Sciences Humaines-* and of the *Electronic Journal for History of Probability and Statistics* on publishing during these 10 years a number of papers I wrote on these themes, Xavier BRY, Henri CAUSSINUS and in the memory of Henry ROUANET who unfortunately died in 2008); from the point of view of social science (Alban BOUVIER, Eric BRIAN, Luc BUCHET, Tom BURCH, Ana DIEZ ROUX, Fatima JUAREZ, Alain GALLAY, Jean-Claude GARDIN, Harvey GOLDSTEIN, Nico KEILMAN, Denise PUMAIN, Jean-Marc ROHRBASSER, Isabelle SEGUY, Herbert SMITH, Eric VILQUIN, Bernard WALLISER, Guillaume WUNSCH); from the point of view of philosophy and methodology (Robert FRANCK, Pierre LIVET, Federica RUSSO); and many other researchers for discussion on these themes, particularly during the session on *Epistemology in demography and sociology*, I organized at the IUSSP (International Union for the Scientific Study of Population) General Conference in 2005.

Two anonymous referees must also be thanked for their very positive and constructive review on the outline of the manuscript.

Many thanks finally to Jonathan MANDELBAUM for his very clever translation of the whole volume.

Contents

General Introduction xiii
    Links Between Probability Theory and Social Science
        at Their Inception xv
    Statistical Inference, Induction, and Social Science xx
    Concordance Between Basic Probability Concepts
        and Social Science xxi
    Overview of Entire Volume xxiv
    General Conclusion xxxiii

Part I   From Probability to Social Science

Introduction to Part I 3

1 The Objectivist Approach 7
    1.1 Objectivist Probability 8
        1.1.1 First Step: A Classical Theory 8
        1.1.2 Second Step: A Frequentist Theory 9
    1.2 Paradigm and Axiomatics of Objective Probability 12
        1.2.1 The Paradigm 14
        1.2.2 The Axioms 17
    1.3 Objectivist Statistical Inference 25
    1.4 Application to Social Science 28
    1.5 Problems Posed by the Objective Approach 38

2 The Epistemic Approach: Subjectivist Interpretation 43
    2.1 Subjectivist Probability 44
    2.2 Paradigm and Axiomatics of Subjective Probability 55
        2.2.1 The Paradigm 56
        2.2.2 The Axioms 57
    2.3 Subjectivist Statistical Inference 67
    2.4 Applications to Social Science 69
    2.5 Problems Posed by the Subjective Approach 79

3	**The Epistemic Approach: Logicist Interpretation**	85
	3.1 Logical Probability	86
	3.2 Paradigm and Axiomatics of Logical Probability	98
	3.2.1 The Paradigm	98
	3.2.2 The Axioms	100
	3.3 Logicist Statistical Inference	112
	3.4 Application to Social Science	113
	3.5 Problems Posed by the Logicist Approach	129
Conclusion of Part I		133
	A Unique Notion or a Multi-faceted Notion?	133
	Inference and Decision-Making Theory	138
	Does Cumulativity Exist in Probability?	144

Part II From Population Sciences to Probability

Introduction to Part II		151
4	**The Dispersion of Measures in Population Sciences**	155
	4.1 From Pascal and Fermat's Wagers to Graunt's Bets	156
	4.2 Introduction of Epistemic Probability in Population Sciences	159
	4.3 Toward an Objectivist Approach in Population Sciences	166
	4.4 Return of Dispersion in the Event-History and Multilevel Approach	170
	4.5 Estimating the Age Structure in Paleodemography	174
	4.5.1 Methods Proposed Earlier	175
	4.5.2 A New, Truly Epistemic Approach	182
	4.5.3 Example of Archeological Application	186
	4.6 Conclusion	189
5	**Closer Links Between Population Sciences and Probability**	191
	5.1 The Framework: The Population and the Individual	193
	5.2 The Object: Not Individual Behaviors but More Abstract Concepts	197
	5.3 The Cross-Sectional Approach	200
	5.3.1 Independence Between Phenomena	200
	5.3.2 Dependence of Characteristics on Society	203
	5.3.3 Problems Posed by the Cross-Sectional Approach	208
	5.4 From a Longitudinal Vision to a Full-Fledged Event-History Approach	209
	5.4.1 An Initially Longitudinal Approach	210
	5.4.2 An Event-History Approach	213
	5.4.3 Problems Posed by This Approach	219

5.5	From a Latent Hierarchical Vision to a Multilevel Approach............	220
	5.5.1 An Initially Hierarchical and Latent Vision	221
	5.5.2 A Contextual Then Fully Multilevel Approach	227
	5.5.3 Problems Posed by the Multilevel Approach........................	230

Conclusion to Part II 233
Intensity of Ties with Probability... 233
Links Between Probability, Social Science,
and Counterfactual Causality 235
Does Cumulativity Exist in the Population Sciences? 238

General Conclusion... 243
Generality of the Use of Probability and Statistics
in Social Science 243
Revisiting Causality in Social Science... 249
Revisiting the Notions of Individual and Levels................................ 252
Predicting Behavior in Social Science 256
Epilogue 259

Glossary 261

References.. 263

Author Index... 295

Subject Index... 303

General Introduction

The notion of chance has always been present in human culture from earliest antiquity, and all peoples have used a wide variety of games in which chance plays a relatively important role (David 1955). For example, traces of the game of knuckle-bones[1] can be found during the First Dynasty in Egypt as early as 3500 B.C.E., and Roman soldiers played it by betting on the sides that would turn up after a throw. There is evidence of dice games in Mesopotamia, Egypt and Babylon dating from the third millennium B.C.E. Games with sticks were played by the Mayas, the Greeks, the Romans, the ancient Bretons, and the Egyptians—along with card games, chess, and others. These games were disseminated either for religious purposes, such as Jewish Talmud (Hasofer 1967; Rabinovitch 1969, 1970), divining among the Greeks, the Romans, and Tibetan Buddhists, or for recreational purposes (David 1955). However, this application of chance was not formalized more rigorously until much later[2] (Kendall 1956) and in gradual stages. Examples include the Latin poem *De Vetula*, possibly written by Richard de Fournival between 1200 and 1250, *Liber de ludo aleae* by Cardano written in approximately 1564 but published only in 1663, a fragment by Galileo Galilei (ca. 1642), and the studies by Pascal and Fermat (1654, 1922). These texts set the stage for the emergence of probability theory as a full-fledged scientific discipline.

The same pattern applies to the study of population and the efforts to enumerate human beings. Population counts were already performed by the Egyptians around 3000 B.C.E., partly to meet labor requirements for the construction of the Pyramids; they were carried out in Mesopotamia during the same period for religious reasons, by Moses in Sinai at God's behest ('Take a census of the whole community of Israelites by clans and families, taking a count of the names of all the males, head

---

[1] Small bones of the tarsus connected to the tibia and fibula. Knuckle-bones of hoofed animals such as sheep and goats have been found in large quantities on archeological sites dating back to at least 40,000 years. The knuckle-bone in these animals is roughly symmetrical; in others such as cats and dogs, it is totally asymmetrical and thus unsuitable for games of chance.

[2] Italian authors from the early fourteenth century to the fifteenth century offered various partial formalizations: see the article by Meusnier (2004).

by head [...] 20 years of age and over [...]', *Numbers*, 1, 2), by the Chinese Emperor Yu or Yao in the Empire of the Center after a great flood in 2238 B.C.E., by the Greeks in sixteenth century B.C.E (Missiakoulis 2010), the Romans, the Incas of Peru, and others (Hecht 1977). Once it had developed an organized structure, a State manifestly needed to count not only its citizens but also its economic resources. Here as well, however, the analysis of censuses and registers occurred much later, when scientists succeeded in measuring and quantifying phenomena that were previously God's secret. The first such analysis was published by Graunt (1662), followed by Christiaan and Lodewijck Huygens (1669, see Huygens 1895; Véron and Rohrbasser 2000) and Leibniz (early 1680s, see Rohrbasser and Véron 2001). The social sciences—such as demography, economics, and epidemiology—could now enter the scene.

A large body of literature has addressed these two broad themes separately: first, the history, methodology, and epistemology of probability and statistics (Gouraud 1848; Todhunter 1865; Matalon 1967; Hacking 1975, 1990; Krüger et al. 1986; Stigler 1986; Porter 1986; Daston 1988; Gigerenzer et al. 1989; Desrosières 1993; Barbin and Lamarche 2004); second, the history, methodology, and epistemology of population and other social sciences (Durkheim 1895; Landry 1945; Granger 1967; Piaget 1967; Franck 1994, 2002; Berthelot 2001; Courgeau 2002, 2003; Martin 2003).

Our purpose here is entirely different. We want to examine the historical connections between those two broad sectors. Analysis and research projects were not carried out independently of one another but, on the contrary, in close interaction.

Pascal, for example, worked on mathematics (*Essay sur les coniques*, 1640), probability theory (*Traité du triangle arithmétique III*, 1654), physics (*Récit de la grande expérience de l'équilibre des liqueurs*, 1648), and philosophy (*Entretien avec Sacy sur la philosophie* and *Les pensées*, 1670). Leibniz worked alternatively on logic, mathematics, probability theory, history, linguistics, law, politics, philosophy, and other disciplines. All these subjects are addressed in his complete works (see the website: http://www.leibniz-edition.de/).

The same is true of many researchers since the seventeenth century, although a greater specialization developed over time. In the twentieth century, for instance, Fisher worked simultaneously on probability theory, statistics, and genetics throughout his life, Keynes on economics and probability theory, and so on. Our aim here is to describe the origin and development of the relationships that have always existed between these disciplines. That is what makes this volume different from its predecessors. In the first part of this General introduction, we illustrate the links that were established between probability theory and the social science at their very inception.

As noted above, the concept of probability arose from the examination of the outcomes of a wide variety of games such as dice and cards. It took shape through a theoretical and mathematical evaluation of the number of possible outcomes, assumed to be equally likely. This *geometry of chance* (*géométrie du hasard*), as Pascal called it, does not suffice in social science, where probabilities cannot be determined in advance. All we can do is perform a certain number of comparable tests and observe a posteriori the number of events occurring in the sample, such as the number of deaths in a population. How can we then use these figures to reconstruct an unknown probability? What significance should we assign to the principles of statistical inference and induction that we can use to infer a probability from an earlier observation of facts?

Moreover, the social sciences have not yet managed to define their specific 'object' and their present state 'may be compared to that of natural sciences in pre-Galilean times' (Granger 1994). The complex and changing life experience that constitutes a 'human fact' still needs to be conceptualized as a scientific object, and we shall try to make some progress toward that goal here. In the second part of this introduction, we address the issues raised by this statistical inference, the problems encountered in social science and possible ways to solve them.

Despite its flowering in the seventeenth century, probability theory was not axiomatized until the twentieth century, with the work of Kolmogorov (1933). However, while the role of axiomatization is to define mathematical beings in formal terms, it does not tell us what entities in nature can be represented by them. For instance Kolmogorov clearly conveys his belief that not every event has a probability (1951):

> Certainly not every event whose occurrence is not entirely determined under given conditions has a definite probability under these conditions

and he asserts his frequentist position (1933). Nevertheless, with slight alterations, his axioms can apply to other approaches to probability theory—for instance, the subjectivist or logicist approaches. It is therefore important to realize that 'probability theory formalizes something that, in a manner of speaking, 'exists' independently; the divergences concern the nature of that 'something' which, according to this approach, is represented by the mathematician's probability' (Matalon 1967).

In the third part of this introduction, we shall examine this axiomatization and the problems encountered in applying it to a universe of experience—in social, biological, or physical science.

The fourth and final part will outline the path followed in this book, so that the readers can locate their position in the overall plan at all times.

Links Between Probability Theory and Social Science at Their Inception

While the investigations by Greek philosophers and mathematicians did not lead them to probability theory or to social science (Granger 1976), their work did enable them to raise the issue of chance and introduce the notion of *justice*, a crucial factor in the establishment of links between probability and social science.

For instance, Aristotle already made a clear distinction between things that 'always occur identically and others [that occur] frequently.'(Physics, 196b). In the Nichomachean Ethics, he writes (III:3):

> And in the case of exact and self-contained sciences there is no deliberation […]; but the things that are brought about by our own efforts, but not always in the same way, are the things about which we deliberate […]. Deliberation is concerned with things that happen in

a certain way for the most part, but in which the event is obscure, and with things in which it is indeterminate.[3]

While he does not succeed in formalizing this probable outcome correctly, the introduction of the notion of justice (fairness) and its formalization led to rules that preceded probability theory by centuries and made it possible. Aristotle defines justice as 'that kind of state of character which makes people disposed to do what is just and makes them act justly and wish for what is just' (Nichomachean Ethics, V:1). He goes on to formalize the notion as follows:

> The just, therefore, involves at least four terms; for the persons for whom it is in fact just are two, and the things in which it is manifested, the objects distributed, are two. And the same equality will exist between the persons and between the things concerned; for as the latter—the things concerned—are related, so are the former; if they are not equal, they will not have what is equal, but this is the origin of quarrels and complaints—when either equals have and are awarded unequal shares, or unequals equal shares.

Aristotle views justice as a critical element in, for example, concepts such as markets and money, which allow contracts between different and unequal persons. He therefore extends the argument by stating:

> This is why all things that are exchanged must be somehow comparable. It is for this end that money has been introduced, and it becomes in a sense an intermediate; for it measures all things, and therefore the excess and the defect—how many shoes are equal to a house or to a given amount of food. (Nichomachean Ethics, V:5)

What ultimately gives justice its full importance in the genesis of the notion of probability is the random contract (Daston 1988).

In his book *Liber de ludo aleae*, written in the mid-sixteenth century but not published until 1663, Cardano invokes Aristotle to define a fair wager:

> Other questions must be examined in a subtler manner, for mathematicians too can err, but differently. I did not want this issue to be set aside, for many people, who have not understood Aristotle, have erred, incurring losses. Thus there is a general rule that requires us to consider the total circuit,[4] and the number of outcomes representing all the ways in which a favorable result can occur, then to compare this number to the rest of the circuit, and lastly to examine the proportion to be used in reciprocal wagers so that they apply to equal terms.[5]

We shall see later how to formalize such a line of argument, which enables us to compute the probability of an event using the notion of circuit.

---

[3] Translation by W.D. Ross, http://classics.mit.edu/Aristotle/nicomachean.html.

[4] Cardano uses the term 'circuit' to denote the set of throws of different dice that can be examined in a given game.

[5] Reliqua ergo subtiliter consideranda; cum etiam in Mathematicis deceptio contigat, sed alia ratione. Volui hoc non latere, quia multi non intellegentes Aristotelem, decipiuntur, & cum iactura. Vna is ergo ratio generalis, vt consideremus totum circuitum, & ictus illos, quot modis contingere possunt, eorumque numerum, & ad residuum circuitus, eum numerum comparentur, & iuxta proportionem erit commutatio pignorum, vt equali conditione certent.

Links Between Probability Theory and Social Science at Their Inception

For Pascal as well, fairness is the concept that enabled him to develop the 'geometry of chance' Indeed, he presented his treatise in the following terms (Pascal 1654):

> ...an entirely new treatise, on a subject hitherto utterly unexplored, namely: the distribution of chance in games that are governed by chance—what is known in French as *faire les partis des jeux* [setting the odds of the game]; the uncertain outcome is so well controlled by the fairness of the computation that each player always receives exactly the amount consistent with justice.[6]

Pascal goes on to show how reasoning allows progress in this area, where experience seems of little use to him:

> And it is there, surely, that we must seek by means of reasoning all the more so as we are less likely to be informed by experience. Indeed, the results of ambiguous chance are rightly attributed to fortuitous contingency rather than to natural necessity. That is why the issue has drifted uncertainly until today. But now, having remained impervious to experience, it has failed to escape the empire of reason. And thanks to geometry, we have reduced it so effectively to an exact art that it partakes of geometry's certainty and has already made bold progress. Thus, by combining the rigor of scientific demonstration with the uncertainty of chance, and reconciling these apparent opposites, it can, drawing its name from both, rightfully claim this astonishing title: *The Geometry of chance.*[7]

In the third section of his *Traité du triangle arithmétique* [Treatise on the arithmetical triangle] (1654), Pascal spells out the prerequisites for reasoning on chance:

> ...the money that players have wagered no longer belongs to them, for they have relinquished their property of it; but, in exchange, they have received the right to expect the share of that money which chance can give them, under the terms they have agreed upon at the outset.

In the third section, Pascal also formulates the two principles that he views as the prerequisites for computing probability:

> The first principle, which is designed to determine how shares should be divided, is this.
> If one of the players finds himself in such a situation that, whatever the outcome, a certain sum accrues to him in the event of loss and gain, without chance being able to deprive him of it, he must not wager it, but take it in its entirety as guaranteed. This is because the wager must be proportional to the chances, and since there is no risk of loss, he must withdraw the entire amount undivided.
> The second principle is this. If two players find themselves in such a situation that, if a player wins, he is entitled to a certain sum, and if he loses, the sum will go to the other

---

[6] Novissima autem ac penitus intentatae materiae tractatio, scilicet de compositione aleae in ludis ipsi subjecti, quod gallico nostro idiomate dicitur *faire les partis des jeux*, ubi anticeps fortuna aequitate rationis ita reprimitur ut utrique lusorum quod jure competit exactè semper assignetur.

[7] Quod quidem eô fortius ratiocinando quaerendum, quò minus ten tando investigari possit. Ambiguae enim sortis eventus fortuitae contingentiae potius quam naturali necessitati meritò tribuuntur. Ideò res hactenus erravit incerta; nunc autem quae experimento rebellis fuit rationis dominium effugerenon potuit. Eam quippè tantâ securitate in artem per Geometriam reduximus, ut certitudinis ejus particeps facta, jam audacter prodeat; & sic matheseos demonstrationes cum aleae incertitudine jungendo, ab utraque nominatinem suam accipiens, stupendum hunc titulum jure sibi arrogat: *aleae Geometria.*

player; if the game is of pure chance and if the chances of winning are equal for both players and therefore the chances of winning are no greater for one player than for the other, if they want to part ways without playing, and reclaim their legitimate shares, they should divide the sum at stake in half, and each should take his half.

Pascal clearly indicates that this is a game of pure chance, i.e., for example, that the dice are not loaded. Using the arithmetical triangle, he generalizes this result to the broader case in which the players break up the game at a time when the first player is missing m shares and the second player n shares. Interestingly, Fermat, who discussed his approach in his correspondence with Pascal on this subject (Pascal 1922), reached the same result by means of a purely combinational method, and this enabled Pascal to conclude:

> I admire your method for wagers, all the more so as I comprehend it very well; it is entirely your own, and has nothing in common with mine, and reaches the same goal but easily. Our [mutual] understanding is thus restored.

In this exchange, Pascal and Fermat were addressing objective probability, for the chances of winning are determined by the fact that the playing tokens have not been tampered with. But Pascal's wager takes the reasoning further and introduces epistemic probability, for unique events, such as the existence of God. In a section of the *Pensées* entitled *Infini rien* [*Infinite nothingness*] (1670), he shows how an examination of chance can lead to a decision of a theological nature. Let us summarize his approach briefly here; we can return to it in greater detail in later sections of this book, when needed. Pascal argues as follows. Consider an individual who hesitates between faith and unbelief, but does not want to rely on the testimony of believers, doctors of the Church or miracles. Pascal begins by stating how the question is formulated absent experimental data:

> And let us say: God is or is not; but to which side shall we lean? Reason is of no avail here. An infinite chaos separates us. A game is being played at the far end of this infinite distance, where heads or tails will turn up. What will you wager?

Pascal shows that we must wager the existence of God, and that a probabilistic approach is possible here:

> Let us weight the gain and loss, wagering tails that God exists. Let us estimate the two outcomes: if you win, you win all, and if you lose, you lose nothing: therefore, without hesitation, wager that God exists. That is admirable.

Here, Pascal examines a hypothesis—the existence of God—and shows that the previous probabilistic argument, which concerned the occurrence of events that could reoccur in identical conditions, remains possible. While we can criticize its premises, this reasoning closely resembles that of game theory, but is based on entirely different arguments.

Let us now examine the situation in social science at the time. The first experiment in social science was, in fact, provided by John Graunt (1662), who submitted his findings to John Lord Roberts, Lord Privy Seal, in these terms:

> Now having (I know not by what accident) engaged my thoughts upon the Bills of Mortality, and so far succeeded therein, as to have reduced several great confused

Volumes into a few perspicuous Tables, and abridged such Observations as naturally flowed from them, into a few succinct Paragraphs, without any long series of *multiloquious Deductions...*

Graunt's approach effectively summarizes many observations by means of clear statistical tables. He uses these mortality statistics to deduce, through probabilistic reasoning, the population of London, formerly estimated at six million by worthy persons:

Next considering, That it is esteemed an even Lay, whether any man lives ten years longer, I supposed it was the same, that one of any 10 might die within one year. But when I considered, that of the 15000 afore-mentioned about 5000 were *Abortive*, and *Still-born*, or died of *Teeth, Convulsion, Rickets*, or as *Infants* are *Chrysoms*, and *Aged*. I concluded, that of men, and women, between ten and sixty, there scarce died 10000 per Annum in London, which number being multiplied by 10, there must be 100000 in all, that is not the 1/60 part of what the *Alderman* imagined.

We shall see later on the errors committed in this reasoning. Suffice it to say here that Graunt's method is still highly approximative and his hypotheses extremely crude. Compiling a true life table would require, at the very least, a series of age-specific probabilities of dying, which are far from constant. This was achieved decades later by Edmond Halley (1693), who set out to estimate the 'Degrees of the Mortality of Mankind' from the bills of mortality and birth of Breslau, a town whose population was less affected by migration than that of Graunt's London.

William Petty (1690) generalized the approach—which he designated as *Political Arithmetic*—not only to demographic issues but also to economic, political, epidemiological, administrative, and other issues in social science:

The Method I take to do this is not yet very usual: for instead of using only comparative and superlative Words, and intellectual Arguments, I have taken the course (as a Specimen of the Political Arithmetic I have long aimed at) to express myself in Terms of Number, Weight, or Measure; to use only Arguments of Sense, and to consider only such Causes as visible Foundations in Nature; leaving those that depend upon the mutable Minds, Opinions, Appetites of particular Men, to the Consideration of others. (Petty 1690)

It is under the label of political arithmetic that the social sciences developed during the seventeenth, eighteenth, and early nineteenth centuries. The rise of political economics began with Petty and de Boisguilbert (1695), followed by Cantillon (1755), Quesnay (1758), and Adam Smith (1776). After Graunt and Petty, demography and epidemiology progressed thanks to Halley (1693), Süßmilch (1741, 1761–1762), and Deparcieux (1746). But the term *demography* did not appear until much later—in its French form of *démographie*—in the title of Guillard's book *Eléments de Statistique Humaine ou Démographie Comparée* (1855). Epidemiology followed a similar path. The term was initially used to denote a medical discipline devoted to large-scale outbreaks of infectious diseases. But it did not emerge as a scientific discipline until the nineteenth century, most notably with the founding of the London Epidemiological Society in 1850.

Statistical Inference, Induction, and Social Science

If the probabilities of successive plays in a game of pure chance can be computed by strictly rational means, in other cases—particularly in social science—they cannot be *a priori* probabilities but only be determined *a posteriori*. Unlike Pascal, who was working on results that he could regard as equiprobable, Graunt had to use empirical observations to deduce probabilities of dying. It is after these observations on human mortality that Pascal's successors tried to generalize the notion of probability.

This involves going back from effects to causes—from empirical observations to the factors that generate them—in order to achieve greater certainty and, above all, greater generality in the analysis. This is known as the problem of statistical inference. After the initial efforts by Jacob Bernoulli (1713) to solve it, the solution was eventually proposed by Bayes (1763). Condorcet and Laplace developed it as the mathematical instrument perfectly suited to social science, where the *a priori* probabilities of causes were always unknown. Let us briefly review the issues raised and solved, which we shall examine in greater detail in the first section of this volume.

Jacob Bernoulli died in 1705, but his book, published by his nephew Nicolas, did not appear until 1713. In it, the author clearly stated the problem of a priori and a posteriori probabilities:

> But, in truth, another path is open to us in our quest for what we are seeking. What we cannot obtain *a priori* can at least be determined *a posteriori*, i.e., we shall be able to extract it by observing the outcomes of many similar examples; for we must assume that, later on, each fact can occur or not occur in the same number of cases as it was previously observed to occur or not occur in similar circumstances.[8]

The problem that Bernoulli is trying to solve is thus indeed complementary to the one raised by Pascal: when we do not know the a priori probability, we must obtain it a posteriori, from the observation of many similar outcomes. However, we are dealing here with objectivist probabilities, where the law of large numbers enables us to confer an objective, non-equivocal status upon the notion of probability. In the process, Bernoulli demonstrates a theorem still known in probability theory as the weak law of large numbers:

> Thus it is this problem that I now propose to solve, after having reflected on it for twenty years: its novelty and great usefulness, combined with its great difficulty, may exceed in weight and value all the other chapters of this thesis.[9]

---

[8] Verum enimverò alia hîc nobis via suppetit, quâ qæsitum obtineamus; & quod *à priori* elicere non datur, saltem *à posteriori*, hoc is, ex eventu in similibus exemplis multoties observato eruere licebit; quandoquidem præsumi debet, tot casibus unumquodque posthac contingere & non contingere posse, quoties id antehac in simili rerum statu contigisse & non contigisse fuerit deprehensum.

[9] Hoc igitur is illud problema, quod evulgandum hoc loco proposui, postquam jam per vicennium pressi, and cujus tum novitas, tum summa utilitas cum pari conjuncta difficultate omnibus reliquis hujus doctrinæ capitibus pondus and pretium superaddere potest.

His demonstration of the theorem is perfectly correct, but he was expecting a fuller result from his investigations. His finding applies to objectivist probabilities whereas his work was intended to apply to subjectivist probabilities, as he clearly states in his treatise.

Jacob Bernoulli accordingly demonstrates that if we know the probability of a phenomenon (assumed to be constant in successive observations), then, when we increase the number of observations, the observed frequency will diverge from its probability by a given quantity, which we can determine with the aid of that number and can set to as low a value as we want.

Bayes managed to go further by proving the opposite theorem, at least in a simple given case. He begins his article (1763) by clearly announcing the problem he intends to solve:

> *Given* the number of times in which an unknown event has happened and failed: *Required* the chance that the probability of its happening in a single trial lies somewhere between any two degrees of probability that can be named.

That is indeed the principle of statistical inference. The approach consists in using the observation of occurrences of an event to draw an inference on the probabilistic distribution responsible for the phenomenon—i.e., to provide an analysis of a past phenomenon, or a prediction of a similar future phenomenon. Throughout this volume, we shall see the various meanings that have been assigned to statistical inference and their links to social science.

Concordance Between Basic Probability Concepts and Social Science

As noted earlier, the notions of chance and of counting populations as well as some of the events they experience have been present in human thought since earliest antiquity. However, the concepts were not refined and initially mathematized until around the seventeenth century—by Pascal and Fermat (1654) for probability and Graunt (1662) for social science. This mathematization should logically lead to a more precise search for the bases on which to build a more robust theory of probability and social science.

Our introduction attempts to outline some of these bases in order to show the concordance or discordance between probability and social science. We shall elaborate on the bases in growing detail throughout the rest of the volume.

As Pascal and Fermat showed, probability could be mathematized, paving the way for probability theory. However, the research on probability focused on concepts that did not fit into the mathematics or the logic of the period: events, proof, randomness, chance, likelihood of an event, expected winnings, and so on—none of these concepts entered into the formalization of social science. Likewise, some of the chosen examples drawn from social science since the very inception of probability theory clearly showed the theory's potential use in fields other than games. Hence the need to define those concepts more precisely in order to use them with greater confidence and ensure that everyone was referring to the same things when applying them.

From the outset, Cardano clearly enunciated the main precondition of equiprobable outcomes, without which there could be no fair wager:

> The most fundamental principle of all in gambling is simply equal conditions, e.g., of opponents, of bystanders, of money, of situation, of the dice box, and of the die itself. To the extent to which you depart from that equality, if it is in your opponent's favor, you are a fool, and if in your own, you are unjust.[10]

This broad notion of equality is thus indeed the bedrock of probability theory, without which it would be meaningless.

Similarly, when Huygens sought to axiomatize probability—in the first true handbook on the subject published in 1657 under the title *On ratiocination in dice games*—he clearly showed that one cannot determine the fair amount of a wager except in a game with even chances:

> ... I start from the hypothesis that in a game, the chance of winning something has a value such that if we possess that value we can obtain the same chance by means of a fair game, i.e., a game that seeks to deprive no-one.

Again, this is a fundamental notion that allowed a reasoned investigation of probability. The notion was taken up by many later authors to serve as a basis for probability theory. Hence the classic definition of probability as: the ratio of the number of positive outcomes to the total number of outcomes provided that these are all equally possible. However, this definition contains a circular element, *equally possible* being an exact synonym of *equally probable*.

For instance, in a game of heads or tails, the only way to determine that the coin is as likely to land face up as face down is to toss it an infinite number of times. In social science, the problem is even trickier, for we must assume that the probability of an event is identical for all individuals in a given population.

This question was directly addressed by Henry (1959) in his discussion of a fundamental issue in demographic analysis:

> A homogeneous cohort may be viewed as consisting of identical individuals whose life histories differ only by chance. We can classify their histories according to the events that characterize them and the dates of their occurrence. This yields a statistical history of the cohort: a given proportion of individuals has experienced a given type of history. Let us now imagine that each individual in the cohort can repeat his or her history indefinitely; the infinite set of histories of each individual could, in turn, be classified according to the same criteria as before; we would obtain a statistical history of the individual. For a homogeneous cohort, the statistical history of the individuals who compose it is identical to the statistical history of the cohort.

---

[10] Is autem, omnium in Alea principalissimum, aequalitas, ut pote colusoris, astantium, pecunarium, loci, fritilli, Aleae ipsius. And quantumcumque declinaueris ab ea aequalitatae aduersum te, stultus es, & pro te iniustus.

However, Henry is then forced to admit that actual cohorts do not consist of identical individuals and that no human group is homogeneous. This finding undermines the analytical methods commonly used in demography, which assume cohort homogeneity or do not address that homogeneity. The author examines the equally theoretical case of a heterogeneous cohort formed by the amalgamation of infinitely large homogeneous cohorts. Once again, we are faced with difficulties similar to those encountered in probability theory when analyzing equally probable outcomes. Henry shows that error can be null only when the cohort is, in fact, homogeneous with respect to the topic studied. We shall return to these issues later.

Another basic notion of probability theory was defined somewhat later by Jacob Bernoulli (1713) and elaborated by Cournot (1843). It involves the case where the possibility of an event may be so close to zero that we may regard it as *physically impossible* or, on the contrary, so close to unity that we may regard it as *physically certain*. In Chap. IV of Part IV of *Ars Conjectandi* (1713), before demonstrating his theorem on the law of large numbers, Jacob Bernoulli clearly states:

> Some new points must be examined here, which may never have occurred to anyone before. We certainly still need to ask ourselves why, after the number of observations increases, there is a greater probability of reaching the true ratio between the number of cases where a given event can occur and the number of cases in which it cannot, so that the probability ultimately exceeds all given degree of certainty...[11]

This notion of certainty or 'moral' impossibility opposed with mathematical impossibility was widely discussed throughout the eighteenth century and in the early nineteenth. It was then revisited more thoroughly by Cournot (1843), who introduced continuity in the measurement of probability. This enabled him to discuss the notions of physical or moral possibility and impossibility:

> *The physically impossible event is therefore the one whose mathematical probability is infinitely small*; and this single statement imparts substance—an objective and phenomenal value—to the theory of mathematical probability.

Let us take the example of a jar containing a single white ball and an infinity of black ones. The probability that a blind agent will extract the white ball is mathematically possible but in fact so small as to be physically impossible. However, the only way to demonstrate this physical impossibility by means of Bernoulli's theorem is to draw an infinity of balls from the jar.

Do we find a similar notion in social science? Again, we can refer to Cournot (1843), who tells us:

> The acts of living, intelligent and moral beings have no explanation, in the present state of our knowledge, and we can boldly proclaim that they can never be explained by the mechanics of geometricians.

---

[11] Ulterius aliquid hic contemplandum superest, quod nemini fortassis vel cogitando adhucdum incidit. Inquirendum nimirum restat, an aucto sic observationum numero ita continuò augeatur probabilitas assequendæ genuinæ rationis inter numeros casuum, quibus eventus aliquis contigere & quibus non contigere potest, ut probabilitas hæc tandem datum quemvis certitudinis gradum superet ...

The notion of probability is therefore the only one applicable to social science, for this second notion of physically impossible event is perfectly suited to human acts. The two disciplines—probability theory and social science—set out to measure and quantify phenomena regarded as secrets of the gods before the seventeenth century: games of chance (such as dice, and cards) and games of life (births, diseases, deaths, migrations, and so on). The means used for these measurements and quantifications will, of course, form the basic theme of this book.

We now reach the twentieth century, in which the axiomatization of probability reached its broadest extension and in which the social sciences sought firmer foundations on which to address human affairs in rational terms.

After a series of more or less fruitful attempts to axiomatize probability (Laemmel (1904), Broggi (1907), Bernstein (1917), von Mises (1919), Slutsky (1922), Łomnicki (1923), Steinhaus (1923), Ulam (1932), Cantelli (1932), etc.), the work of Kolmogorov (1933) is now regarded by most probability theorists as the most consummate foundation for the science. We shall examine its basic principles in greater detail throughout this book, and point out the links between that axiomatization and the way in which we can interpret that formalization. Despite near-general acceptance of the axioms, controversies over the nature of this calculation and its possible interpretation persist in barely muted form. We shall therefore need to examine in greater detail how the different approaches view social science, in order to assess their validity in that field.

In social science, we are still a long way from axiomatization, and 'the transformation of the complex and changing life experience that constitutes the human fact into a scientific object—even in those of its aspects that are commonly recognized as public—remains problematic' (Granger 1994). We shall therefore need to examine in detail the multiplicity of viewpoints adopted on human facts over time in order to identify the operation that may enable us to reconstruct them in all their complexity. For this reconstruction, probability may prove essential.

Overview of Entire Volume

This volume will be structured as follows:

Part I From Probability to Social Science

Introduction to Part I

Depending on the historical period examined and the authors, the number of alternatives theories of probability is very variable and ultimately leads us to distinguish three broad types: objective probability, subjective probability, and logical probability. The last two categories can be grouped under the heading of epistemic probability.

Chapter 1 The Objectivist Approach

Classical probability theory relied from the outset—as early as Aristotle—on the notion of fairness. In 1654, Pascal referred to it for the purpose of defining a fair wager. But the notion, fully applicable to games of chance, did not hold up when transposed to the social sciences. These needed to assume that the now unknown probability of a demographic event—or, more generally, a social event—nevertheless existed, and remained the same throughout the period observed. This led to the notion of frequentist probability. The nineteenth-century debates over its validity showed that it cannot be applied to all feelings of uncertainty. The approach is suited to only a small number of social phenomena—particularly demographic ones.

The paradigm of objective probability had to reconcile the notions of equipossibility and physical impossibility. While the first was not specific to objective probability, the second proved indispensable, contrary to what later happened for epistemic probability. Objective probability is confined to events that can repeat themselves in identical conditions. Therefore, we cannot speak of the probability that a proposition, unique by nature, is true.

A proper search for axioms, however, did not become possible until after the establishment of set theory and axiomatics in the late nineteenth century. Setting aside many other attempts, we describe in greater detail two main types of axiomatization of probability, which were to formalize the two notions of paradigm. The first, introduced by von Mises in 1919, defined the notion of *collective* as the origin of probability. But many authors questioned the notion's consistency, undermining von Mises's axiomatics. In the end, it was the second type, introduced by Kolmogorov in 1933, that won the acceptance of most authors working on objective probability.

At this point, it is important to see how to apply objective probability to the statistics supplied by the physical and social sciences: this is known as the problem of statistical inference. The aim is to make the best use of the incomplete information available in order to move from data on a given phenomenon to the prediction of a similar phenomenon in the future. But, as the notion of 'an objective probability that a proposition is true' is meaningless, all we can estimate here is the probability of obtaining the observed sample if the hypothesis underlying the prediction is met.

We give some examples of applications of this approach to the social sciences. In developing political arithmetic, Graunt and Arbuthnott still used the notion clumsily. Another application concerns epidemiology, with the analysis of the effects of inoculation to prevent smallpox. Likewise, in sociology, Durkheim sought to identify social phenomena stripped of all extraneous elements by using the method of concomitant variations, i.e., a regression method.

This approach raises various problems. For example, while it allows a proper analysis of the outcomes of games with no cheating, it cannot determine whether a player is cheating or not. Similarly, the statistical inference made possible by objective probability is imperfectly suited to the study of decision-making. And it is suitable for analyzing only a small proportion of social phenomena.

Chapter 2 The Epistemic Approach: Subjectivist Interpretation

To apply probability calculus to the greatest possible number of feelings of uncertainty, however subjective they may be, we must abandon the notion of frequency—the foundation of objective probability—and hence the notion of physical impossibility. In 1713, Jacob Bernoulli envisaged what is now called a direct approach, which actually takes the probability of the studied event as a given. In 1763, Bayes solved the problem of the inverse approach, which assumes not only that the probability is unknown, but that its very existence is hypothetical. This leads to the notion of epistemic probability, which becomes fully subjective when one takes the view that it can be defined only for a specific individual, and not for an event as in the objective approach. As a result, the scope of application is substantially enlarged. For instance, we no longer need to assume the lack of cheating, for this probability is also defined in situations where players cheat, and the probability that a proposition is true now has a clear meaning.

The subjective-probability paradigm must rely on notions that differ from those underlying the objective approach. The notion of *coherence* in individual behavior must be reconciled with the notion of utility of winning for the individual. Coherence means that the reasoning of individuals must not contain any intrinsic contradiction, even as they are free to adopt any probability value that they prefer for an event. The notion of *utility*, introduced by Daniel Bernoulli in 1738, represents the subjective value of the stakes and will depend on each individual's condition. We can complete this paradigm by introducing the notion of *belief*, which is not probabilistic but allows the formalization of a psychological level outside the forecasting domain, and that of *plausibility* in order to reintroduce probability.

We must now apply a set of axioms to characterize the choice made by a rational individual faced with an uncertainty situation. Here as well, many axiomatizations have been proposed and we shall describe only the main ones. In 1931, de Finetti showed that a set of personal opinions, if it satisfied certain axioms, could be represented by a numerical measure. His axioms specified the notion of coherence. Savage completed them in 1954 by introducing the notion of utility, which arithmetizes the preference relationship between actions. Interestingly, the resulting quantitative probability satisfies Kolmogorov's axioms. Some criticisms of the axioms led to modifications introducing the notion of belief, which exists independently of the notion of probability examined in this volume. We shall therefore give only a brief presentation of it: Suppes in 1974 and Shafer in 1985 proposed axiomatizations incorporating two probabilities; Smets, in 1990, proposed an axiomatization that did not even include the concept of probability.

The objectivist approach offered only a partial solution to the problem of inference by twisting its meaning. By contrast, the subjectivist approach provided a perfectly clear answer. Using a *prior distribution*[12] and a data set, it allows an

---

[12] We need to distinguish here the term *prior*, which denotes any information beyond the immediate data and even used to express our ignorance, from the term *a priori*, which denotes a proposition, whose truth can be known independently of experience (Jeffreys 1939; Jaynes 2003).

estimation—under certain conditions—of a *posterior distribution* that predicts a future phenomenon. To ensure this outcome, the notion of *exchangeable* events, introduced by de Finetti, becomes indispensable.

We give examples of applications. The first concerns the combination of testimonies and is applicable in jurisprudence, artificial intelligence, and other areas. This problem has been addressed by many researchers over several centuries: the earliest solution used results found by Hooper in 1699; the latest uses Smets's theory of 1990. In our second example, the notion of exchangeability is applied to educational-science data for the purpose of drawing a correct statistical inference.

The approach is open to several criticisms. Psychological experiments have shown that, depending on how events are described, the subjective probabilities actually chosen by individuals do not necessarily meet the coherence principle. Although subjectivists reply that they study rational choices, the psychological problems posed by actual choices remain a fundamental issue. Moreover, an individual cannot always make choices transitively or even decide which choices to make: in such cases, his or her feelings of uncertainty cannot be represented by subjective probability. We also examine the criticisms of Savage's axiomatics by Allais in 1953 and show that the attempted modifications of his axioms cannot adequately explain all the phenomena connected to the choice paradox. The subjectivist approach seems too closely tied to individual psychology. Could a more logical yet still epistemic approach offer a means to avoid such criticisms?

Chapter 3 The Epistemic Approach: Logicist Interpretation

While a subjective probability is defined only for a given individual, a logical probability must be definable in the same manner for all individuals. For this, rather than start from the notion of personal odds for each individual, we must return to Pascal's notion of fair odds: when an individual wagers on a random event, fair odds yield a zero loss or zero expected gain. Yet fair odds will always reflect a degree of belief and are therefore applicable to all situations involving uncertain events, such as subjective probabilities.

The logical-probability paradigm introduced the logical notion of *consistency*, which specifies the required relationship between a proposition and the information available. Subjective probability depends on the individual. By contrast, logical probability, when obtainable in different ways, must yield the same result. It must also use all the information available for defining it. To this end, it incorporates the notion of *entropy* proposed by Shannon in 1948. Lastly, its focus is not on repetitive events, as in objective probability, or a single event, as in subjective probability, but on propositions made about events.

At this point it is useful to provide an axiomatics of the logic of propositions, introduced by Boole in 1854. It forms a basis for describing the main axiomatics of logical probability. The axiomatics proposed by Jeffreys in 1939 was initially rejected by most probabilists, philosophers, and statisticians of the time, but came to be recognized as highly innovative. However, without the notion of entropy, introduced later, Jeffreys was led to question the uniqueness of the choice of the prior probability. In 1961, Richard Cox showed that it was possible to derive the rules of probability from two axioms independent of the notion of set. One could thus use Kolmogorov's axioms, applying them now not to sets but to propositions. However, these axioms contain implicit conditions that van Horn later spelled out in order to make them more comprehensive. Similarly, while Cox effectively introduced the notion of entropy, it is Jaynes (2003) who showed more clearly how to use it to estimate a distribution of prior probabilities under different information scenarios.

For its application to social science, the epistemic approach concentrated on the incomplete information available on a phenomenon in order to draw inferences on the outcome of future experiments, using the consistency condition. It would no longer consider the personal probability that different individuals may choose, but those that they should choose on the basis of information shared by all. Statistical inference and probability would then form an inseparable whole.

We give examples of the use of logical probability in social science. The first example, from demography, is Laplace's application to the masculinity proportion at birth in 1781. The second is an application to legal science, which we illustrate with a wide-ranging review of results from 1785 to 2003, from Condorcet to Jaynes, via Laplace, Quetelet, Poisson, and others.

We conclude with a discussion of problems posed by this approach. The first is that impossibility and logical necessity are incompatible with the notion of zero probability for certain events when they can actually occur. But this criticism, which would be valid for an Aristotelian deductive logic, does not apply to a logic of plausible reasoning. The second problem is the difficulty, in certain cases, of finding a single prior distribution, although in many other cases we can deduce a non-informative distribution directly from the distribution of observations. This leads to a more general problem of dependency between the language used to pose a problem and the prior probability that can be deduced from it. We offer some solutions, but it is important to realize that the problem is inherent in all forms of epistemic probability, whether logical or subjective.

Conclusion to Part I

We begin by setting the three different approaches described in the preceding chapters in the context of the history of probability. The classical theory of probability that prevailed from the mid-seventeenth century to the first half of the nineteenth century was a unified theory in which the three aspects were closely linked: the probability of an event was simultaneously objective (considering its long-term frequency when it could be measured), subjective (considering the degree of our belief in its occurrence), and logical (considering the notion of fair odds). This type of probability was used in all fields, particularly the social sciences. In the first half of the nineteenth century, many criticisms led specialists to prefer the objective approach, which soon established its dominance for reasons that we discuss. In the

1930s, Kolmogorov's axiomatization of objective probability was swiftly followed by an in-depth examination of subjective and logical probabilities, although this did not result in their immediate adoption. They did not regain a stronger position until the second half of the twentieth century. However, they did not loosen the grip of objective probability—particularly in the social sciences, where it prevails to this day. We conclude with a methodological reflection on this revival of subjective and logical approaches, which leads us to examine if there is some cumulativity in probability.

Part II From Population Sciences to Probability

Introduction to Part II

We now examine the development of population sciences to show their methodological ties with probability throughout their history. While we cannot discuss all the social sciences—our work is not an encyclopedia—we show, when possible, that some methods used in this field are also suited to many other social sciences. We can thus extend the conclusions of these chapters beyond the specific field of population sciences.

Chapter 4 The Dispersion of Measures in Population Sciences

The aspect of probability that played a crucial role in the history of population sciences pertains to the *dispersion* of measures, either around their mean value, called *rate* (first sense of 'dispersion'), or as a function of other characteristics of the population studied (second sense). We devote particular attention to the use of statistical regression methods.

From the outset, Graunt's wager on the probability of dying is based on other hypotheses than Pascal's wager on the outcome of a game. Whereas Pascal can assume without too much difficulty that the odds are fair, it is far harder for Graunt to assume that the probability of dying is identical for all members of a population. Although the only information available to him was the number of observed deaths, he nevertheless chose that course in order to establish political arithmetic by positing an identical probability for all persons between ages 10 and 60. We show his errors, and how other researchers with access to fuller data improved his estimate by demonstrating that one should regard mortality as a function of age. Moving in the other direction, the introduction of the law of large numbers allowed Nicolas Bernoulli to refute Arbuthnott's argument on the distribution by sex of births in London from 1629 to 1710.

At the beginning of the nineteenth century, Laplace's application of the multiplier method, which allows a transition from observed births to the total population, supplied an estimate of the French population in 1782 within precise limits, confirmed by modern studies in historical demography. One might have thought that the regression methods developed by Gauss between 1795 and 1809 would have provided applications of interest for population sciences, but it took nearly a century for the methods to be used in *cross-sectional* (period) analysis.

The reason for the delay is that in the nineteenth century, at the same time as the abandonment of epistemic methods in probability, the dissemination of exhaustive censuses in Europe led to the rejection of Bayesian methods in population sciences. There was no longer any point in calculating the variance of a rate once the variance had become so insignificant. For dispersion in the second sense (see above), after a long reflection described in next chapter, *cross-sectional* analysis was finally able—in the late nineteenth century—to use the aggregated regression models to study the effect of different characteristics on rates. However, the advent of *longitudinal* analysis at the end of World War II, by introducing the time lived by the individual, no longer allowed the use of regression methods in the absence of a theory that could perform these regressions throughout an individual's life.

Such a theory did take shape in the 1970s—driven, in fact, by the social sciences. By the early 1980s, it was being used in population and other social sciences. Known as the *event-history* approach, it was first developed by David Cox in 1972 and Aalen in 1975 in an objectivist framework. More recently, it has been adapted to the Bayesian framework, which allows a better integration of all the information relevant to the topic of study. In both cases, the approach reintroduces the notions of variance and regression model, now applied to the flow of time. It was later extended by a *contextual*, then *multilevel* approach. These make it possible to avoid the *ecological fallacy* (a risk with aggregate models) and the *atomistic fallacy* (a risk with event-history models) once individuals' living environments have been properly taken into account.

We conclude this chapter by presenting a very recent study conducted by Caussinus and Courgeau (2010, 2011) in paleodemography. The study shows that it is possible to estimate the age structure of a past population for which no age measurements exist but for which proxy indicators are available. After a detailed criticism of methods proposed in the past, we show that only a fully Bayesian approach allows a correct estimate of the age structure and its dispersion from samples of a few dozen observed individuals.

Chapter 5 Closer Links Between Population Sciences and Probability

We now look at how the complex experience of a human lifetime has become a better-defined object for population sciences, while losing some of its complexity in exchange.

The notions of *population* and *individual* formed the basic framework of population sciences. While Plato and Aristotle managed to address some aspects of both notions, we show why they did not succeed in establishing a science of population. The concept of population did not take shape until the seventeenth century with the notions of 'comprehension' and 'extent' (étendue) introduced by the logicians of

Port Royal. In today's language, we would speak instead of the *intension* of the term population, which establishes its properties and characteristics, and its *extension*, which consists of the set of individuals who satisfy these properties. To define the concept of individual, we must replace its unlimited and unknowable character of the observed individual—fully recognized by Aristotle—and reduce it to a small set of aspects that can be addressed by a science. We thus arrive at the notion of abstract individual, called the *statistical individual*, whose characteristics we describe.

The object of population sciences is not the study of births, deaths, and migration flows that concern the members of a population, but indeed the study of their fertility, mortality, and mobility, measured by their probability. This clearly establishes close links between probability and population sciences, and the reason for their near-simultaneous emergence in the seventeenth century. Ultimately, population sciences were to study the changes to the population caused by the above-mentioned events. Depending on the perspective from which the changes have been viewed, different approaches have been applied.

The first perspective was the *cross-sectional analysis*, which prevailed from the earliest days up to the end of World War II. It holds that the social facts of a given period exist independently of the individuals who experience them and that they are explained by the various characteristics of the society to which the individuals belong. As early as 1760, Euler framed the independence hypothesis and defined the notion of stationary or stable population. The methods to study forms of dependence were developed later. We follow the path taken by several authors throughout the nineteenth century, and describe how they eventually showed—with the aid of the notion of correlation—that the least-squares method, used to study astronomic phenomena, was also suited to the social sciences. Durkheim applied it to demographic and social data at the end of the nineteenth century. Under this approach, the statistical individual became, in fact, a group of individuals defined by their age and various characteristics. Their aggregated behavior was observed in specific units such as geographic regions.

But these methods posed a number of problems—examined in detail here—which led researchers in population sciences at the end of World War II to incorporate personal 'lived time' into their approach. This was a two-stage process.

The initial *longitudinal analysis* observed the life of a cohort over time, and determined what would be the frequency of the studied phenomenon and its time distribution in the absence of disturbing phenomena. The approach assumed that the studied phenomena and disturbing phenomena are *independent*, and that the studied cohort is *homogeneous*. These assumptions overcame some of the objections to cross-sectional analysis. In the longitudinal approach, the statistical individual still consists of a group of homogeneous individuals, but they are tracked over their entire lives instead of being observed at a given point in time.

This approach, however, raised new problems, of which the most important were: (1) the impossibility of studying—as in cross-sectional analysis—the effect of various characteristics of the population on the probability of the studied events and (2) the impossibility of determining whether the condition of independence between phenomena is effectively met.

The consequent introduction of an *event-history analysis* solved these difficulties. Event-history methods originated in the martingale theory elaborated by Doob in 1953, which allows the study of ever more complex stochastic processes. To answer questions posed by demographers in particular, Aalen developed stochastic counting processes in 1975. They provided a solid probabilistic foundation for the analysis of life histories that began to take shape in demography in the early 1980s. The event-history approach could now properly integrate an analysis of *dependence* between events and an analysis of the *heterogeneity* of populations, thereby solving the difficulties encountered in longitudinal analysis. Statistical individuals were now assumed to follow an identical complex random process, whose parameters could be estimated with the aid of a sample of observed individuals. We also show that this approach, initially developed in an objectivist context, was extended to the epistemic approach.

As before, we review some of the criticisms of this approach, notably the problems raised by unobserved characteristics and by the existence of other aggregation levels, which frailty models attempt to incorporate.

First, we discuss a hierarchical, latent vision that covers a wide spectrum of models described briefly here, with a more detailed examination of frailty models. These assume an underlying distribution of individual probabilities and try to estimate it. However, the distribution is unknown, whereas only one model exists that can be estimated without observed heterogeneity: as a result, an infinity of distributions will fit observed data identically.

We also discuss the introduction of epistemic models, with fuller details on the artificial neural network method. It should be noted, however, that such models supply a 'black box' for effectively predicting a given distribution, without actually explaining the phenomenon studied.

We therefore turn now to a *contextual analysis*, followed by a fully *multilevel analysis*. Contextual analysis incorporates both individual and group characteristics into event-history models. This avoids two fallacies: (1) the *ecological fallacy* that an aggregate-level study can generate, and (2) the *atomistic fallacy* inherent in a pure event-history analysis. However, contextual analysis ignores potential intra-group dependence between individuals, which may produce overly narrow confidence intervals. Multilevel analysis overcomes this drawback by introducing random effects at group level in addition to individual variance.

To conclude, we present some of the objections to the approach, most notably that it fails to take into account the mechanisms for moving from more aggregated levels to the individual level. Only a new paradigm could allow advances in this field.

Conclusion to Part II

Our study has thus shown that nearly all population studies paradigms display such close ties with probability as to make it impossible to separate the two disciplines: population study is the application of probabilistic concepts to populations. We have also shown that, throughout their history, the social sciences have used the successive approaches to probability to address specific issues. For instance, in the late eighteenth and early nineteenth centuries, Laplace and Duvillard applied logical probability to a number of demographic subjects. By contrast, the late nineteenth century and the first half of the twentieth century saw the triumph of objective probability with the use of census data and aggregate regression methods. While the objectivist approach prevailed in the early days of event-history analysis, the epistemic approach later became important owing to the smallness of the populations observed. The same is true for the multilevel approach.

We also examine shortly here the role of counterfactual causality in social science and give some arguments against its use. Lastly we try to show how some cumulativity is possible in population sciences.

General Conclusion

Our general conclusion summarizes the main findings of our study, emphasizing what can and what cannot be generalized from population sciences to the other social sciences. While we cannot explore all the vaster implications of this issue—our book is not an encyclopedia—we suggest some ways of gaining a clearer picture of the situation.

Accordingly, we examine in greater detail the links between sociology or artificial intelligence and probability in order to understand their limits and to see the alternatives to statistical logic.

This leads us to discuss in more details the problem of causality in probability and social science and to go further than the counterfactual approach previously discussed. The role that mechanisms play in social sciences seems very important to explore simultaneously with their multilevel character.

We move on to various questions to which our book has provided only partial answers, and we suggest various approaches to supplement those answers.

For instance, we have solved the delicate problem of individual cases by means of the notion of statistical individual. This allows the introduction of many time-dependent individual characteristics into an event-history analysis where, initially, all individuals were equally likely to experience the event. True, there will always be unobserved characteristics capable of influencing the phenomenon, and an effect specific to each individual: his or her frailty. The more general problem is thus the transition from the individual to the population, under these various conditions. We show the formal relationships that link the parameters of an analysis of event histories at individual level and population level. We discuss recent approaches introducing more complex stochastic processes.

Lastly, we discuss the problem of forecasting in the social sciences, which implies the use of probability. We show the importance of using epistemic methods to solve the forecasting problem, with particular reference to the results obtained by microsimulation methods.

All the recent examples given in the conclusion show the enduring relevance of the methodological problem addressed in our work—a problem whose history we have recounted from its seventeenth-century origins to the present.

Part I
From Probability to Social Science

Introduction to Part I

In Part I, we will explore the path that runs from probability to statistics and social science. Before we begin, a cautionary note: it would be vain to assign a historical order of precedence to the three fields, as Good (1956) has shown so clearly for probability and statistics:

> Fermat and Pascal, and the other authors mentioned, started the mathematical theory of probability in order to explain the results of some statistics obtained experimentally, so that it could be contended that statistics came first. But since we have dated probability by mathematical probability, the reasonable question is 'When did mathematical statistics start?' … Apparently, then, the *mathematical* theory of statistics started at least sixty years later than that of probability.

In other words, the answer to the question depends on the definitions given for probability and statistics, and cannot be provided without bringing them into the discussion. In our view however, as noted below, Graunt's book (1662) marks the de facto establishment of a full-fledged statistical theory. That theory, therefore, appeared well before the date given by Good. However, it follows the advent of probability (Pascal 1654a), so Good's basic argument remains correct. The interval has simply shortened from 60 years to 8. In this 'chicken-or-egg' situation, we cannot determine which came first.

The same is true of social science, which we can view as preceding or following probability and statistics. Accordingly, under a certain definition of social science, we can say that when Plato and Aristotle described the foundation of a society, they were already engaged in population science by defining the component groups of a population (for more details, see Chap. 5), well before Pascal and Fermat introduced probability. Yet we can also argue that a truly statistical approach to population did not appear before John Graunt (1662), and this time it followed the introduction of probability.

To underscore this duality, Part II of our book will examine the opposite path from social science to probability and statistics—in the specific case of population science, for our scope is not encyclopedic.

As noted in the General Introduction, probability made it possible to reason about chance, i.e., about uncertain events. For more than three and a half centuries now, probability theory has developed in many different directions whose number, as well, is hard to specify. For Laplace (1814), the analytical theory of probability is unique and governs all our knowledge, even of mathematics, for the principal means of arriving at the truth—induction and analogy—are based on probability. Good (1971), on the contrary, believes that one can distinguish between 46,656 varieties of Bayesian probability, to which he adds the non-Bayesian variety proposed by von Mises (1942). Good arrives at his figure by considering that there are 11 facets to probability, each with 2, 3 or even 4 categories, yielding a total of $2^4 \times 3^6 \times 4 = 46,656$ categories. It is not useful to elaborate on the categories here, since some may be empty—as Good himself recognizes.

Between these extremes, we shall focus on three broad types of approaches: the objective approach, the subjective approach, and the logicist approach—the second and third of which can be regarded as epistemological. The first approach, which is purely empirical, concerns events that can recur in identical conditions. The second approach, which is purely subjective, is concerned with all types of feelings of uncertainty—however subjective they may be; it takes individual opinions on these events as a starting point to develop a coherent theory. The third approach, based on an extension of logic, also examines feelings of uncertainty but begins with the notion of fair wager in order to develop a consistent theory. As the second and third approaches consider the total set of feelings of uncertainty, we may regard them as 'epistemological' in the sense that Hacking (1975) uses the term to describe a knowledge-related entity. Here, however, we prefer to maintain the distinction between subjective and logicist, which we view as essential.

We shall add to the discussion the theories that resemble the two approaches, but without lingering on those that are of little value in social science (such as Popper's *propensionist* interpretation of the probability of isolated events (1983), which he notably applies to quantum theory and Werner Heisenberg's uncertainty equations), or on those (such as Zadeh's theory of fuzzy sets (1965, 1978)) that diverge too sharply from the notion of Boolean tribe, a characteristic of the forms of probability examined here.

It is not enough to describe the main theories related to the three approaches. We must also try to identify the underlying paradigm and axiomatics. The object of probability theory was initially elaborated via paradigms that sought to better define the object using abstract models whose elements, such as 'coherence' and 'consistency', must be defined more precisely. The probabilists then attempted to place their science on firmer foundations, i.e., to find the most suitable axiomatics for it. These axiomatics are now well established, but sometimes exhibit differences depending on the approach adopted. However, as we shall see in Part II, while some social sciences have tried to follow this path, most are still a long way from having fully axiomatized their discipline. Economy may be the field that has gone furthest in that direction (see, for example, the discussion between Mongin (2003), Agliardi (2004), and Armatte (2004)), for it is the most mathematized of the social sciences. However, in our opinion, we cannot say that this axiomatization has entirely succeeded. We can also say that population sciences have not yet succeeded in axiomatizing their field, although one can define manifestly different paradigms in the discipline.

To connect these types of probability to existing statistics, we must take a closer look at what is known as statistical inference. The purpose of inference is to make the most appropriate use of the incomplete information that statistics give us on a specific phenomenon. Inference consists in verifying the possible hypotheses on the behavior of the population studied and in estimating the population's underlying characteristics. Again, we shall see the very different meanings that the various approaches to probability can assign to inference. The notion of statistical inference plays a critical role in the application of probability to social science.

In these conditions, how should we apply the different forms of probability and the resulting varieties of statistical inference to the social sciences—for example, demography, economics, sociology, education sciences, and legal science? We illustrate the application of a type of probability to these disciplines without dwelling on their conceptual implications and the reasons that have driven authors to use probability. Here however, we examine succinctly different social sciences, while in Part II we will examine population sciences in more detail.

Lastly, it is obvious that each approach to probability, while offering solutions to certain problems posed by the others, has its own limitations and is open to criticism. In the conclusion to Part I, we seek to transcend these criticisms by offering a broader view of the different approaches to probability. After discussing their emergence over the centuries, we offer an overview of issues raised by statistical inference in social science.

Chapter 1
The Objectivist Approach

We have already noted that, from its inception, probability displayed two totally distinct facets: (1) objective probability, which concerns observed facts and relies on the frequencies of such events as rolls of the dice, human deaths, and human births; (2) epistemic probability, which concerns knowledge and relies on the assessment of degrees of belief in the truth of propositions as diverse as the existence of God, miracles, and a defendant's guilt. We shall examine each form separately here, bearing in mind that one did not precede the other, but that both emerged simultaneously. As we shall see, they imply a fuller, more encompassing approach: the logicist approach.

We have chosen to begin with the objectivist approach, for it seems to be more attuned to the initial concerns of the social sciences—particularly demography. From the outset, these disciplines addressed statistical and frequentist issues with the aim of measuring human phenomena such as births and deaths rather than topics of a more subjective kind such as law, jurisdiction, and certainty.

Moreover, the first theoretician to truly axiomatize probability, Kolmogorov, was an utterly convinced objectivist. He based the application of his probability theory on the existence of systems capable of infinite repetition. He sought to apply probability not to the largest possible number of feelings of uncertainty, but only to the feeling of uncertainty regarding the occurrence of events liable to be repeated in identical conditions. However, as discussed later, his purely formal axioms are open to a subjective interpretation as much as to a logical interpretation—with major differences, of course, in the concepts used.

By contrast, Kolmogorov fully realized that his axioms could have non-probabilistic interpretations, in a wide variety of research fields. Indeed, many quantities that have nothing to do with probability satisfy these axioms: standardized masses, lengths, areas, volumes, and so on—in fact, everything that falls within the scope of measure theory. That is the specific characteristic of successful axiomatization, which can apply to totally different systems, whether natural or human.

1.1 Objectivist Probability

Let us first try to see how this notion of probability took hold over time before discussing its paradigm and axioms, the inference it involves, and its application to demographic issues.

1.1.1 First Step: A Classical Theory

As already noted in our introduction, the notion of justice lies at the root of probability, and we would add objectivist probability here, for that is exactly what Cardano already meant in the mid-sixteenth century, when he spoke of a just wager:

> I am equally capable of casting a one, a three or a five as I am of casting a two, a four or a six. Consequently, bets are placed in keeping with this equality if the die is fair, and, if it is not, the bets are more or less proportional to the divergence from true fairness.[1]

He clearly sets out an equal number of alternatives, sustained by the vision of a fair die.

Let us examine more closely how to formalize in mathematical terms Aristotle's rule for a fair wager using Cardano's terms, quoted in our introduction.

Let us suppose that a player bets a sum x against another player who bets a sum y. The ratio of the number of positive outcomes for the first player to the total 'circuit' is in fact his/her probability of winning p, while that of the second player is $1 - p$. The game is, in this case, fair if the expected gains for both players are identical, i.e.:

$$\underset{\text{first player's expected gains}}{y \cdot p - x \cdot (1-p)} = \underset{\text{second player's expected gains}}{x \cdot (1-p) - y \cdot p}$$

which we can rewrite as: $\dfrac{x}{y} = \dfrac{p}{1-p}$

This is tantamount to stating that the ratio of bets must be identical to the ratio of the chances of winning, which is exactly the rule given by Cardano. This is also consistent with Aristotle's rule, where the 'shares' constitute the wagers and the 'justice' of individuals denotes their respective chances of winning.

Obviously, to make those wagers, we need to know the chances of winning, without which no reasoning is possible. Thus, as already noted, when Cardano refers to the basic principle of games, he clearly indicates the impossibility of reasoning when we do not know those chances. This confirms the general hypothesis formulated in the early days of probability calculus, namely, that the game is fair.

---

[1] Tam possum proiicere unum tria quinque, quam duo quatuor sex. Iuxta ergo hanc aeqalitatem pacta constant, si alea sit iusta; & tanto plus, aut minus, quanto a vera aequilitate longius distiterit.

For instance, if we play heads or tails, we must assume that the probability of the coin's coming up heads is one-half, and that of coming up tails is also one-half. We thus assume that the coin is perfectly fair—a heroic assumption. The same reasoning applies to a die for which we assume a 1/6 probability of landing on a given side, to a deck of cards for which we assume a 1/52 probability of drawing a given card (for a 52-card deck), and so on. That is how Pascal, Fermat, Christiaan Huygens, Jacob Bernoulli, de Montmort, and many other authors reasoned in the earliest days of probability calculus, by introducing ever more complex games, but keeping this hypothesis consistently in mind. Only by framing it could they determine the probabilities of the various possible outcomes. This initial approach forms what is known as *classic* probability theory.

This hypothesis of 'a fair game' or 'equally possible events' would lead us, when analyzing events other than games, to a definition of probability as a branch of psychology, for probability would be the only way to decide, in this case, whether the hypothesis is true or false. But objective probability concerns itself not with what people believe, but with what they should think of objective facts.

1.1.2 Second Step: A Frequentist Theory

We must therefore try to define probability by means of an objective property of the objects examined. Let us consider the classical definition of probability as the ratio of the number of favorable outcomes for an event to the total number of equally possible outcomes. This possibility is not measurable in the same way as a length or duration. The expression 'equally possible' is merely a synonym for 'equally probable'. As a result, the definition is circular: it merely reduces the general case in which the probabilities of several possible events are different to the specific case where all the probabilities have the same value.

Moreover, if we want to generalize these results to other phenomena—for example, demographic phenomena—or even to the situation where we do not know whether the coin is perfectly fair in the game of heads or tails, we need to posit new hypotheses. As we do not know the number of favorable outcomes and so cannot compute their ratio to all possible outcomes, the occurrence or non-occurrence of an event in a single trial can hardly tell us about its probability. We need to repeat the trial many times and observe the ratio of the number of trials that effectively produced the phenomenon studied to the total number of trials.

For this operation to be meaningful, we must define more precisely the phenomena to which it can apply. This approach was fully elaborated by the logician Venn (1866), known for his representation of sets by means of a simple diagram that can be applied to display various probabilistic operations. The approach 'combines individual irregularity with aggregate regularity.' Let us examine his argument in greater detail.

For Venn, the main error of certain earlier probabilists—and not the least eminent among them: Laplace, for example, is to have applied probability theory to events for which it was not applicable. Venn argues that the notion of series is the most essential for deciding whether probability theory applies to a given event or not, and the theory makes no sense unless it is linked to that notion. But he then needs to define with precision what he means by series. Venn's demographic example clarifies the term.

Let us consider the statement: 'Some children will not reach age 30.' If we view the statement as a logical proposition, the notion of series is utterly foreign to it. If, instead, it is a proposition that we can express in numerical form by replacing the term 'some' with a given proportion, then it is hard not to speak of a series any longer. This is not, however, the same thing as stating that if we observe a certain number of children, we shall observe this exact proportion of deaths before age 30. What it does mean is that if we observe a growing number of children, the proportion of deaths observed will tend toward that limit. The problem then becomes

> to see whether, with a steady increase in the number of observations, there is a growing likelihood of obtaining the true proportion between the number of cases in which an event can occur, and the number of cases in which it cannot, so that this likelihood will exceed the degree of certainty that we want (Jacob Bernoulli 1713, chapter IV).[2]

The underlying hypothesis is that this probability, while not calculable a priori as in games, exists and remains identical over time for the event studied. Bernoulli demonstrates this proposition—which in fact is anything but self-evident—in Chapter V of his book.

We shall now no longer need to assume that we can determine probabilities *a priori* but that they exist, even if we do not know them. For example, in a coin-toss exercise, the observation of a large number of persons tossing the same coin will enable us to estimate the probability of heads or tails, which may no longer be 50:50 if the coin is not fair. We shall also be able to estimate the probability of dying at a given age from the observation of a very large number of individuals in a population, even though there is no combinatory calculus enabling us to estimate it. Lastly, this approach leads to the notion of geometric probability, which means that instead of counting equally likely outcomes, we measure the extension of their surface area in a geometric space. But the probability remains a ratio of favorable outcomes to total outcomes—this time, a ratio of areas rather than of whole numbers (Cournot 1843).

We can thus assign an objective and non-equivocal status to the concept of probability when dealing with events liable to occur in identical conditions, during repeated trials. Normally, the trials should be indefinitely repeatable in order to allow the definition of an objective probability. While that is indeed the case of the coin in the previous example—provided that its wear over time leaves the probability unchanged—the same is not true of deaths observed at a given moment in a finite population. In the latter case, we need an additional hypothesis, namely, that this

---

[2] Inquirendum nimirum restat, an aucto sic observationum numero ita continuó augeatur probabilitas assequendae genuinæ rationis inter numeros casuum, quibus eventus aliquis contingere & quibus non contingere potest, ut probabilitas hæc tandem datum quemvis certitudinis gradum superet.

1.1 Objectivist Probability

finite population is extracted from an infinite population for which the probability of dying at a given age is always the same. Fisher follows this approach (1922b), when he defines probability as follows:

> It is a parameter which specifies a simple dichotomy in an infinite hypothetical population, and it represents neither more or less than the frequency ratio which we imagine such a population to exhibit.[3]

We can accordingly perform an empirical estimation from the objectivist standpoint, with a degree of certainty determined by the size of the sample observed. This is known as frequentist probability theory.

As noted earlier, in the approach defended by Venn (1866), many events are not open to probabilization. For example, he has this to say about social science:

> Many of the events which occur to human beings cannot be repeated at all, or not often enough to secure in the case of one individual any sufficient statistical uniformity.

Thus, in the objectivist approach:

> We can say nothing about the probability of death of an individual even if we know his condition of life and health in detail. (von Mises 1957)

In this case, we can only speak about the probability of dying in a population as large as we wish. Likewise, to speak of the probability of an inherently unique event or, more generally, of the probability that a proposition is true makes no sense for an objectivist. The event must form part of a series in which it is merely one of an infinity of elements.

As we shall see later, this frequentist probability theory was elaborated on a systematic scale by von Mises (1919, 1928, 1932), who named such series a *collective*:

> A collective is an infinite sequence of experiments whose results are represented by certain points in a space with r dimensions, the correspondence between the results and the order of the experiments meeting the two following conditions:
>
> 1. Let A be a random portion of the characteristic set; the ratio $\frac{n_A}{n}$ of the number n_A of such experiments among the first n whose results belong to A to the total number n of experiments tends toward a specific limit, with n rising infinitely;
> 2. Let A and B be two non-empty portions of the characteristic set without common points, n_A and n_B the numbers of such experiments among the first n whose results belong to A or B respectively; the limits p_A and p_B do not both vanish. If we select a place on the $(n_A + n_B)$ experiments such that only n'_A and n'_B remain, the ratio $\frac{n'_A}{n'_A + n'_B}$ tends toward a limit equal to the limit of the ratio $\frac{n_A}{n_A + n_B}$ or to $\frac{p_A}{p_A + p_B}$. With both conditions met, the limit is 'called the probability of a result belonging to set A in the collective examined' (von Mises 1932).

---

[3] While Fisher may be viewed as a frequentist in most of his writings, he gave a different definition of probability at the end of his life, noting that 'no sub-set may be recognizable having a fraction possessing the characteristic differing from the fraction P of the whole' (Fisher 1960). This concept of probability is generally regarded as unclear and has been little used since (Savage 1976).

Kolmogorov followed a different approach in his book (1933), where he set out a solid axiomatics of probability. In a footnote, however, he indicated that:

> In laying out the assumptions needed to make probability theory applicable to the world of real events, the author has followed in large measure the model provided by Mr. von Mises.

In a later article (1951), he added that he believed many events have no probability in the objective sense:

> The assumption that a definite probability (i.e. a completely defined fraction of the number of occurrences of an event if the conditions are repeated a large number of times) in fact exists for a given event under given conditions is a hypothesis which must be verified or justified in each individual case.

In our section on the axiomatics of objective probability, we shall take a more detailed look at the contribution of von Mises and Kolmogorov in this area.

The philosopher of science Popper (1982) also defended this position. For quantum mechanics, he proposed a propensionist interpretation of probability, which is a refined version of the frequentist position. He began by noting:

> For a long time, it was believed (and many eminent mathematicians and physicists still do) that we could take a system of subjectively interpreted probabilistic premises *and then draw objective statistical conclusions from these subjectivist premises. That, however, is a grave blunder.*

Later, he added: 'The error was carefully analyzed by Richard von Mises and also by me', to arrive at the following conclusion: 'the error is quite clear: from premises regarding degrees of belief, we can never reach a conclusion regarding the frequency of events.'

In Chap. 2, we shall see how the subjectivists respond to this criticism. Popper did, however, indicate (1959, 1983) that von Mises's axiomatics was met by a number of objections and that there was a need to elaborate a *propensionist* approach. The latter preserves the objective idea that probabilities are estimates of statistical frequencies observed in long real or virtual sequences. But it takes into account the fact that these sequences are defined by the way in which their elements are generated. These probabilities will thus depend on the generation conditions, and they can change when the conditions change (Popper 1983). This approach is particularly useful for studying stochastic processes found in nature—such as the radioactivity of certain sources—and it may be suitable for understanding probabilistic theories in physics. We shall not discuss it further here, as it is less relevant to social science (for more details see Suppes (2002a, b)).

In sum, the objectivist or frequentist approach to probability has come to prevail in the thought processes of many researchers, including statisticians and probabilists, in both the physical and social sciences.

1.2 Paradigm and Axiomatics of Objective Probability

Let us begin by specifying what we mean by 'paradigm' and 'axiomatics' before describing how these concepts emerged and developed in the field of objective probability.

1.2 Paradigm and Axiomatics of Objective Probability

First, although Masterman (1970) identified 21 different meanings of the term 'paradigm' in Thomas Kuhn's work (1962), Kuhn himself eventually isolated two main meanings:

> On the one hand, it stands for the entire constellation of beliefs, values, techniques and so on shared by the members of a given community. On the other, it denotes one sort of element in that constellation, the concrete puzzle-solutions which, employed as models or as examples, can replace explicit rules as a basis for the solution of the remaining puzzles of normal science. (Kuhn 1970)

Here, we propose a slightly different approach from those offered by Kuhn—an approach that actually addresses the following question: how does one move from experienced phenomena to the scientific object as defined by the philosopher Granger (1994)? For Granger:

> the complex life experience grasped in the experience of sensitive things has become the *object* of a mechanics and a physics, for example, when the idea was conceived of reducing it to an abstract model, initially comprising only spatiality, time, and 'resistance' to motion.

Granger further recognizes that the content of this object is not explicitly and broadly defined at the outset. For instance, sciences such as physics and biology perform successive elaborations of their objects, as illustrated by the transition from Newton's physics to Einstein's general relativity. Likewise, probability spelled out its object by means of successive paradigms, each of which specified its own distinct relationship between observed phenomena and the scientific object (Courgeau and Franck 2007). This notion will suit us well here for our examination of probability and population sciences.

Once we have identified the paradigm of a science, it is useful to formalize the science more fully by means of an axiomatics. Let us spell out what we mean by this term.

We begin with the period that gave birth to modern science and probability, and the rules established by their founders: Galileo (1613), Bacon (1620), Descartes (1647), Newton (1687), and others. We shall not explore their approach in detail here. Franck (2007) has provided a very clear description of it, and his main conclusions are as follows:

> Descartes teaches us that Euclid's axioms are not self-evident truths that need no demonstration; nor are they postulates, contrary to a common assertion. These axioms have indeed been demonstrated, admittedly not by means of logical deduction, but by analyzing the properties of geometric figures.

Descartes (1647), for instance, writes: 'Euclidean axioms have been deduced from the properties of geometric figures, contrary to what geometers pretend to believe.'

This thesis was recently revived, most notably by McKinsey, Sugar, and Suppes, in 1953. In 2002, taking measure theory as an example, Suppes showed that

> [a]n analysis of how this passage from the qualitative to the quantitative may be accomplished is provided by axiomatizing appropriate algebras of experimentally realizable operations and relations.

These studies led to the development of the semantic approach to theories (van Fraassen 1980; Suppe 1989). Similarly, the analysis of the empirical properties of objective probability is what allowed a definition of its basic axioms.

1.2.1 The Paradigm

From the outset, the theory of objective probability was fully rooted in the empirical phenomena that it aimed to address, such as games and population science's events. Gauss, for instance, counted the number of aces that chance allotted to him in his daily games of whist. We can even say that, although it is not possible to perform an infinity of trials in a game, the experiment can be conducted and confirmed a very large number of times, but this number will always be finite. By contrast, if these phenomena do not produce frequencies broadly converging toward a value that is, in principle, unknown, then they do not fit into the framework of the theory.

The study of games highlighted a first key principle for this axiomatics. By observing a large number of dice games, card games, and so on, we can determine the number of different outcomes—provided, of course, that the games are not biased, for example by loaded dice or marked cards. As discussed earlier, that is the principle used by Cardano to estimate his chances of winning. Likewise, Jacob Bernoulli (1713) observed that

> the number of outcomes is known from the dice; indeed, there are as many outcomes as there are sides, and all equally predisposed, since because of the similarity of the sides and the even weight of the die, there is no reason why one side should tend to turn up more than another, as would occur if the sides were shaped differently, or if one part of a die were made of heavier material than another.[4]

In the same vein, de Moivre (1711) introduced equally possible[5] events to define probability. A few years later (1718), he spelled out two essential notions to consider in analyzing probability: independent events and dependent events:

> Two Events are independent, when they have no connection one with the other, and that the happening of one neither forwards nor obstructs the happening of the other. Two Events are dependent, when they are so connected together as the Probability of either's happening is altered by the happening of the other.

He goes on to define the notion of compound probability:

> the Probability of the happening of two Events dependent, is the product of the Probability of the happening of one of them, by the Probability which the other will have of happening, when the first is considered as having happened.

---

[4] Ita ex. gr. noti sunt numeri casuum in tesseris; in singulis enim tot manifestè sunt quot hedrae, iique omnes æquè proclives; cùm propter similitudinem hedrarum & conforme tesseræ pondus nulla sit ratio, cur una hedrarum pronior esset ad cadendum quàm altera, quemadmodum fieret, si hedræ dissimilis forent figurae, aut tessera una in parte ex ponderosiore material constaret quàm in altera.

[5] eæque faciles.

1.2 Paradigm and Axiomatics of Objective Probability

He also uses, without defining it precisely, the notion of total probability: if two events are independent, the probability of either happening is equal to the sum of the respective probabilities of one happening and the other happening.

Should we include all these concepts in the paradigm or, on the contrary, can some be deduced from the others? To answer this question, we must examine the essential preconditions for the emergence of the concept of probability.

When the dice are loaded or when the probability of an event cannot be determined in advance—as is the case in demography—we must see which hypothesis enables us to calculate a probability for these events. To do so, the hypothesis that each successive draw is always equally possible (equipossibility) suffices to make the concept of probability usable. In this case, if the probability of a draw is unknown, but if we know that all the draws are equipossible, then Bernoulli's theorem cited above will apply, allowing us to estimate the probability. For instance, if a die is loaded, we can estimate probabilities differing from one-sixth for each side; in demography, we can estimate a probability of dying, marrying, migrating, and so on. In the words of de Montessus (1908):

> The term 'probability' merely indicates that, for a large number of trials, the ratio of the number of occurrences of an event to the number of tests will converge precisely toward the ratio of the number of positive outcomes to the number of possible outcomes: and nothing more.

We can thus state a portion of the paradigm of the calculus of objective probability. That is what Poincaré (1912) calls *petitio principii* ('begging the question'):

> how can we recognize that all outcomes are equally probable? A mathematical definition is not possible here; we shall have to formulate *conventions* in each application, to say that we regard given outcomes as equally probable.

Once these conventions are framed, the concepts of compound or total probability become theorems for Poincaré. Interestingly, as we shall see later, Kolmogorov (1933) took the opposite stance, introducing the concept of total probability as one of his axioms. Some of his contemporaries, instead, preferred to choose compound probability as one of their axioms (Lévy 1937; Ville 1939). But, in our view, this *petitio principii* seems to be the true foundation of the paradigm of the concept of objective probability.

Another important concept for objective probability is physical possibility or impossibility—known in the eighteenth century as 'moral certainty'. Jacob Bernoulli (1713) had already showed that, when the number of observations increases, the probability that the measured frequency converges toward the event's probability tends toward unity. He noted that this probability may be viewed as a moral certainty: 'which is a reasonable practice in civil life, where moral certainty is viewed as absolute certainty'.[6] Moral certainty was the subject of considerable discussion in

---

[6] quod sane in usu vitæ ciilis, ubi moraliter certum pro absoulte certo habetur.

the eighteenth century, particularly by D'Alembert (1761a) in his *Réflexions sur le calcul des probabilités*:

> Indeed, we must distinguish between what is *metaphysically* possible, & what is *physically* possible. The first class contains all things whose existence is in no way absurd; the second comprises all things whose existence not only has nothing absurd about it, but even has nothing too extraordinary about it, & [nothing] that does not belong to the daily course of events.[7]

Cournot (1843) gave the clearest definition of this concept of physical impossibility. For this, he shifted his focus away from fair games—which always offer a countable number of possible outcomes—toward the possibility of increasing to infinity the number of concomitant causes generating a given event. This enabled Cournot to examine events whose probability can be infinitely small from the outset. In his Chapter IV, 'On chance – on physical possibility and impossibility', he wrote:

> *the physically impossible event is therefore the one whose probability is infinitely small;* and this remark alone gives substance, an objective and phenomenal value to the theory of mathematical probability.

A physically impossible event may be envisaged as mathematically possible, but in fact it never occurs. What Bernoulli had shown for games by increasing the number of trials is just as valid for events that have an infinity of causes.

Hadamard (1922) stated the two basic concepts on which probability is based: totally equivalent events and totally impossible events. This is an exact reformulation of the concepts of equipossibility and physical impossibility. Lévy (1925, 1937) spelled out the role of both concepts in fuller detail. The concept of equipossible events offers a foundation for probability theory, but does not enable us to distinguish between objective and epistemic probability. By contrast, the concept of physically impossible events allows us to characterize objective probability with precision:

> A very improbable event is therefore *practically* equatable with an impossible event; we shall describe it *as practically impossible*, or *nearly impossible*; its opposite is *nearly certain*. (Lévy 1925)

Bernoulli's law of large numbers offers an illustration: we must expect the difference between frequency and probability to be all the smaller as the number of observations increases. 'We must expect' effectively means here that the opposite is practically impossible.

Thus, with the aid of the two concepts of equipossibility and physical impossibility, we can define the paradigm on which objective probability is based.

---

[7] Indeed, Fréchet (1951) remarks that this principle, while attributed to Cournot, 'seems to have been already stated more or less clearly by D'Alembert'. But we should note that D'Alembert, while a great mathematician, made a number of errors in his reasoning on probability (Bertrand 1889; Delannoy 1895; Maupin 1895). However, as we shall see later, D'Alembert's occasionally subjective stance in his reasonings gave rise to some of the criticisms directed against objectivist probabilists.

1.2.2 The Axioms

Lastly, we must identify the more specific axioms on which this discipline is founded. While Euclid had defined the axioms of geometry about 33 centuries ago, the axioms of probability were not clearly set out until the twentieth century. It was Hilbert—in his famous list of the 23 problems to be addressed in the twentieth century, which he presented at the International Congress of Mathematicians in Paris in 1900—who set the sixth problem as follows:

> The investigations on the foundations of geometry suggests the problem: to treat in the same manner, by means of axioms, those physical sciences in which mathematics plays an important part; in the first rank are the theory of probabilities and mechanics. (Hilbert 1902)

Before axiomatizing probability, however, we need to elaborate a theory and an axiomatics of sets and set measurement, known in Hilbert's day as additive function of abstract sets. The events of interest to the probabilist—such as the tosses in a heads-or-tails game, the draws from a jar, or the occurrence of demographic events—are neither geometrical or arithmetical elements. By contrast, the concept of 'set' allows us to capture these events, and the concept of 'set measurement' allows us to measure them. Each can be viewed as an element of a set, and we shall see that operations on these sets indeed apply to more complex events whose probability we want to estimate.

Space precludes a detailed discussion here of the concept of set, which has been extensively debated. Suffice it to recall that the theory was largely elaborated by Cantor (1874), who was also interested in probability. In an earlier paper (1873), Cantor had called for a debate in which the validity of his calculations could be established with exactitude. Set theory raised a number of paradoxes, later resolved thanks to a more precise and more general axiomatics (axioms of the ZFC theory[8]).

To develop an axiomatics of objective probability, we do not need to examine the general concept of set but only the concept of measurable set. Three operations on these sets will suffice for our purposes: (1) and (2), union and intersection applied either to a finite number or to a countable infinity of elements in the set; and (3) complement. Such a set is called closed with respect to these operations if, by applying it to a finite or countable number of its elements, we still obtain an element of the same set. A σ-algebra or tribe is a closed set with respect to the three operations above. In fact, if we axiomatize this concept of tribe, we simply need to posit the axiom of closure with respect to the finite or countable union of sets and the axiom of closure with respect to the transition to the complement: from this, we deduce closure with respect to the finite or countable intersection. If Ω is the set

---

[8] The acronym stands for the Zermelo-Fraenkel theory, formulated with the axiom of Choice.

examined, $\mathscr{B}$ a collection of sub-sets of Ω, and $\cup$ the operation to unite the sets, the axioms are as follows:

1. The empty set and Ω are elements of $\mathscr{B}$
2. If B is an element of $\mathscr{B}$, its complement is an element of $\mathscr{B}$
3. If B_n is an element of $\mathscr{B}$, then $\cup B_n \in \mathscr{B}$

These axioms of set theory were developed from the observation of a large number of countable sets that meet all the conditions above but cannot be demonstrated. They then served as a basis to demonstrate theorems on these sets and formalize probability theory.

From the notion of σ-algebra, Borel (1898) was able to establish a measure theory that is inseparable from Lebesgue's theory of integration (1901): the integral of a function of a numeric variable is a measure of the area bounded by the x-axis and the curve plotting the function. While Borel confined himself to Euclidean spaces, Fréchet (1915) generalized the approach from n-dimensional spaces (Radon 1913) to any abstract set.

Such a measure is a function that associates each element A of an σ-algebra with a value $\mu(A)$ that is a positive or infinite real number. The conclusively recognized axioms that allowed a definition of this measure are as follows:

1. The empty set, $\emptyset$, has a null measure.
2. The measure is σ-additive: if B is the union of sets $B_1, B_2,...$, disjoined two by two, then the measure $\mu(B)$ is equal to the sum $\sum_{i=1}^{\infty}\mu(B_i)$.

This definition of measure is in fact based on many earlier definitions (such as Jordan's measure, Cauchy-Riemann integral, and Stieltjes integral) and it remedies some of their defects. As we can see, the notion of measure has evolved over time, as have its axioms. However, while the empty set has a null measure, it is not necessarily the only set: we shall see how valuable this result is when studying practically impossible events. The axioms would later be needed for probability theory, as Lévy (1936) showed so well.

Interestingly, most mathematicians responsible for advances in measure theory worked on probability at the same time, clearly demonstrating the common ground between the two. Borel wrote *Éléments de la théorie des probabilités* (1909) and *Le hasard* (1914). Fréchet wrote a book entitled *Généralités sur les probabilités. Variables aléatoires* (1937). Lévy published two volumes: *Calcul des probabilités* (1925) and *Théorie de l'addition des variables aléatoires* (1937).

This brief overview of set theory and measure theory shows that it was impossible to axiomatize probability without the groundwork described above, for probability is based on the concepts embodied in both theories. We can now see how to move from the many observations made since the seventeenth century in fields where objective probability applies (such as games of chance, demographic events, and Brownian motion) to the foundations of this science: induction, advocated by Francis Bacon (1620), aims to discover and demonstrate the axioms on which probability is based.

1.2 Paradigm and Axiomatics of Objective Probability

We shall not give a detailed description of all the axiomatics proposed for probability: for more details, see Shafer and Vovk (2005, 2006). In particular, we shall not examine the attempts made by Laemmel (1904), Broggi (1907), Bernstein (1917), and others before 1919, the year von Mises published his essay on axiomatization. But we shall discuss in detail the axiomatics of von Mises and the criticisms that it raised. This will give us a better idea of the requirements that an axiomatics must meet in order to be accepted by the majority of researchers, as occurred with Kolmogorov's axiomatics.

The first objective of von Mises (1919, 1928) was to axiomatize the notion of collective, mentioned earlier. At that stage, probability was merely an attribute of the collective: the probability of an element of a collective is simply the element's limit frequency in the collective. Let us briefly recall the meaning von Mises gives to the concept of collective: 'A collective is an infinite sequence of experiments whose results are represented by certain points in an r-dimensional space' (von Mises 1932). He specifies that without the notion of collective, one cannot speak of the probability of an event. He goes on to give the two founding axioms of probability theory, in the simple case of a sequence of n trials repeated with one coin, where x is the ratio of the number of 'heads' results to n:

1. The first axiom of probability theory holds that if one continues the trials by increasing n to infinity, the ratio x tends toward a determined limit p.
2. The second axiom holds that if we make a place selection and if we take into account only the chosen elements when calculating the ratio x, the limit remains invariable.

It is easy to generalize these axioms to the case where the results of the trials are represented by points in an r-dimensional space. Von Mises is more specific about what he means by 'axioms':

> In these axioms use is made of general experience; they do not, however, state directly observable facts. They delineate the subject of the theory; all theorems are but deductions from the axioms, i.e. tautological transformations [...]

Von Mises therefore seems to have come fairly close to our view of what axioms should be, but we shall see that the validity of his axioms has been disputed.

From these two axioms, von Mises demonstrated the main results of probability theory, which became its theorems. Without discussing these demonstrations, we shall examine the axioms in greater detail in connection with the paradigm and with the criticisms directed at them.

To begin with, the axioms do not use the equipossibility hypothesis that formed part of the probability-theory paradigm. As von Mises put it (1932):

> No doubt the expression 'equally possible' is merely a misnomer for 'equally probable,' so that the expression contains a partial *vicious circle*; it simply reduces the general case where the probabilities of several possible events are different to the special case where they all have the same value.

Its axiomatics, on the other hand, enables us to show that 'the classic statement on the ratio of the number of positive outcomes, etc., will obtain the legitimate status of a special theorem.'

Likewise, the physical-impossibility hypothesis is not viewed as an axiom but as a consequence of a theorem on the summing of n collectives. These axioms therefore allow us to identify more relevant elements—according to von Mises—than those constituting the paradigm, so that we can define unequivocally a more robust theory, all of whose theorems can be deduced from the axioms.

Unfortunately, these axioms were the target of many justified criticisms, which eventually led to Kolmogorov's axiomatics. Let us examine these criticisms.

An initial series of criticisms challenged the consistency of the theory of collectives[9] and entailed a more specific definition of the second axiom, also called irregularity axiom (*Regellosigkeitsaxiom*). The term reflects the fact that the sequences complying with this axiom cannot be represented by a formula or rule. Hence, we cannot speak of a 'limit' for these sequences. For example, Kamke (1932) examines the simple case in which collective x induces probabilities $P(0) = P(1) = \frac{1}{2}$. Let us now consider, independently of this collective, the increasing sequence $\{n_k\}_{k \geq 1}$ and let us extract selection $x_{n_1}, x_{n_2}, \ldots$ from the collective x. If all place selections made on x are allowed (axiom 2), then there will be at least one choice such that $x_{n_k} = 1$ for all k and another such that $x_{n_k} = 0$. As a result, x cannot be a collective, for these values differ from $\frac{1}{2}$, counter to expectations. The demonstration is valid for all more complex cases. In sum, if arbitrary selection rules are allowed, then there can be no collective.

We must therefore restrict these overly general sequences to more narrowly defined sequences, but they must be sufficiently numerous to form a set with a non-null measure. We shall not go into the details of the discussions on this topic (see, in particular, Copeland 1928, 1936; Tornier 1929; Wald 1936; Church 1940). However, none of the proposed solutions allows the definition of a general probability theory, as von Mises himself admitted (1928):

> It is not possible to build a theory of probability on the assumption that the limiting values of the relative frequencies should remain unchanged only for a certain group of place selections, predetermined once and for all.

Indeed, whatever the group chosen, we can always find sequences of places that lie outside the group yet comply with the second axiom as initially posited. The most recent studies on the definition of random sequences (Martin-Löf 1966) actually lead to a totally fresh interpretation of von Mises's theory: 'Such an interpretation of the randomness axiom is of course anathema to von Mises' (van Lambalgen 1987).

---

[9] This consistency is essential for the *constructivist* mathematical school, of which Gauss, Borel and Lebesgue are the best-known representatives: for them, mathematical objects exist only if there is a precise method that tells us how to construct them. By contrast, for the *formalist* school, of which Moritz Pasch and David Hilbert were the most famous representatives, the lack of contradiction in a system of axioms is a sufficient precondition for accepting that system. Despite siding with the constructivists, von Mises was not unduly troubled by these criticisms. He actually claimed that 'collectives are in a sense 'the rule,' whereas lawfully ordered sequences are 'the exception'.'

1.2 Paradigm and Axiomatics of Objective Probability

A second series of criticisms concerns the combination of axioms 1 and 2 (Waismann 1930): they argue that it is inadmissible to apply the mathematical concept of convergence in axiom 1 to a sequence that, by virtue of its definition in axiom 2, can be subject to no mathematical rule or law. The notion of mathematical limit is only a property of the law that defines the sequence.

Next, for his own theory, von Mises borrowed many elements of game theory, notably the fact that, whatever the players' strategies, their final fortunes can never exceed their initial fortunes (on average). The strategies are effectively equivalent to the replacement of the full sequence of observations with a random sub-sequence. However, Ville (1939) began by showing that, given a countable set of selection rules, there is always one collective that reaches its limit at infinity without fluctuating around the limit, i.e., it remains consistently greater than or equal to the limit. Such a collective cannot be equated with the intuitive notion of the game of heads or tails. Moreover, Ville showed the existence of strategies, called martingales that cannot always be represented by selections of sub-sequences. He thus demonstrates a theorem of martingale theory that 'one cannot imagine demonstrating […] by means of the theory of collectives'.

Von Mises also rejected from the field of his theory a number of studies that addressed the concept of probability and were even closely tied to its axiomatics. For instance, Borel showed the paradoxical properties of what he termed absolutely normal numbers (1909). Taking segment [0, 1] supplied with Lebesgue's measure and developing a random number in this segment in base two, Borel demonstrated that nearly all real numbers in the segment are absolutely normal. Such sequences indeed appear to be random, as they hold up to many classic stochasticity tests. Even more important, Borel provided an explicit construction of such a sequence. However, von Mises (1932) refused to recognize such sequences as collectives, which, in his view, are the object of probability calculus in the usual sense:

> If we only supposed, for a sequence of numbers forming a collective, the restricted irregularity that defines a normal number, we could no longer state that no martingale, no game system was possible.

This position is surprising, given that von Mises's collectives are a step in the definition of random sequences that began with Borel (1909), continued with the collectives of von Mises (1919) and Wald (1936), the studies by Ville (1939), and Church's random sequences (1940), and extended to the work of Martin-Löf (1966): for more details see the excellent synthesis made by Dellacherie (1978).

Lastly, von Mises stated that the concept of random variable used by Cantelli and Fréchet is not 'a special class of independent variables, but rather a special class of *functions*, namely, distribution functions' (1932). But Fréchet (1938) believed that von Mises unduly restricted his theory's scope of application:

> No probabilist will refuse to admit that the collectives defined by Mr. De Misès [sic] are particularly interesting sequences, which accordingly deserve the fullest attention. Likewise, in function theory, it is quite legitimate to focus on derivable functions, provided one does not assume that all functions are derivable.

This reading downplays von Mises's contribution to special sequences that are far from exhausting all conceivable sequences in probability theory.

All these difficulties encountered by collectives show that the concept is far from perfect for axiomatizing objective probability. Indeed, this axiomatization has now been completely abandoned. In the words of Suppes (2002a):

> the von Mises relative-frequency theory ultimately seems to fall between two stools. On the one hand, it is a very awkward and unnatural mathematical theory of probability as compared to the standard measure-theoretic approach [...]. Yet despite its technical complexity, it fails to provide an adequate theory of application for dealing with finite sequences.

There were other, later attempts, for example by Steinhaus (1923) and Cantelli (1932, 1935), which in many respects anticipated Kolmogorov's axiomatics (1933). We shall leave them aside here in order to examine the latter in detail. Its validity was eventually recognized by most objective probabilists, and—unlike von Mises's axiomatics—it proved immune to criticism.

Kolmogorov begins his book with the five founding axioms of probability theory, which use all elements of set theory and measure theory. For this purpose, he considers a set Ω of elementary events and a collection of sub-sets $\mathscr{B}$ of Ω; the elements of set $\mathscr{B}$ are called random events.[10] Kolmogorov then introduces the five axioms:

I. $\mathscr{B}$ is a field of sets.
II. $\mathscr{B}$ contains the set E.
III. To each set A in $\mathscr{B}$ is assigned a non-negative real number $P(A)$. This number $P(A)$ is called the probability of the event A.
IV. $P(E)$ equals 1.
V. If A and B have no element in common, then $P(A + B) = P(A) + P(B)$.

He adds a sixth axiom, called continuity axiom, which is useless when $\mathscr{B}$ is finite, but which, when $\mathscr{B}$ is infinitely countable, is independent of the five others:

VI. For a decreasing sequence of events $A_1 \supset A_2 \supset ... \supset A_n \supset ...$ of $\mathscr{B}$, for which $\bigcap_{n=1}^{\infty} A_n = \emptyset$ the following equation holds: $\lim_{n \to \infty} P(A_n) = 0$.

Today, the triplet $(\Omega, \mathscr{B}, P)$ is called a probability space. Kolmogorov shows that this system of axioms is consistent, i.e., no statement in it can be both true and false, but it is not complete, i.e. some statements may be neither true nor false 'for in various problems in the theory of probability different fields of probability have to be examined' (Kolmogorov 1933).

The sixth axiom is indispensable only for infinite probability spaces: in Kolmogorov's own view, 'it is almost impossible to elucidate its empirical meaning, as has been done, for example, in the case of axioms I-V'. In his opinion, infinite probability spaces can occur only as idealized models of real random processes. We can clearly recognize his distinctive concept of objective probability.

---

[10] For consistency with the notations already used in this chapter, we have modified those used by Kolmogorov.

1.2 Paradigm and Axiomatics of Objective Probability

Indeed, Kolmogorov devotes an entire section of his book to spelling out the links between the world of experience and the axioms, and the empirical procedure that he has used to deduce his axioms. He recognizes that if one works in certain conditions, which do not need to be described here, then:

a) One can be practically certain that if the complex of conditions is repeated a large number of times, n, then if m is the number of occurrences of event A, the ratio m/n will differ very slightly from $P(A)$.
b) If $P(A)$ is very small, one can be practically certain that when the conditions are realized only once, the event A would not occur at all.

He shows that the proposed axioms can be empirically deduced from these premises.

Condition a) is equivalent, in the finite case, to von Mises's axiom 1, which, as we have seen, replaces equipossibility with the frequency of the event. Condition b) is a strong form of Cournot's principle. An event A with zero probability will practically never occur in an individual trial. This does not mean, however, that it cannot occur in a sufficiently long series of trials. For instance, in a fair game of heads or tails, let us consider the theoretical set of all games with an infinite number of tosses. That is possible under Kolmogorov's axiom VI, although he regards this case as 'ideal'. All games with a finite number of tails versus an infinite number of heads—or, on the contrary, a finite number of heads versus an infinite number of tails—will constitute a null set, i.e., a set of events with zero probability. Of course, when the number of games is finite, the games that produce a very small number of heads versus a large number of tails—or the opposite—will have a very low probability. The probability of one tail versus $(n-1)$ heads will thus be $\frac{n}{2^n}$, a value that tends quickly toward zero when n increases. With the zero-measure concept, we can thus clearly specify the meaning of the paradigm that states that a highly unlikely event is virtually the same thing as an impossible event.

That is indeed the paradigm of objective probability formulated above, which Kolmogorov now chose as the basis of his axioms.

Kolmogorov has used, in a large measure, the work of von Mises, but he no longer resorted to the concept of collective. On the other hand, he introduced the substantial contribution of set theory and measure theory by showing the perfect parallel between set theory and the theory of random events. In a 1939 letter to Fréchet (Shafer and Vovk 2005), he noted that a theory based on the concept of finite collectives would rely on notions that cannot be defined in purely formal terms. By contrast, Wald's work, based on infinite collectives, may be defined in a non-contradictory manner, but in this case the relationship to experience is not of the same nature as in other axiomatized theories. Far more than that, Kolmogorov's axiomatics made it possible—along with the concept of recursive function developed by logicians in the 1930s—to mathematize the intuitive notion of effectively negligible set; it also allowed Martin-Löf to provide an exact characterization of random sequences, which are in fact collectives as defined by von Mises (Dellacherie 1978).

As early as 1934, the mathematician Feller[11] was commenting enthusiastically on Kolmogorov's book:

> The calculus of probabilities is constructed axiomatically, with no gaps and in the greatest generality, and for the first time systematically integrated, fully and naturally, with abstract measure theory.

At the 1937 Geneva Conference on the Theory of Probability (Wavre 1938–1939), the best probabilists and mathematicians of the time compared the axiomatics of von Mises and Kolmogorov. They included Bernstein, Cantelli, Cramér, Dœblin, Feller, de Finetti, Fréchet, Glivenko, Heisenberg, Jordan, Kolmogorov, Lévy, von Mises, Neyman, Pòlya, Slutsky, Steinhaus, and Wald. While neither von Mises nor Kolmogorov participated in the event, both submitted papers.

In his opening presentation, Fréchet attacked von Mises's approach:

> Here, therefore, is how we see the distribution of roles in probability theory. After observing as a practical fact that the frequency of a fortuitous event in a large number of trials behaves like the measure of a particular physical constant attached to this event in a particular category of trials—a constant that we can call 'probability'—we deduce, through reasonings of less than absolute rigor, the laws of total and compound probability and we verify them in practice. The possibility of performing this verification means that the looseness of the reasonings used to induce these laws ceases to be of any consequence. Inductive synthesis stops here. We now match these realities (riddled with experimental errors) to an abstract model, the one described in the axiom set, which—unlike the axioms of Mr. de Misès [sic]—offer not a constructive definition of probability, but a descriptive definition [.]

In contrast, Fréchet congratulated Kolmogorov for having put into practice an axiomatics created by Borel by adding countable additivity to classic probability. Most conference participants—apart from Wald, who presented a demonstration of the existence of collectives—endorsed Fréchet's verdict, rejecting collectives and recognizing that Kolmogorov had correctly axiomatized the classic theory of objective probability.

This axiomatics became standard in many objective-probability textbooks that followed Kolmogorov's book. We shall see later that its strength was also recognized in other approaches to probability, whose axiomatization bears at least a partial resemblance to Kolmogorov's.

Let us now examine how this objective approach to probability makes it possible to define the principles of statistical inference that can be associated with it.

1.3 Objectivist Statistical Inference

Before describing the application of objective probability to social science, we must address the complex issue of the interpretation of statistical inference. First, what exactly do we mean by 'statistical inference'?

---

[11] Author of a major work on probability theory (Feller 1950, 1961).

1.3 Objectivist Statistical Inference

Only in exceptional cases do researchers have all the information they would require to reach a definitive conclusion. They therefore need to make the most of whatever information is available. The purpose of statistical inference is to try to reach the most robust conclusion possible by making the best use of this incomplete information on a given phenomenon.

Also, as the data cannot say more than what they are, the transition from incomplete premises to a conclusion that is as specific as possible cannot be achieved without recourse to principles that are inherently extrinsic to the data. Consequently, the inference principle that we shall use will always be somewhat arbitrary, and will inevitably depend on our conception of the nature of knowledge. In particular, the objectivist conception of probability will place a number of constraints on this inference.

Chapter 2 will examine in greater detail the use of the Bayesian theorem to perform this inference under the epistemic-probability hypothesis. For the moment, suffice it to say that the theorem posits prior probabilities of observing the sample under different possible hypotheses, in order to estimate the posterior probability of each hypothesis. The theorem is easy to demonstrate using Kolmogorov's axioms, provided that the prior probabilities have been defined. Now we have already indicated that, for an objectivist probabilist, the 'probability' of a hypothesis is a meaningless notion: it is impossible to define unambiguously a trial that can be repeated in an identical manner and that would produce one of two outcomes: either 'the hypothesis is verified or the hypothesis is not verified.' Moreover, under pain of being dragged into a regression *ad infinitum*, we must introduce the prior probabilities at a certain point—probabilities whose origin cannot be empirical. As a result, when we work on objective probability, it is out of the question to use the Bayesian theorem for drawing an inference from incomplete data.

For this purpose, we must therefore frame the issue of statistical inference in terms more restricted than before. Let us consider a sample of observations, among the set of all those potentially observable in a wider population of observations. That is the population we want to capture in its entirety, but in practice we have only the sample observed. We must therefore extrapolate some of our findings on this sample to the total population.

Statistical inference involves two procedures. First, we need to verify the validity of several hypotheses on the population studied: this procedure is called hypothesis testing. Second, we need to estimate the population's characteristics and conduct statistical tests to see if their effect is significant. The two procedures are actually very similar and closely interlinked: both aim to draw conclusions about the total population from information on the sample alone.

To this end, let us first examine the approach used by Pearson (1900), Student [Gosset] (1908a, b) and Fisher (1923, 1935). We want to determine if a given factor influences the phenomenon under study or not. We shall estimate parameters linking the factor to the phenomenon. At this point, a question arises: can we explain the values of the estimated parameters by chance alone, or does the factor studied also play a role? This 'Type I error' is the one we commit when wrongly rejecting the hypothesis that observations can be explained by chance alone. The authors listed above devised hypothesis tests to verify whether the factor does or does not influence the phenomenon examined.

We can interpret the tests in strictly frequential terms. Suppose a population in which the hypothesis we want to test is true. Let us assume that we draw a large number of samples at random from this population, in the same conditions as the sample already selected. Some of these samples will be very rare, others far more frequent. If the probability of the sample drawn is too low—say, under 0.05—we shall reject the hypothesis. The solution consisting of a large number of draws—not actually performed, but supposed—does indeed enable us to work with probabilities open to a frequential interpretation. At this point, we are no longer examining the probability of a hypothesis, but only the probability of obtaining a particular sample, if the hypothesis is true.

In the wake of the authors above, Neyman and Pearson (1928, 1933a, b) observed that another type of error can occur at the same time, which the 'Type I error' leaves aside. This other type of error, called 'Type II error' is the one incurred by wrongly rejecting the opposite hypothesis, namely, that the observations cannot be explained by chance alone. We must estimate both types of error to obtain a more robust conclusion: when we guard against one, we necessarily increase the probability of the other, if the information remains the same. However, we can see that this second risk is far more complex to analyze, for the contrary hypothesis actually comprises an infinity of possibilities of deviations from chance: strictly speaking, therefore, we should compute an infinity of type-two errors. That is why probabilists very often simply assign a low value to type-one error, setting aside the 'Type II error'. In any event, accepting a hypothesis after subjecting it to a statistical test does not mean that we declare it to have been verified, but only that we choose to act as if it were.

Often, the reasoning that we have just used to obtain a frequentist statistical inference from observed data is interpreted incorrectly. Let us take the statement that the 95% confidence interval for an unknown parameter, θ (such as the mean age at first childbirth, in the French 1920 birth cohort, estimated from a representative sample of that cohort), lies between two values θ_1 and θ_2. This appears to indicate that the parameter has a 95% probability of lying in that interval. But that is incorrect, for we can apply the interval only to the parameter's estimation and not to the parameter itself, which is unknown.

We would actually want to answer the following question: what is the probability that the unknown parameter lies in a given interval? But in this case we can only state that, if we draw many samples of identical size and if we build such an interval around the mean of each sample, then we can expect that 95% of the resulting confidence intervals will contain the unknown parameter. That is an answer to a far more complex question than the first, which seemed much clearer and does not actually exist in frequentist theory. The question is the following: if we draw a large number of different samples, N, what is the probability that the unknown parameter is contained in a certain number of the samples, n? As the analysis is often confined to a single sample, we conclude that the question makes little sense.

1.3 Objectivist Statistical Inference

Let us go further and try to establish more specifically what objectivist statistical inference can demonstrate. Suppose we want to see whether a given factor influences the phenomenon studied or not: for example, we want to determine if the fact of being a farmer influences a given sub-population's probability of migrating. The question then becomes whether the differences between the estimated probabilities of migrating for farmers and the rest of the population can be explained by chance alone or whether they diverge significantly. If these probabilities prove to be different at a preset limit, for example at a 1% 'significance level', then we can conclude that they do diverge, since the result observed is hardly likely to have been obtained by chance. We therefore interpret these probabilities here in frequential terms, by imagining a population larger than the one observed, from which we can draw many sub-populations at random, including the one observed. If the probability of the observed sample is too low, we shall reject the tested hypothesis. This finding is consistent with our earlier statement: objectivist statistical inference makes it possible to test the probability of obtaining the observed sample—if the hypothesis is true—but not the probability of the hypothesis itself, which is either true or false (Matalon 1967).

The above is also consistent with Fisher's position (1956):

> This fundamental requirement [of no recognisable subset] for the applicability to individual cases of the concepts of classical probability shows clearly the role of subjective ignorance, as well as that of objective knowledge in a typical probability statement. It has been often recognised that any probability statement, being a rigorous statement involving uncertainty, has less factual content than an assertion of a certain fact would have, and at the same time has more factual content than a statement of complete ignorance. The knowledge required for such a statement refers to a well-defined aggregate or population of possibilities [,] within which the limiting frequency ratio must be exactly known. The necessary ignorance is specified by our inability to discriminate any of the different subaggregates having different limiting frequency ratios, such as must always exist.

This quotation generalizes the difficulty of defining an interval in which an unknown parameter lies, by indicating that frequentists are unable to discern with sufficient clarity the different sub-populations with different limiting frequencies.

Thus, when we work on a sample of a larger population, the results of a statistical analysis of these data enable us, under certain hypotheses, to draw an inference about the behavior studied with regard to a member of the population who lies outside the sample but displays some of the characteristics observed. We must assume that the member resembles the individuals in the sample possessing the characteristics and that he or she belongs to the same subgroup, which cannot be broken down in greater detail either. It is hard to see how to incorporate this individual into such a sub-population.

The situation is even more complex when we are working not on a sample but on the total population. We must assume that this observed population is itself a sample of a 'super-population' (Royall 1970) from which we have been able to draw only one selection, which we therefore need to examine. This is indeed what we do when we try to project the population into the future on the assumption that its behavior will remain identical to its observed behavior. The demographic methods of population projection and micro-simulation use this hypothesis (van Imhoff and Post 1997, 1998).

However, there is no obvious reason why it should hold, and, as the population changes over time, the hypothesis soon becomes irrelevant.

In such cases, statistical inference is possibly only through a postulate of ignorance, i.e., stating that some things are unknown but that the validity of the argument implies that they should not be (Fisher 1958). Later, we shall see that the epistemic approach enables us to address this inference more clearly.

1.4 Application to Social Science

This probabilistic reasoning was applied very rapidly to social science—in particular, to demographic issues—as early as the seventeenth century. However, our aim is not to provide an overview of the application of probability to social science, but rather to give an idea of how the process occurred.

The first use of probability calculus in games of chance, by Pascal and Fermat, foreshadows the application of probability to economics, which, centuries later, would draw on game theory. For instance, von Neumann and Morgenstern (1944) proposed a classic game theory that rested on a few simple principles governing confrontation between players. Economic behavior was accordingly viewed as the choice of a tactic in a situation where the set of possible tactics can be determined. This approach led to many advances in economics, but we shall not discuss it in further detail here.

In our introduction, we noted the use of probability calculus in Graunt's *Natural and political observations* (1662), which prefigured demography and epidemiology.[12] His book is on the objective side of probability. As Hacking (1975) said: 'Graunt's *Observations*, [.], is entirely dedicated to demography and the analysis of stable frequencies'. For example, he tried to compute the probability of dying within a year from the probability of dying within 10 years. Although Henri VIII had introduced parish registers of burials, baptisms, and marriages as early as 1538, Graunt's book was the first to use statistics from them in order to construct chronological tables—with the greatest possible accuracy—of the numbers of persons having experienced these events and to analyze those numbers. Graunt elaborated a method for determining population size from the registers alone, using an indirect calculation method (later called multiplier). He also tried to compile a life table, although

---

[12] Ever since the *Observations* appeared, it has been claimed that their true author was William Petty. Petty himself claimed authorship when applying for a political office in Ireland. Some observers, such as Le Bras (2000), use this argument to prove that demography never was and never will be a science—contrary to the thesis advocated by Graunt's supporters. Rather, because of the possibility that Petty might have founded it, demography should be viewed as a political instrument in the hands of political authorities. Le Bras's contentions—particularly on the determination of the number of deaths—and his attacks on some Graunt specialists hardly allow us to take his demonstration seriously (for more details, see Reungoat 2004).

his observations did not enable him to do so, for he lacked data on the distribution of the population and deaths by age.

However, Graunt's implementation of probabilistic reasoning remained far from perfect. Let us begin by examining in greater detail how he applied it to calculate the probability of dying within a year from the probability of dying within 10 years (see text quoted in the Introduction).

Contrary to Hacking's presentation[13] (1975), this probabilistic reasoning is still far from totally correct. Let us consider, as Hacking does, a population of N persons. If we assume a constant annual probability of dying, written q,[14] the number of survivors in 10 years will indeed be, as Hacking says, $N(1-q)^{10}$, equal here to $\frac{1}{2}N$ according to Graunt's first hypothesis. As a result, the annual probability is $q = 1 - \sqrt[10]{0,5} = 0,067$ and not $\frac{1}{20} = 0.05$ as claimed by Graunt ('I supposed it was the same (an even lay), that one of any 10 might die within one year') and Hacking.[15] Hacking's reasoning actually assumes an equal number of deaths, d, each year and not an equal probability of dying each year. The resulting number of survivors within 10 years is $N-10d$, hence $\frac{10d}{N} = \frac{1}{2}$. But in that case, $\frac{d}{N}$ is not an annual probability of dying.

From the probability above, the London population can be estimated at $N = \frac{10000}{0.067} \approx 150000$. Graunt made a new error by calculating the population with a multiplier of 10 and not 20, as his prior calculation showed him. He obtained a population of 100,000, an estimate 50% lower than the previous one, whereas he should have found a population of 200,000.

It is also interesting to compare this estimate with the one that Graunt gave later in his book, starting from age six, to construct the life table given below on the basis of a 10-year probability of surviving close to 5/8. This led to an annual probability of dying of $q' = 1 - \sqrt[10]{0.625} = 0.046$—again, a different value from the two earlier estimates.

How did Graunt construct the age-specific life table that he proposed later in his book? His first figure was based on the observation 'of 100 quick Conceptions about 36 of them die before they be six years old'.[16] As for the following six proportions

---

[13] Hacking offers the following argument: 'Graunt assumes a uniform death rate, that is, that there is a constant chance p of dying in a given year. If the chance of living 10 years is 0.5, consider a population of size N. The number who survive the first year is $N(1-p)$. The number who survive the second is $[N(1-p)-pN(1-p)]$ or $N(1-p)^2$. The number who survive 10 years is $N(1-p)^{10}=0.5 N$. Now let q be the chance that at least one man in a group of ten dies in a given year; then $1-q$ is the chance that no one dies. This is just $(1-p)^{10}$, which, solving the above equation is 0.5. So, as Graunt says, q is also 0.5'.

[14] We use the standard demographic notation for a probability (quotient), q, which Hacking writes p.

[15] Leibniz (1675) gave the correct solution to this problem.

[16] 'quick Conceptions': 'live births' in modern English.

Table 1.1 Graunt's life table

Viz. of 100 there dies within the first 6 years	36
The next 10 years, or *Decad*	24
The second *Decad*	15
The third *Decad*	9
The fourth	6
The next	4
The next	3
The next	2
The next	1

Source: Graunt 1662

of deceased, he based their determination on 'six mean proportional numbers between 64, the remainder, living at six years, and the one, which survive 76'. This is tantamount to calculating the terms of a geometric progression of root 54 and ratio 5/8 (see commentaries by Vilquin in French translation of Graunt's book, 1997, page 106). He rounds off the figures and arrives at the life table, given in Table 1.1.

As we can see, this table is still far from a true life table, for it is based on hypotheses too crude to be valid. In fact, as noted above, its calculation is based on a constant 10-year probability of dying, of 3/8, leading to the rounded percentages of deaths between ages 6 and 66. At no point does Graunt discuss the validity of this hypothesis.

We may therefore conclude that Graunt's probabilistic reasoning was still highly uncertain and that its demographic hypotheses are equally debatable.[17] As Lodewijk Huygens wrote in a letter to Christiaan Huygens in 1669 (Huygens (1895), correspondence 1666–1669), when they were seeking to estimate mean length of life at different ages: 'I admit that my determination of ages is not totally correct, but there is so little to say that it [*i.e., the error*] is no way considerable, and even less so as the English table, on which we rely, is not of the utmost accuracy either [.]'.

The astronomer Halley (1693) went further in the quest for a more satisfactory life table. He recognized some shortcomings of Graunt's initial calculations. The population at risk was lacking, ages of death were unknown, and immigration to London and Dublin was substantial.[18] Halley decided to use the data for Breslau, a

---

[17] Let us again note his misuse of his own table, when he confuses the number of deceased persons and living persons. He gives the percentage of individuals aged between 16 and 56 as 34%, which is in fact the percentage of deaths. According to his life table, the percentage of living persons is in fact 41%.

[18] Halley's exact words are as follows: 'But the Deduction from those Bills of *Mortality* seemed even to their Authors to be defective: First, In that the *Number* of the People was wanting. Secondly, That the *Ages* of the People Dying was not to be had. And Lastly, That both *London* and *Dublin* by reason of the great and casual accession of Strangers who die therein, (as appeared in both, by the great Excess of the *Funerals* above the *Births*) rendered them incapable of being Standards for this purpose; [.]'

1.4 Application to Social Science

Table 1.2 Halley's first table

7 .	8	9 .	҃14	. 18	. 21	. 27	. 28 .	.	35 ҃.
11 .	11 .	6 .	5½ . 2 . 3½	5	6	4½ 6½	9 .	8 . 7	.7 .

36 .	42 .	45		49 54 .	55 . 56	.	63
8 .	9½ 8 .	9 .	7 . 7 .	10 11 .	9 .	9 . 10 .	12

	70 71 .	72	77	81	84 .	90 91 .	
9½	14 9 .	11 9½	6 .7	.3	.4 .2	.1 . 1 .	1 ⸺

98 .	99 .	100 .
0 .	⅖ .	⅗

Source: Halley 1693

town with far less migration (1,238 births per year versus 1,174 deaths). This enabled him to formulate the underlying hypothesis of a stationary population, later elaborated by Euler (1760). Admittedly, Euler did not use the term, but he did clearly note that 'if every year as many children are born as men die, the number of all men will always remain the same, & there will then be no multiplication'.

From the data over the 5-year period 1687–1691, Halley provided the table of age-specific deaths per year in Breslau: the upper lines show the ages and the lower lines show deaths (Table 1.2).

When no age is shown above a number of deaths, we should assume that this annual figure indicates persons who died between the ages of the previous and following columns. Halley noted that 348 newborns died in their first year and 198 children died between ages one and six. In other words, he aggregated certain ages and therefore smoothed his results, without giving reasons for doing so or for his choice of ages.

Our first observation is that the table is incomplete: there are no deaths between ages 50–53 and 92–97, despite the fact that the figures are based on a table of data by annual age.

Halley then estimated an age-specific life table, from birth to 84; he reconstructed the Breslau population by 7-year age groups, whose sum—34,000—gives an estimate of the total population (Table 1.3).

This table, which should have been derived directly from the previous one, exhibits differences, for Halley again smoothed the data without saying how he did so. For a stationary population, from an estimated 1,238 annual births (n_0), we can determine the population at age x by recurrence, taking $n_x = n_{x-1} + \overline{d}_x$, where $\overline{d}_x$ is the smoothing estimator of deaths between ages $x-1$ and x.

Halley did not use the term 'probability', but he also computed the odds that a person of a given age will live another year:

> […] if the number of Persons of any Age remaining after one year, be divided by the difference between that and the number of the Age proposed, it shews the odds that there is, that a Person of that Age does not die in a Year.

Table 1.3 Halley's second table

Age. Curt.	Perſons.	Age. Curt.	Perſons	Age. Curt.	Perſons	Age. Curt.	Perſons	Age. Curt.	Perſons	Age. Curt.	Perſons
1	1000	8	680	15	628	22	585	29	539	36	481
2	855	9	670	16	622	23	579	30	531	37	472
3	798	10	661	17	616	24	573	31	523	38	463
4	760	11	653	18	610	25	567	32	515	39	454
5	732	12	646	19	604	26	560	33	507	40	445
6	710	13	640	20	598	27	553	34	499	41	436
7	692	14	634	21	592	28	546	35	490	42	427

Age Curt	Perſons.	Age. Curt.	Perſons	Age. Curt.	Perſons	Age. Curt.	Perſons	Age. Curt.	Perſons	Age. Crit.	Perſons
43	417	50	346	57	272	64	202	71	131	78	58
44	407	51	335	58	262	65	192	72	120	79	49
45	397	52	324	59	252	66	182	73	109	80	41
46	387	53	313	60	242	67	172	74	98	81	34
47	377	54	302	61	232	68	162	75	88	82	28
48	367	55	292	62	222	69	152	76	78	83	23
49	357	56	282	63	212	70	142	77	68	84	20

Age.	Perſons.
7	5547
14	4584
21	4270
28	3964
35	3604
42	3178
49	2709
56	2194
63	1694
70	1204
77	692
84	253
100	107
	34000
	Sum Total.

Source: Halley 1693

In fact, this table was based on annual age-specific data supplied by Neumann (Graetzer 1883; Rohrbasser 2002; Bellhouse 2011). We can therefore reconstruct Halley's source table and compare the numbers of annual deaths calculated from different sources: Neumann's table and Halley's death function and life table (Fig. 1.1).

We can see that Halley's first table reflects an initial smoothing of Neumann's data, but uncertainty persists on how Halley proceeded and on the difference between deaths and survivors. Today, we would apply one of two procedures: (1) an aggregation by 5-year intervals—3–7, 8–12, 13–17, and so on—as ages ending in 0 or 5 are favored (which is the case here); (2) moving averages on 5-year intervals.

Moreover, we can fill the gaps in Halley's age-specific mortality table for ages 50–53 and 90–97 from Neumann's table[19] or even the life table. This yields 9.75 deaths and 1 death per year respectively.

Lastly, while the population may seem stationary, it is only approximately so, and modern demographers reject Halley's estimate. For instance, Henry (1957) took an example of the use of Halley's method to note:

> At a time when historical demography is enjoying a revival, we thought it would be useful to give an example of the errors that an obsolete method [Halley's] may entail.

By estimating infant mortality from the ratio of deaths in the first year to total deaths enumerated,[20] Halley would find 265 per 1,000, whereas if he divided the

---

[19] Ignoring Neumann's table, Jaynes (2003), who describes Halley's work in detail, regrets that Halley did not supply his data in more detailed form. However, despite Neumann's detailed table, it is clearly impossible to reconstruct Halley's tables without additional hypotheses.

[20] Interestingly, Neumann distinguished between stillbirths and deaths occurring before the age of 1 year. This would have made it possible to determine separate probabilities of stillbirth and infant mortality.

1.4 Application to Social Science

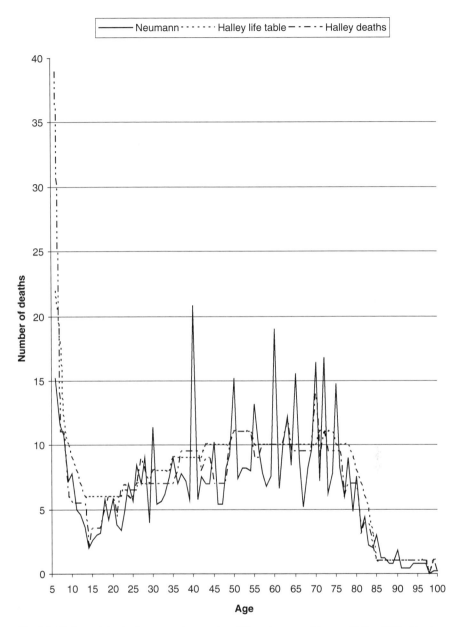

Fig. 1.1 Estimated deaths from age six up, using different sources (Source: Halley 1693; Graetzer 1883; Bellhouse 2011)

first-year deaths by total births, he would obtain a rate of 233 per 1,000—in other words, a 20% overestimation with the first method. By using this method, he would unfortunately be unable to estimate any mortality rates at older ages, because migration exerts an unmeasurable influence.

We can therefore conclude that Halley's method—while an improvement on Graunt's rough procedure—did not allow him to obtain an accurate estimate of mortality in the late seventeenth century.

In fact, Halley, like most researchers in the seventeenth century and the first half of the eighteenth century, had only birth and death statistics at his disposal, which are inadequate for constructing a proper life table. He was missing populations at risk. It was not until 1766 that the Swedish astronomer Wargentin gave a true life table—thanks to the fact that his country compiled population registers, which provide figures for populations at risk, and death registers, which give the numerators of the rates or quotients to calculate. The censuses introduced in the nineteenth century made it possible to generalize life-table compilation.

The eighteenth century saw the rise of political arithmetic in Europe, which led to an ever fuller knowledge of its inhabitants and demographic phenomena. Without discussing this development in detail here, let us highlight some aspects that are most relevant to probability.

Arbuthnott (1710) introduced the first statistical test, applying it to the hypothesis that the number of male and female births is identical. Let us examine more fully the calculations he performed for this purpose. He began with a table listing the annual numbers of baptisms of boys and girls in London from 1629 to 1710. In his article, he sought to demonstrate two propositions, expressed here in modern terms:

1. It is not by chance that the number of baptisms of boys and girls is roughly the same.
2. It is not by chance that the number of baptisms of boys exceeds that of girls, in a constant proportion.

We shall see that his demonstration of the first is incorrect, whereas that of the second leads to an accurate statistical test.

To demonstrate the first proposition, he began with an argument resembling that of Pascal by using the numbers of the arithmetic triangle: for example, if two children are born and the probability of the birth of a boy or girl is one-half, then there is one chance that two boys will be born, one chance that two girls will be born, and two chances that a boy and a girl will be born. When the number of births increases, the number of chances that there will be as many boys as girls divided by the total number of births becomes very small.[21] Naturally, Arbuthnot recognized that the equality of proportion between boys and girls is not mathematically verified but that, even in this case, he noted that it is highly unlikely not to obtain extreme

---

[21] Arbuthnot wrote: 'in the vast Number of Mortals there would be but a small part of all the possible Chances, for its happening at any assignable time, that an equal Number of Males and Females should be born'.

1.4 Application to Social Science

values occasionally. His reasoning was incorrect, as Nicolas Bernoulli promptly showed in a letter to Montmort of October 11, 1712 (de Montmort 1713). Bernoulli's argument is very clear:

> I was forced to refute this argument, & to prove the high probability that the number males and females occurs each year between limits even more narrow than those observed for 80 years in a row.

Indeed, it is not enough to remain vague and invoke the possibility of an extreme result. We must calculate the probability that, after performing a large number of draws, we will obtain the extreme result. We can then verify that the probability of an extreme result can be regarded as physically impossible although mathematically possible, as noted above. Taking the same example as Arbuthnott, Nicolas Bernoulli, writing to Montmort on January 23, 1713 (de Montmort 1713), asserted that

> the probability that among 14,000 children the number of males will be neither greater than 7,363, nor smaller than 7,037, will stand in a ratio greater than $43\frac{58}{100}$ to 1 to the probability that the number of males will fall outside these limits.

In other words, when the number of observations increases, the probability of overstepping narrow boundaries around their mean value decreases sharply. Nicolas Bernoulli added that his uncle Jacques Bernoulli had already demonstrated a similar proposition in his treatise *De Arte Conjectandi*, i.e., the solution to the problem that we described at the start of this chapter.

By contrast, Arbuthnott's demonstration of his second proposition is far more robust as it closely resembles a statistical test.

Arbuthnott states the problem thus:

> A lays against B, that every Year there shall be born more Males than Females: To find A's Lot or the Value of its Expectation.

Arbuthnott formulates his hypothesis as a wager. From his data, the probability of A is obviously below 1/2 for each year observed. Let us suppose, however, that it is 1/2 for a given year. If this hypothesis remains the same for all 82 observation years, its probability will be $(1/2)^{82}$. As a result, we can reject hypothesis A of equiprobability of male and female births. Therefore, this is indeed a statistical test. Arbuthnott does add that he attributes these observed proportions to Divine Providence, but on that point we part ways with him, and refer interested readers to Brian and Jaisson's book, *The descent of human sex ratio at birth* (2007).

The other application of objective probability that we shall now examine concerns epidemiology. It consists of Daniel Bernoulli's analysis (1760) of the effects of inoculation to prevent smallpox. Interestingly, Bernoulli uses Halley's table, discussed above, to 'distinguish, in total mortality, that due to smallpox at all ages'.

It is important to bear in mind that there were no vaccines at the time, and that smallpox was a scourge. Bernoulli wrote:

> A long series of observations shows that smallpox takes away one-thirteenth or one-fourteenth of each generation; [...] and that it takes away one-eighth or one-seventh of those who are ill with it.

However it was said that, as early as the eleventh century, the Chinese practiced inoculation by placing the person to be immunized in contact with the pus of an ill person. The practice spread to countries near the Caspian Sea, particularly Circassia (Voltaire 1734), then to England in 1721 and France in 1723, but Europe did not effectively adopt inoculation until 1750. Meanwhile, a sharp dispute erupted between advocates and opponents of the practice, leading Daniel Bernoulli to perform a full-fledged exercise in probabilistic modeling.

From the outset, Bernoulli stated very clearly the two questions that he would consider in defining his hypotheses:

> The first issue is the risk that people who have never had smallpox incur every year of catching it; the second is the risk of dying from it at different ages when people catch it.

He then showed, with the aid of observations, that his hypotheses were confirmed by 'our notions on this illness & by the results of all the calculations concerning this foundation'. The first hypothesis is that 'as long as one has never had smallpox, one continually runs the same risk of catching it. The second hypothesis is that the risk of dying from smallpox, when one is attacked by it, may well be, in the same year, identical at all ages'.

Having enunciated these hypotheses, Bernoulli could begin to model the phenomenon (a process that we shall not discuss here). This led him to estimate the parameters of his model from the data, which he posited as known. From Halley's table, he deduced the proportion of individuals who had or had not caught smallpox at each age, and the proportion of persons who had died of smallpox and other diseases as well, at each age from 0 to 24. He then calculated (1) the potential size of the population at each of these ages, if no one died of smallpox, and (2) the outcome if persons had a one-in-200 chance of dying after inoculation. Probabilistic arguments led him to conclude, for the latter scenario, that

> the danger of inoculation reduces the average length of life by only one month and twenty days; & notwithstanding this danger, the gain is still three years on twenty-six years and seven months, which is the average length of life in the natural state.

Section 1.5 below provides greater details on the debate triggered by Bernoulli's study—a controversy in which D'Alembert played a very active part.

Let us now move to the most recent period. Population censuses, which appeared in the eighteenth century, came into general use in Europe during the nineteenth century. Their adoption, coupled with vital statistics, would change the use of probability. By collecting exhaustive data on the total population at a given moment, demographers, for example, could now implement an objectivist approach for studying very large numbers of individuals. The variance of estimated probabilities became so weak that demographers would no longer even calculate them. Let us illustrate this with the example of the size of the French male cohort reaching age 60 in 1962 (Pressat 1966): given the number of deaths between ages 60 and 61, D_{60}, and the number of 60-year-olds, $N(60)$, the annual probability of dying is estimated at $q_{60} = \dfrac{D_{60}}{N(60)} = 0.0234$, or 23.4 per 1,000, and the author does not even calculate its variance, which is:

$$Var(q_{60}) = \dfrac{(N(60) - D_{60})D_{60}}{N^3(60)} = 0.000000085, \text{ or } 8.5 \text{ per } 100 \text{ million.}$$

There are only two cases where demographers will need to take these variances into account: (1) the determination of probabilities for shorter periods (for example, monthly ratios) (Hoem 1983), as the number of persons experiencing the event will be much lower, even if measured exhaustively; (2) the analysis of survey data, for topics not covered in censuses or population registers.

Similarly, the sociologist Durkheim (1895), when he sought to identify and isolate what he called social facts, observed:

> But statistics gives us the means to isolate them. They are indeed quantified, not without exactitude, by the birth rate, the marriage rate, and the suicide rate, i.e., by the number obtained by dividing the average annual total of marriages, births, and voluntary deaths by the number of persons old enough to marry, procreate, and commit suicide.

These statistics, which encompass all special cases without distinction, make it possible to neutralize individual circumstances and so to highlight social phenomena, stripped of all extraneous elements. To link these social facts together, Durkheim recommended the method of concomitant variations—in essence, a regression on aggregate data (Courgeau 2004a, 2007a). He clearly stated his position on the subject (see for more details 5.3):

> For [*the method*] to be demonstrative, we do not need to strictly exclude all the variations that differ from the ones we are comparing. The mere parallel pattern of the values assumed by both phenomena, provided that the pattern has been identified in a sufficient number of sufficiently diverse cases, is the proof that a relationship exists between them.

His applications of the method—for example, to suicide rates among Protestants and non-Protestants in Prussia (Durkheim 1897)—show that he is examining a sufficiently long period (1883–1889) to work with significant numbers of suicides and is covering the population of all 14 Prussian provinces. As in the demographic example given earlier, Durkheim worked on the rates, without concerning himself with their variance.

But the possibility of infinite repetition of observations does not exist even in studies using exhaustive data, such as cross-sectional and longitudinal analyses in demography (with data from censuses and comprehensive population registers) or Durkheim's method for explaining social facts in sociology. The reason is that if we try to extend the observation over time, the population will not remain identical and the probabilities of the events observed will thus change. As we shall see in Chap. 5, the introduction of event-history and multilevel analysis methods was to raise this same question even more acutely.

The time has come, therefore, to take a more detailed look at the main problems posed by the objective approach.

1.5 Problems Posed by the Objective Approach

In this chapter, we have sometimes noted various problems encountered in the introduction or application of the objective approach. In this section, we shall try to identify the main problems more specifically and to show that, in fact, they are often interlinked.

We already mentioned D'Alembert in presenting the concept of physical impossibility, but we did not discuss his criticisms of the concept of objective probability, which we shall now examine.

It is important to bear in mind that when objective probability, in the wake of Venn, was regarded as the only valid form of probability, many probabilists violently attacked D'Alembert for his errors in resolving certain probability problems. At the start of his chapter on D'Alembert, Todhunter (1865) wrote:

> This great mathematician is known in the history of the Theory of Probability for his opposition to the opinions generally received; his high reputation in science, philosophy, and literature have secured an amount of attention for his paradoxes and errors which they would not have gained if they had proceeded from a less distinguished writer.

Similarly, Joseph Bertrand, in his preface to *Calcul des probabilités* (1889), comments:

> D'Alembert's mind, usually balanced and subtle, went totally askew when addressing probability calculus.

However Delannoy (1895) showed that D'Alembert's errors in probability were not, as Maupin (1895) thought, omissions but that he emphasized those points strongly enough for us to regard them as genuine reasoning errors.

While some of these criticisms are entirely appropriate, others—leveled by objectivists—take on a new dimension when viewed from an epistemic standpoint.

For instance, in an initial article on probability (1761a), D'Alembert examined whether, if the same event occurs several times in a row, there is an equal probability that heads or tails will turn up in the following throw. He states:

> Therefore the more often *tails* will have occurred consecutively, the more likely *heads* will turn up on the following toss.

An objectivist may reasonably conclude that D'Alembert had committed a true error. But in a later article (1768a), he spelled out what he meant by those terms:

> If chance alone decides the event, *tails* cannot turn up, in my view, many times in a row; that seems proved to me by the reasons I have given above and elsewhere. Therefore if tails turn up many times in a row, say, a hundred times, it is a sign that some special cause is at work to bring up *tails* rather than *heads*; hence we may reckon that, if the cause persists, *tails* will turn up on the one hundred and first toss [...]

This shows more clearly that D'Alembert was reasoning in terms of epistemic probability here. Observing past events, he offered a probability for the outcome of the next toss that took those past events into account. Here, the observation led him to conclude that the player was cheating because the coin was loaded, and so

1.5 Problems Posed by the Objective Approach

D'Alembert responded accordingly. We can see why such reasoning is unacceptable for an objectivist probabilist: to speak of the probability of a person cheating in a game is meaningless to the objectivist probabilist in probability calculus, even if the notion has a meaning in ordinary language. It is difficult to provide an unambiguous definition of a trial that can be repeated identically, and whose outcome would be either 'the player is cheating' or 'the player is not cheating.'

We mentioned the second problem in our section on statistical inference. After conducting a campaign in 1934–1936 against the misuse of the correlation coefficient, Fréchet sent his colleagues at the *International Institute of Statistics*, in 1948, a problem regarding an assessment based on a sample. He obtained 16 different types of response, showing that the leap from inadequate premises to a conclusion cannot be taken without recourse to arguments extraneous to probability.

Later, Kendall (1963), discussing the controversies on statistical inference in his obituary of Fisher, noted that

> a man's attitude towards inference, like his attitude towards religion, is determined by his emotional make-up, not by reason or mathematics.

Despite this, in their quest to define the epistemological bases of their practice, statisticians and probabilists have attempted to spell out—if not justify—the inference principles that they advocated. However, for the objective approach to probability, this justification is nearly impossible because of the many epistemological problems raised by the establishment of a theory of statistical tests aimed at drawing an inference on an observed phenomenon from observed data.

For instance, let us suppose that we perform a statistical test, in the objectivist manner, to determine if an estimated parameter may be regarded as significantly different from zero. One might conclude that the rejection of this hypothesis at the 5% limit, for example, indicates a 5% probability of the hypothesis being verified. However, as noted earlier, the objectivist approach cannot address the probability of a hypothesis.

In fact, what we can state here is that, if we perform a large number of identical trials, then the parameter studied would differ from zero in only 5% of the trials. But as a rule, especially in social science, the trial can be repeated in identical conditions only a few times, and often not at all. This answer is therefore very different from what we should expect in a statistical test.

If we cannot test a hypothesis, we must therefore content ourselves, in objective probability, with testing the probability of obtaining a given sample if the hypothesis is true. Statistical inference is defined here in a narrower sense, so as to avoid speaking of the probability of a hypothesis.

The third problem is the application of objective probability to social science. While objective probability does seem applicable to games of chance and to the measurement of stable physical phenomena—such as the speed of sound and light—over a very long period, we may question its suitability for human phenomena, which are far less predictable and stable.

It is therefore curious to note that probability and social science were born nearly simultaneously and that probability was applied to the study of human phenomena from the outset, as Graunt showed us. Arguably, there is considerable evidence that, at the time, the belief in a divine order made it possible to assume that human phenomena were as stable as physical phenomena. For instance, Arbuthnott (1710) concluded his earlier-quoted study by stating that it was not by chance that the number of male births exceeded the number of female births, in constant proportion, and that this proved the existence of a *Divine Providence*. Similarly, Süssmilch (1741, 1761–1762) widened his observation to many demographic phenomena for the sole purpose of revealing the existence and nature of the *Divine Order*. Studying mortality (1761–1762), while showing that 'the difference between towns and villages lies in the manner of feeding, habits, and the way of life', he concluded: 'If habits and ways of life were similar everywhere—as nature is—mortality would be so as well.'

Süssmilch brought in mythological thought to offset the lack of an explanation for the observed regularities. But his approach was rapidly superseded in the social sciences by a more thorough investigation of the economic, political, religious, social, and other causes of these phenomena.

Social scientists were forced to admit the close and complex links between these various aspects of human life and their extreme variability across time and space.

How, in these conditions, could one continue to apply the objective approach, which supposes that such phenomena can be repeated identically across time and space?

We have already noted how cross-sectional analysis, by using exhaustive census data, and longitudinal analysis, which adds the equally exhaustive data from population registers, allowed population scientists to set aside the study of variance and confine themselves to studying the mathematical expectations of the rates and probabilities used. But once event-history analysis sought to explore in greater detail the effect of a large number of characteristics on the behavior studied, it became necessary to deploy the full probabilistic arsenal in order to conduct this research. In particular, variance became a central topic again (Courgeau and Lelièvre 1989, 1992, 2001). Multilevel analysis (Courgeau 2003, 2004a, b, 2007a), which additionally introduces a large number of regions or units of different levels, naturally intensified the need for epistemic analysis, for the use of an objective approach to probability became a pressing issue.

To conclude, let us examine D'Alembert's eleventh paper (*mémoire*) (1761b) on the inoculation of smallpox, discussed by Daniel Bernoulli (1760) and already described earlier.[22]

First, D'Alembert did not reject Bernoulli's conclusions, and he ended the first part of the text by stating that

> these advantages are real for those who will undergo it [inoculation] with appropriate precautions; we must therefore carefully avoid halting or delaying its spread.

---

[22] D'Alembert returned to the subject in his twenty-third and twenty-seventh '*mémoires*' (1768a, b).

1.5 Problems Posed by the Objective Approach

However, he did point out that this approach made it possible to measure 'the advantage that the State can draw from inoculation but hardly addressed the advantage that individuals may hope to obtain from it'. In support of this statement, he noted that individuals

> see inoculation as an instant and imminent peril of losing their lives in a month, & smallpox as an uncertain danger, & whose place in a long life cannot be assigned.

Thus D'Alembert did draw a contrast between an objective approach to probability—as we defined it in this chapter—and a subjective approach that would seek to apply probability calculus to the largest possible number of feelings of uncertainty, however subjective they may be.

The time has come, therefore, to engage in a deeper examination of the other side of probability. We shall thus be able to consider propositions that are more subjective than events repeatable to infinity in identical conditions. In other words, we must turn to the realm of epistemic probability.

Chapter 2
The Epistemic Approach: Subjectivist Interpretation

Unlike the objectivist approach, which applies probability calculus to events liable to repeat themselves in identical conditions, the epistemic approach seeks to apply probability calculus to the largest possible number of feelings of uncertainty, however subjective they may be. The goal, however, is not only to extend the scope of application of the calculus by taking into account the most diverse categories of human knowledge, especially of non-frequential origin. The epistemic approach also aims to change the mode of reasoning and the premises on which objective probability is based, starting from a very different paradigm—as we shall see later.

As already noted, both approaches were used practically from the introduction of probability calculus. But two very different paths were soon taken to explore epistemic probability.

Pascal had already used an initial subjectivist approach in the section of the *Pensées* entitled *Infini rien* ('Infinite nothingness') (1670), cited in our General introduction. The object of Pascal's wager was no longer a repeatable event as in a game of pure chance, but a unique event: the existence of God. Thus he could no longer rely on the constancy of the chances of coming up heads in a coin-toss game, for example. His wager had to be made in the absence of all prior experience. This is, of course, an extreme approach, for probability theory usually deals with situations where the future is uncertain, and where data on prior experience are available. Ramsey (1926), de Finetti (1937), Savage (1954), and other authors elaborated on this approach in the twentieth century.

By contrast, Leibniz took a distinctly logicist approach when seeking the origin of probability in logic. In *New essays on human understanding* (written around 1703 but only published in 1765), he wrote:

> I believe the study of the degrees of probability would be very valuable and we do not yet have such a study, and this is a serious shortcoming in our logic textbooks.

Hence the need to try to develop an alternative to classic deductive logic—an alternative introduced by Keynes (1921), Jeffreys (1939), Richard Cox (1961), and others in the twentieth century. Chapter 3 will discuss this development. Here, we shall focus on the more subjectivist approach.

We need, therefore, to define probability not only for objective events, such as a dice toss or a card draw, but also for totally subjective events, such as the fact that a player cheats or any other hypothesis. As a result, we can no longer rely on the frequency of identically defined events, since the event may be non-repetitive. After describing the origins of this approach, the present chapter identifies its underlying paradigm and axioms, discusses the resulting inference, and gives examples of applications to social science.

2.1 Subjectivist Probability

In the seventeenth and eighteenth centuries, probabilists had not yet drawn the sharp divide between objective and epistemic probability. That was accomplished in the nineteenth and twentieth centuries, when Venn, von Mises, Fisher, and others rejected the use of probability calculus for totally subjective events. For this reason, many of the seventeenth- and eighteenth-century authors quoted in this chapter also appeared in Chap. 1, and some will return in Chap. 3.

As early as the seventeenth century, a number of authors proposed the application of probability calculus to the degree of certainty of a legal ruling, an insurance policy, an annuity, a witness's testimony, and so on. Arnauld and Nicole (1662), for instance, speak of the judgment 'that we make with regard to the truth or falsehood of human events', particularly when we 'consider them in the time to come'. To judge the truth of such an event, they said:

> we must take heed of all the attendant circumstances, both internal and external. I use the term 'internal circumstances' to refer to the circumstances inherent in the fact itself and 'external circumstances' to those that concern the persons whose testimony leads us to believe that fact.

Arnauld and Nicole accordingly used this approach to examine the belief in miracles as well as notarized deeds, future contingent events such as the positive or negative outcome of diseases, future events in a war, the loss of one's life or property, fear of thunder, and so on. For instance, to judge whether a contract signed by two notaries has been ante-dated, Arnauld and Nicole observe that:

> of one thousand contracts, there are nine hundred ninety-nine that are not ante-dated; it is thus incomparably more likely that this contract that I see is one of the nine hundred ninety-nine than the one in a thousand that may be ante-dated.

If we also know the honesty of the notaries who signed the contract, we can be very certain that it is not ante-dated. But if, instead, we know that it might have been in these notaries' interest to falsify the date, the circumstance lessens the degree to which we can believe that the document is properly dated, even though we cannot

2.1 Subjectivist Probability

conclude with certainty that the contract is ante-dated. This marks the appearance of the epistemic interpretation of probability as a *degree of certainty*. In their last chapter, Arnauld and Nicole even supply a measure for roughly estimating the probability of a risk such as dying by thunder:

> but if it is only the danger of death that fills them with their extraordinary fear, it is easy to show that this is unreasonable. It would be an exaggeration to say that one in two million people is killed by a thunderstorm; there is scarcely any kind of violent death less common.

Thus the authors greatly minimize an excessive fear displayed by a very large number of people.

At the same time, Arnauld and Nicole believe that, in making a decision, we should consider not only the degree of certainty of an event, but also what they describe as its advantage (*avantage*), now called its utility. Taking the example of a lottery, they write:

> That is what attracts so many people to lotteries: winning, they say, twenty thousand *écus* for one *écu*, is that not a most advantageous thing? Everyone believes he will be that happy person who will collect the jackpot; and nobody will reflect on the fact that if the jackpot is, say, twenty thousand *écus*, each person will perhaps be thirty thousand times more likely not to win it than to win it.

To decide on the best course of action, we must therefore take into account not only the degree of certainty of a gain but also its advantage. This marks the emergence of the link between probability and utility—one of the basic concepts of subjective probability.

In an unsigned paper, Hooper (1699) went further and introduced human testimony into the *credibility* calculus. He gave the example of the calculation required to assess concordant accounts by several witnesses:

> if the First Witness gives me $\frac{a}{a+c}$ of Certainty, and there is wanting of it $\frac{c}{a+c}$; the Second Attester will add $\frac{a}{a+c}$ of that $\frac{c}{a+c}$; and consequently leaves nothing but ... $\frac{c^2}{(a+c)^2}$.

Hooper was thus able to determine the degree of credibility of concomitant information, and to show that it tends toward unity when the number of witnesses increases. Conversely, he showed that the credibility for information that is not concomitant but sequential will tend toward zero when the number of witnesses increases. He notes that:

> I therefore suppos'd to have $\frac{a}{a+c}$ of Certainty from the First Reporter; I shall have from the second $\frac{a}{a+c}$; from the third $\frac{a^3}{(a+c)^3}$.

In a later section, we shall discuss the interpretation of these results with the aid of the *belief function*, introduced by Smets (1988) and others.

Jacob Bernoulli (1713) offered a more detailed reasoning on degrees of certainty. For him, chance does not reside in the external world, for he clearly states:

> All things that exist or are done under the sun, in the past, present, and future, always carry the greatest certainty in themselves and objectively so.[1]

Rather, chance resides in the imperfection of our grasp of this world:

> An opinion of certainty that we can voice is not identical for all things, but varies in many ways, toward either a greater or lesser degree. ... The other things cannot be measured by our mind except imperfectly, more or less precisely depending on their probability, which teaches us that something is, will be, or has been[2]

Bernoulli goes on to state that 'probability is indeed the degree of certainty, and it differs from certainty as the part does from the whole'.[3] He is therefore describing a full-fledged epistemic probability, a degree of certainty or belief, which he will try to measure—a process that he calls 'to conjecture' (*conjicere*).

To perform such a measure, he posits nine axioms that such a conjecture must satisfy. Without describing them all in detail here, we shall present the fourth, as it provides the bridge between 'universal things' and conjectures on specific events. This shows us how strongly Bernoulli believed that probabilities are applicable not only to objective events—such as those occurring in games of chance—but above all to subjective events of any kind. His demographic example is most eloquent:

> For instance when we search, in the abstract, how much more likely it would be for a youth of twenty to outlive an old sexagenarian, rather than the latter outliving the former, there is nothing you can take into account apart from their difference in age and their years; but when the discussion specifically concerns young man Peter and old man Paul, you need once again to pay careful attention to their particular constitution and their likings, which determine how the two take care of their health; for if Peter is more ill, if he indulges in passions, if he lives an intemperate life, it is conceivable that Paul, despite his older age, may yet be able to contemplate a longer life expectancy.[4]

In the classic demographic approach using objective probability, the only criteria for distinguishing between any two members of a population were their ages and their age gap. Bernoulli's example above implies that the classic approach ceases to apply when we examine two specific persons many of whose other characteristics—

---

[1] Omnia, quæ sub sole sunt vel fiunt, præterita, præsentia, præsentia sive futura, in se & objectivè summam semper certitudinem habent.

[2] Certitudo rerum, spectatata in ordine at nos, non omnium eadem is, sed multipliciter variat secundùm magis & minus. ...Cætera omnia imperfectiorem ejus mensuram in mentibus nostris obtinent, majorem minoremve, prout plures vel pauciores sunt probabilitates, quæ suadent rem aliquam esse, fore aut fuisse.

[3] Probabilitas enim is gradus certitudinis, &ab hac differt ut pars à toto.

[4] Ita cùm qæritur in abstracto, quantò sit probabilius, juvenem vigenti annorum senem sexagenario fore superstitem, quàm verò hunc illi, præter discrimen ætatis & annorum nihil is, quod considerare possis; sed ubi specialiter sermo is de individuis Petri juvenis & Pauli senis, attendere insuper opportet ad specialem eorum complexionem & studium, quo uterque valetudinem suam curat; nam si Petrus sit valetudinarius, if infectibus indulgeat, if intepemperanter vivat, fieri potest, ut Paulus, etsi ætate provectior, optima tamen ratione longioris spem vitæ concipere valeat.

2.1 Subjectivist Probability

apart from age—are very familiar to us. Significantly, Bernoulli indicates that the characteristics are to be included in the analysis 'only if they can be acquired.'[5] If so, they will improve our estimation of the chances that one individual will outlive the other. A few pages later, Bernoulli describes how, from trials performed on people who resemble one another as closely as possible, we can extract more specific information on a given person's probability of survival:

> if, for example, in a test conducted on three hundred men resembling Titius, of identical age and constitution, you observed that two hundred of them had already died before the exact age of ten, you could conclude more surely that Titius is twice as likely to die before age ten as he is of living beyond that limit.[6]

Bernoulli therefore believed that, by testing a large number of individuals (here, 300) he could obtain a rough estimate of the unknown subjective probability that an individual (here, Titius) will survive beyond age ten.

While noting that no mortal will ever be able to determine the number of diseases, accidents, and other potential causes of human death, Bernoulli argues that, by observing a large number of similar cases, we can extract this probability with a precision proportional to that number. In fact, he acknowledges that Arnauld and Nicole (1662) already proposed this method, but he elaborates it, first by positing what would later be called the *principle of insufficient reason*,[7] then by seeking to estimate what we now call a confidence interval. Bernoulli notes that, to estimate a probability: 'All cases are equally possible, i.e., each can occur as easily as any other;'.[8]

He can thus assign an epistemic probability to a fact about which he knows the various arguments for or against its existence. Later, we shall examine the extensions of this principle and the criticisms that have been voiced against it.

Elaborating on the estimation of a confidence interval, Bernoulli assumes from the outset that the probability of the event studied is known to the author but not to the experimenter:

> in a given urn I place three thousand white tokens and two thousand black ones, these numbers being unknown to you, and to determine the number by experiment you remove one token after another (replacing each token as you remove it, before choosing the next one, so that the number of tokens in the urn remains constant) and you observe how many times a white token comes out and how many times a black one comes out.[9]

---

[5] si modo haberi possunt.

[6] si ex. gr. facto olim experimento in tercentis hominibus ejusdem, cujus nunc Titius is, ætatis & complexionis, observaveris ducentos eorum ante exactum decennium mortem oppetiisse, reliquos ultravitam protraxisse, satis tu colligere poteris, duplo plures casus esse, quibus & Titio intra decennium proximum naturae debitutm solvendum sit, quàm quibus terminium hunc transgredi possit.

[7] This designation allows the principle to be contrasted with Leibniz's *principle of sufficient reason*, which posits that for each fact there exists a sufficient reason to explain why it occurs and not another. Keynes, who was dissatisfied with the term, renamed it the *indifference principle*.

[8] omnes casus æquaè possibiles esse, seu pari facilitate evenire posse;

[9] pono in urna quadem te inscio reconditos esse ter thousand calculos albos & bis thousand nigros, teque eorum nyumerum experimentis exploraturum educere calculum unum post alternum (reponendo tamen singulis vicibus illum quem eduxisti, priusquam sequentem eligas, ne numerus calculorum in urna minuatur) & observare, quoties albus & quoties ater exeat.

It is precisely with respect to this probability—unknown to the experimenter—that Bernoulli then determines what we now call a confidence interval 'between two limits, which we can reduce as much as we want'.[10] Using current notations, if p is the unknown probability, he can calculate the number of observations n needed to obtain a confidence interval ε such that the estimated value, $\hat{p}_n = \frac{m}{n}$ (where m is, for example, the number of draws of a white token divided by the total number of trials n), lies within the interval $[p-\varepsilon, p+\varepsilon]$. This does indeed assume the initial hypothesis of our imperfect grasp of a world that is totally deterministic. An increasingly precise observation of that world should enable us to reveal all of its mechanisms and—returning to the previous example—to compute with a growing accuracy the probability that Titius will live beyond 10 years. But as the experimenter, in this case, does not know the reference value p, the confidence interval thus determined is of little use to him.

Bernoulli's theorem allows what we call a *direct approach* to probability—which is, in fact, the one adopted by his predecessors—and allows an accurate quantification of probability. The approach assumes that the probability of the event studied is known, and shows how through successive trials the estimated frequency tends toward that probability. One example is fair games, where we can determine *a priori* the probability of the various outcomes considered. By contrast, the approach is not applicable to subjective phenomena.

The problem that constitutes the *inverse approach* is the one that Bernoulli thought he had solved by exploring an ever greater number of cases. However, Leibniz challenged the solution by noting, for example, that new diseases could spread—making the estimation of mortality not perfect but variable in the future (Bernoulli and Leibniz 1692–1704). This time, all we know is the sample observed. Not only is the population from which it is drawn unknown, but its very existence is a hypothesis. In such circumstances, can we estimate the probability of the event studied? It is this question that Bayes (1763) tackled from a specific angle and that later investigators, such as Condorcet and Laplace, tried to generalize to more complex cases.

Beforehand, let us see how the notion of utility—already glimpsed by Arnauld and Nicole (1662), and one of the bases of subjective analysis—gradually developed from the *Saint Petersburg paradox*[11] in the eighteenth century.

Nicolas Bernoulli, in a letter to Montmort (1713) is the first to have formulated the following problem:

> A promises to give an écu to B, if with an ordinary die he turns up a six at the first throw, two écus if he turns up a six at the second, four (2^2) écus if he turns it up at the third throw, eight écus (2^3) if he turns it up at the fourth, & so on. The question is: what is B's expectation.

A simple calculation shows that we need to gamble an infinite sum to ensure a fair game. But any sane individual will refuse to play such a game because the initial

---

[10] binis limitibus conclusam, sed qui tam arcti constitui possunt, quam quis voluerit.

[11] This paradox owes its name to the fact that Nicolas Bernoulli's cousin Daniel Bernoulli published a paper on the problem in the *Commentaires de l'Académie des Sciences de Saint-Pétersbourg* in 1738.

2.1 Subjectivist Probability

wager is too high. This situation is called the *Saint Petersburg paradox*. The problem was revisited by Daniel Bernoulli (1738), who believed that not all persons can use the same rules to assess a game. To allow for this fact, he introduced the concept of utility (*emolumento*), which he described as resting on:

> the value estimated not from the price of the thing but from the *utility* that each person can derive from it. The price estimated from the thing itself is the same for all; the *utility* depends on each person's circumstances.[12]

Utility introduces a probability that differs for each individual. The concept later proved very important in the theory of subjective probability as well as in economic and financial theories of risk aversion. Daniel Bernoulli posited the following rule:

> By multiplying the utility of each lot by the number of outcomes in which it is obtained, summing the products and dividing by the total number of outcomes, we will unfailingly obtain a mean *utility* and the gain associated with this *utility* will be equal to the value of the risk in question.[13]

Mean utility thus provides a measure of the risk. This notion was later profitably used in the subjective approach, most notably by Savage (1954), who recognized his debt to Daniel Bernoulli in no uncertain terms:

> it is Bernoulli's formulation together with some of the ideas that were specifically his that became popular and have had widespread influence to the present day.

Let us now examine the highly innovative approach introduced by Bayes. At the very start of his paper, he states the problem clearly:

> *Given* the number of times in which an unknown event has happened and failed: *Required* the chance that the probability of its happening in a single trial lies somewhere between any two degrees of probability that can be named.

He thus sets out to predict the occurrence of a trial on the basis of a finite number of similar trials, which can be very small.

Bayes begins with the simple case of a flat square table. First, he casts a ball W, which determines an initial stopping point. He then casts a ball O, n times; the ball lands m times to the right of W. This yields an estimation of the frequency $\hat{p}_n = \dfrac{m}{n}$. Bayes assumes that W's position on the table is uniform. We shall not go into the details of his demonstration, which ultimately shows that the probability that W's projection on one side of the square lies in the interval $[\hat{p}_n - \varepsilon, \hat{p}_n + \varepsilon]$ is equal to[14]:

$$\frac{2(n+1)!}{m!(n-m)!} \hat{p}_n^{\,m} (1-\hat{p}_n)^{n-m} \varepsilon$$

---

[12] *valor non is aestimandus ex pretio rei, sed ex emolumento,* quod unusquisque inde capessit. Pretium ex re ipsa aestimatur omnibusque idem is, *emolumentum* ex conditione personae.

[13] Cum *emolumenta* singula expectata multiplicantur per numerum casuum, quibus obtinetur aggregatumque productorum dividitur per numerum omnium casuum, obtinebitur *emolumentum* medium, and lucrum huic emolumento respondens aequivalebit sorti quaesitae.

[14] Bayes, in fact, seeks the more complex probability that the sought-for probability lies in an interval $[b, f]$. He thus obtains an integral relative to $\hat{p}_n$, between b and f, of the equation below divided by 2.

Bayes, therefore, effectively obtains an interval around the estimated probability $\hat{p}_n$, in which the sought-for probability must lie, while Bernoulli built an interval around the unknown value of p. This time, the interval is perfectly usable by the experimenter.

Bayes then generalizes his result in a scholium in which he considers a random event whose probability he does not know, but on which he has performed n trials, resulting in m positive outcomes. Bayes notes:

> concerning such an event I have no reason to think, that, in a certain number of trials, it should rather happen any one possible number of times than another.

It is indeed the number of trials producing the event, not its unknown probability,[15] that Bayes regards as being uniformly distributed. He can thus regard the previous formula as equally valid for random events.

Laplace, in his 1774 paper[16] on the probability of causes, generalized this principle of inverse probability to any given number of different causes:

> If an event can be produced by a number n of different causes, the probabilities of the existence of these causes given the event stand with respect to one another as the probabilities of the event given these causes, and the probability of the existence of each is equal to the probability of the event given this cause, divided by the sum of all the probabilities of the event given each of these causes.

We can express this principle more concisely. Let E be an observable event and $\{C_1, C_2, \ldots, C_n\}$ the set of its causes. Let us assume that we know the probabilities of E for each cause, C_i. If we view all the causes as equally likely, the probability of C_i, knowing E, is:

$$p(C_i|E) = \frac{p(E|C_i)}{\sum_{j=1}^{n} p(E|C_j)} \tag{2.1}$$

---

[15] A number of authors (Pearson 1920; Fisher 1956; Hacking 1965) have not properly understood this hypothesis advanced by Bayes; they believe that it consists of Laplace's *principle of insufficient reason*, which we shall examine later. Laplace's principle states that, when no information exists, it is the unknown probability that we must regard as uniformly distributed. In this case, we would also be unaware of any monotonic function of the unknown probability, which would yield different results (see Stigler (1986) for a fuller discussion of this hypothesis).

[16] Interestingly, Laplace does not seem to have been aware of Bayes's work at that date, for the introduction to his paper (written by Condorcet) does not mention Bayes. By contrast, 4 years later (1778), Laplace's introduction by Condorcet quotes Bayes and Price, who published his results in the *Philosophical Transactions*.

2.1 Subjectivist Probability

That is exactly what Laplace demonstrated (1783a), clearly designating the hypothesis that all the causes are equally possible (this hypothesis was already mentioned in note 15 as the principle of insufficient reason[17]):

> we shall obtain the probability of a cause, determined from the event, by dividing the probability of the event, given this cause, by the sum of all the similar probabilities.

In the same paper, he generalized the formula to the case where the events C_i are not equally possible but have different probabilities:

> If the values of x, regarded independently of the observed result, are not all equally possible, but if their probability is expressed as a function z of x, we need only replace y by yz in the preceding equations; this is the same as assuming that all values of x are equally possible, and considering the result observed as comprising two independent results, whose probabilities are y & z.

If we write this function $p(C_i \mid I)$, where I is the information available on the events, the previous equation applied to the present case becomes:

$$p(C_i \mid E, I) = \frac{p(E \mid C_i) p(C_i \mid I)}{\sum_{j=1}^{n} p(E \mid C_j) p(C_j \mid I)} \quad (2.2)$$

However, Laplace himself clearly noted the constraints to which both formulas are subjected (1814):

> But that [*the first equation discussed here*] assumes that the various cases are equally possible. If they are not [*i.e., if the second equation applies*], we shall begin by determining their respective possibilities, whose fair evaluation is one of the most delicate points of the theory of chance.

He visibly confined himself to indicating the limitations without actually offering a means to escape them. In fact, Laplace had no principle for determining these *prior* probabilities in the case where the information available does not allow us to make the various outcomes equally possible. Admittedly, his determinism led him to believe that probability is related partly to our ignorance and partly to our knowledge, but that all phenomena can eventually be explained in full.

Poisson continued on this path by applying epistemic probability calculus to judging crimes (1837), as did Bienaymé (1838). However, afterwards many statisticians rejected this method, sometimes violently, recommending a strictly objective approach to probability calculus (Ellis 1849; Boole 1854; Venn 1866). In particular, they refused Bayes's formula, which assumes a uniform prior

---

[17] This hypothesis is therefore different from Bayes's hypothesis, namely, that it is the number of trials leading to the event that is regarded as uniformly distributed and not its probability. Many authors criticized Laplace's hypothesis (Edgeworth 1885a; Fisher 1922a, 1956), arguing that other monotonic distributions of p, for example $\frac{1}{2} Arc \cos(1 - 2p)$, could be equally suitable and yield different results. We shall discuss the hypothesis in greater detail at the end of the chapter.

probability in order to estimate from observations the probability of the event studied. In fact, what they rejected as strictly meaningless was the probability of a hypothesis—here, the hypothesis of a uniform distribution—as well as the notion of probability as a description of a state of our knowledge. For these critics, the term 'probability' merely designates the frequency of an event in a given trial, as we showed in Chap. 1.

This rejection lasted nearly a century, although some researchers continued to support Laplace's position (Pearson 1920, 1925) without, however, using it in specific applications. Jeffreys (1939), for instance, had this to say about Pearson:

> The anomalous feature of his work is that though he always maintained the principle of inverse probability, and made this important advance, he seldom used it in actual applications, and usually presented his results in a form that appears to identify a probability with a frequency.

Toward the 1930s, a number of authors revived this epistemic argument, which was now channeled toward the subjectivist approach (by Ramsey, de Finetti, Savage, and others) or the logicist approach (by Keynes, Jeffreys, Carnap, Jaynes, and others). As noted earlier, this chapter will discuss the subjectivist approach, while Chap. 3 will focus on the logicist approach.

In an essay published after his death at 26, Ramsey (1931) laid the foundations of a subjective approach to probability, while Kolmogorov axiomatized objective probability in 1933. Ramsey introduced the notion of *degree of belief* in the occurrence of an event and sought to formalize the way in which people may define belief when making decisions in uncertain situations. But it was de Finetti who defined the true principles of the subjectivist approach.

For this purpose, he considered a relationship of subjective comparison between different events and sought the conditions that would enable us to transform the comparison into a measure of probability (de Finetti 1931a). For example, in a tournament between several teams, in order to calculate the probability of victory that a person may assign to either team, de Finetti showed the need to assume that the person is coherent in his or her choices. De Finetti saw this coherence condition as the sole principle from which he could deduce all of probability calculus:

> this calculus accordingly emerges as the set of rules with which a given person's subjective assessment of the probabilities of various events must comply in order to avoid a fundamental contradiction between them. (de Finetti 1937)

It is important to understand that, for de Finetti, an event is always unique, and that if he is examining several trials, he will speak of trials for a single phenomenon. Likewise, people are free to adopt whatever assessment of probability they prefer, provided that the assessment satisfies the property of coherence.

Let us now see in greater detail how he defines subjective probability on the basis of this notion, which is common to all subjectivists, although authors differ slightly as to what they mean by coherent behavior. For the sake of clarity, De Finetti introduces the notion of wager and looks for what a perfectly rational person would be willing to accept concerning the occurrence of a certain event. The probability p attributed by a person to an event E is defined as the ratio of the sum that

2.1 Subjectivist Probability

(s)he would be willing to bet, pS, to win a sum S, if E occurs. Using this definition and the coherence condition, de Finetti shows that the probability of a random event must be non-negative and less than unity. He generalizes this result to a random number of incompatible events, in order to show that the sum of their probabilities must equal unity. He goes on to the determination of conditional probability, and—again for consistency's sake—deduces the Bayesian theorem from conditional probability.

De Finetti thus shows that subjective probability is suited to a wide range of events to which objective probability did not apply, such as sports results, weather events, and political events. But subjective probability is also appropriate for events for which objective probability applies. For example, in the theory of games of chance,

> the symmetry characteristics displayed by the various 'possible outcomes' can force our mind to view them as equally possible, but not to *impose* such an evaluation of probability by logical means. (de Finetti 1937)

For instance, if we toss a loaded coin, the observation of successive trials will allow us to adjust our assessment of the chances of coming up heads.

Another important component of this theory is the notion of 'equivalence', now called 'exchangeability'.[18] We shall examine it in greater detail in the section on subjectivist statistical inference. Let us note here that it makes it possible, for example in a game of heads or tails, to replace an unclear definition of 'independent events with fixed but unknown probabilities' with that of 'equivalent events' (de Finetti 1937), now called exchangeable events. This notion applies to the assessment of the probabilities of individual events and allows a clear response to statistical inference, as we shall see later.

Lastly, de Finetti offered an axiomatization of subjective probability, which we shall examine in the next section. This work was continued by Savage (1954), who linked subjective probability to game theory (von Neumann and Morgenstern 1944) in economics. Savage extended de Finetti's studies by taking fuller account of the behavioral aspects of decision-making. He introduced a utility function, which reflects a person's risk aversion, and developed a methodology for maximizing mean utility. Suppes and Zanotti (1982) supplemented these axioms by constructing a measure of probability for a single distribution.

None of these subjective theories challenged the additivity principle for probabilities of independent events, which was also valid for objective probability. A generalization of this approach is found in the Dempster-Shafer theory of degrees of credibility, which specifically waives this condition.

The Dempster-Shafer theory was foreshadowed by Hooper's rules of credibility of successive or concurrent information (1699), described at the beginning of this

---

[18] In his 1937 article, de Finetti used the term 'equivalent events', but the designation was not kept in later publications dealing with the concept. Today, the term 'exchangeable events' is used in all of the literature, even though some authors do not regard it as perfect.

chapter. Shafer (1986) shows that the Bayesian interpretations of these rules by Gärdenfors et al. (1983) are open to criticism and that an interpretation in terms of degrees of confidence is the only way to understand them properly. The same applies to Lambert's generalization of these rules (1764), which led Dempster (1967) to define the notions of upper and lower probability[19] and the combination rule for confidence functions.

Starting from an initial space X where a classic measure of probability exists, Dempster assigns to each element of the space one or more elements of a new space, S. He shows that we can define an upper probability and a lower probability for any element T in S—probabilities whose various properties he describes. He then shows that these notions apply when we try to combine different sources of information on a given phenomenon. Suppes (1974) uses these results to offer a better axiomatization of subjective probability, examined in the next section.

Following in this path, Shafer (1976) argues that this belief should be measured on two levels and not on a single one, as additive theories assume: that of belief or support, and that of plausibility, where these degrees of confidence are used in decision-making. This ties in with earlier work by Choquet on the theory of capacities (Choquet 1953; Shafer 1979). The postulate that propositions need to be well defined—i.e., that we must be able to determine if a proposition is true or false—is not applied here, for there may exist well-defined propositions to which no degree of belief is assigned.

The difference between probability functions and belief functions can now be stated clearly:

> Whereas probability functions assume belief is apportioned to the points of the frame Θ, belief functions allow basic probability numbers (or mass numbers) to be assigned to whole sets of points in Θ without further subdivision. (Yager and Liu 2007)

With these mass numbers we can define the belief for a set A, $Bel(A)$, as the sum of all the masses of subsets B of this set:

$$Bel(A) = \sum_{B | B \subseteq A} m(B)$$

and the plausibility of the same set, $Pl(A)$, as the sum of all the masses of the sets B that intersect A:

$$Pl(A) = \sum_{B | B \cap A \neq \Phi} m(B).$$

The relationship between the subjective probabilities of an event A and its opposite, $\bar{A}$, under the state of knowledge C:

$$p(A|C) + p(\bar{A}|C) = 1$$

---

[19] In fact, this notion predates Dempster's work. It was notably formulated by Good (1962), Cedrik Smith (1961, 1965), and Fishburn (1964).

is replaced by a relationship between A's plausibility, $Pl(A)$, and the credibility of its opposite, $Bel(\overline{A})$:

$$Pl(A) + Bel(\overline{A}) = 1$$

but there is no longer a general direct relationship between $Pl(A)$ and $Pl(\overline{A})$. When multiple states exist, they must be combined according to Dempster's orthogonal rule. According to the latter, in a two-state situation, we can write:

$$(Bel_1 \oplus Bel_2)(A) = \frac{\sum_{B \cap C = A} m_1(B) m_2(C)}{1 - \sum_{B \cap C = \Phi} m_1(B) m_2(C)}$$

where m_1 and m_2 are the weights calculated on the two independent bodies of evidence noted 1 and 2. The rule is easy to generalize to any number of states.

The theory can now serve to take account of imprecise information and the ignorance of certain probabilities. It is freed from the additivity constraint for probabilities of independent events in classical subjective theories. From imperfect knowledge and data, it constructs the most consistent set possible.

It does, however, impose a complex rule for combination of beliefs (Dempster's orthogonal rule) without clear theoretical justification. A number of authors have sought to justify this rule (Dubois and Prade (1988), Gacôgne (1993)) by means of various simpler notions. However, it was Smets (1988, 1991) who, starting with a conditioning rule that was easier to axiomatize than the combination rule, ended up with a transferable belief model not tied to any interpretation or probabilistic hypothesis. This is what distinguishes Smets's theory from those of Dempster (1967) and Shafer (1976), which, although also based on belief functions, remain linked to a subjectivist probabilistic model (Smets 1994). As this approach departs from probability theory while generalizing it, we shall provide only a succinct description of Smets's approach in the next section. By contrast, we shall not discuss the theory of *fuzzy sets* (Zadeh 1965, 1978; Dubois and Prade 1988). It diverges even further from probability theory, as fuzzy sets no longer form a Boolean tribe, but lead to weaker structures.

In sum, the subjectivist approach to probability, after having been largely rejected during the second half of the nineteenth century and the first half of the twentieth century, enjoyed a revival of interest by the mid-twentieth century and is now commonly used by many researchers.

2.2 Paradigm and Axiomatics of Subjective Probability

As in Sect. 1.2, let us begin by trying to isolate the paradigm underlying the subjectivist approach before examining in greater detail the various axiomatizations that have been offered for it.

2.2.1 The Paradigm

As already noted, subjective reasoning must abandon any notion of event frequency in order to define as generally as possible the probability of the largest number of feelings of uncertainty, however subjective they may be. In this case, subjective probability will be derived from a reasoning that is valid for a given person and not—as in the objectivist approach—for the observed event. However, to avoid the arbitrariness that this situation would engender, we need to lay down conditions for the person's reasonings so that the probability can be used and calculated in the largest possible number of cases considered.

Subjectivists start with the notion of coherence of behavior, of which we have already spoken, in order to calculate the degree of probability that a person will assign to an observation. The authors who have addressed the subject offer slightly different definitions of coherent behavior, but we shall not examine those differences in detail here. It will be the first basic notion in the subjectivist paradigm. From that notion, we shall be able to demonstrate a certain number of classic properties of probability—such as the theorem of total probability and the Bayesian theorem—that now apply to subjective probability.

The second fundamental notion is the utility expected by a given person. We saw how Arnauld and Nicole (1662), then Daniel Bernoulli (1738), introduced it to denote the fact that a person's assessment of probability may vary according to circumstances. It is therefore important to estimate a subjective probability by taking into account the expected utility of the gain considered, i.e., the subjective value of the wager. For instance, the perceived utility of an identical gain may be greater for a poor person than for a rich person. Likewise, a wealthy prisoner who possesses 2,000 ducats but needs 2,000 more to buy back his freedom will assign a greater utility to a gain of 2,000 ducats than another, less wealthy player. This notion, which allows us to determine a preference relationship between different actions, was used by von Neumann and Morgenstern (1944) in economics, and later by Savage (1954) to axiomatize subjective probability.

However, when we want to explore the subjectivity of probability further, we need to ask questions about the origin of the feeling of belief. Coherence no longer suffices. Nor does the notion of expected utility, which is the foundation of subjective probability and allows decision-making amid uncertain events. But if we are only interested in the credibility of events, outside of all decision-making, we need to move out of the probabilistic framework to address the feeling of belief in an event, when a person has a set of information—naturally, an incomplete set—on a specific proposition.

In particular, this situation renders obsolete the principle of insufficient reason—which consists, in the absence of information on the various outcomes of an event, in assigning equal probabilities to them. The problem is that the principle does not enable us to take total ignorance properly into account. Consequently, if we assign a non-null belief to all simple outcomes of the event, then we should attribute a double belief to all outcomes comprising two simple and mutually incompatible outcomes. However, there is no reason for these two beliefs to differ, and the only logical solution is to assign a null belief to each simple or compound outcome. By contrast, a single belief is attributed to the combination of all these outcomes, which is the only proposition to be verified. This third element of the paradigm therefore leads not to a probability in the strict sense, as before, but to a characterization of the propositions by two numbers: belief, a new and non-probabilistic notion, and plausibility, which reintroduces subjective probability and is thus not actually new. In other words, only new information will enable us to assign a non-null belief to a given set of event outcomes, but will still leave a null belief in each outcome taken separately.

Later, we shall discuss the axioms that allow the definition of belief functions. These therefore precede the notion of probability, but from them we can deduce probabilities when we need to make a decision in the presence of such events.

2.2.2 The Axioms

We must now examine how to axiomatize this coherent behavior in the face of uncertainty—and that is where divergences between subjectivists appear. As in Sect. 2.1, we shall not recount the history of the axiomatization of subjective probability, but simply outline the main stages.

Once the bases of set theory had been firmly established, the quest for axioms for subjective probability followed very closely on the search for axioms for objective probability. In a 1926 article entitled *Truth and probability* (published posthumously in 1931), Ramsey sketched out a program for an axiomatics, but unfortunately he did not have the time to elaborate on it before his premature death at age 26. It was de Finetti (1931a) who effectively showed that, by complying with a certain number of axioms, a set of personal opinions on probabilities could be represented by a numerical measure. Let us examine his axioms.

De Finetti considers a specific event about which we do not know if it will happen or not. He shows that our doubt regarding its occurrence can be measured by degrees of probability if the following axioms are fulfilled (de Finetti 1937):

1. an uncertain event can only appear to us to be: (*a*) as probable as, (*b*) more probable than, and (*c*) less probable than another;
2. an uncertain event will always appear more probable to us than an impossible event and less probable than a certain event;
3. an event E' cannot appear more probable than another event E'', when we deem E' more probable than a third event E, itself viewed as more probable than E'' (transitive property);
4. the inequalities remain in the logical sum: if E is an event incompatible with E_1 and with E_2, then $E_1 + E$ will be more or less probable than, or as probable as, $E_2 + E$ depending on whether E_1 is more or less probable than, or as probable as, E_2.

This is tantamount to defining a binary relationship for the event set, often noted $\succeq$, which constitutes a weak-order relation among the events. De Finetti had to add a fifth axiom for what he called subordinate events, which we now call *conditional events*. He designated the three-value logical entity $E'|E''$ [20] as: true, if E'' and E' are true; false, if E'' is true and E' is false; null, if E'' is false. This fifth axiom for subordinate (i.e., conditional) probabilities is accordingly enunciated as follows:

5. if E' and E'' are contained in E, $E'|E$ is more or less probable than, or as probable as, $E''|E$ depending on whether E' is more or less probable than, or as probable as, E''.

While taking a system of purely qualitative axioms as a starting point to define a structure of qualitative probability, de Finetti thus succeeded in obtaining a quantitative measure of that probability. He then showed that his structure satisfies various classical properties of probability, such as the theorem of total probability, the theorem of compound probability, and the Bayesian theorem. We should note, however, that a probability complying with de Finetti's coherence condition does not need to satisfy the σ-additivity condition, discussed in Chap. 1, which was used to define an objective probability. The only condition that needs to be met is finite additivity. But we can preserve coherence by moving to the limit when studying a sequence of probabilities whose number tends toward infinity.

Thanks to these axioms, we can spell out the coherence paradigm. The first axiom states that we are examining events for which we must be able to say that one is more or less probable than, or as probable as, the other. The second effectively curtails the freedom of opinion of a person who wants to remain coherent: the person cannot regard an event as more probable than a certain event and less probable than an impossible event. The third axiom specifies how this coherence satisfies transitivity: a person cannot hold an event E' to be more probable than an event E, itself more probable than a third event E'', and regard this third event as more probable than the first. The fourth axiom states that inequalities between probabilities can be compounded, in the same manner as logical additions of incompatible events. The fifth axiom enables us to deduce the theorem of compound probabilities, which could not be shown with the first four. On the other hand, de Finetti's axioms make hardly any allowance for the utility paradigm.

De Finetti does not incorporate the wager concept into his presentation of the axioms, but he does indicate that it offers a clearer way to make the concept of probability understandable. In fact, he defines the degree of probability as follows:

Let us assume that a person must evaluate the price p for which he would consent to exchange the possession of any given sum S (positive or negative), contingent upon the occurrence of a given event, E, for the possession of the sum pS; we shall say by definition that this number p is the measure of the degree of probability attributed by the person in question to the event E, or, more simply, that p is the probability of E (depending on the person considered; this specification may actually be implied if there is no ambiguity). (de Finetti 1937)

---

[20] Actually, de Finetti noted this event $\dfrac{E'}{E''}$, but we shall use standard probability notations here.

This definition—under the assumption that the person is coherent—leads to a psychological approach to the concept of probability, already recommended by Ramsey (1931) but criticized by many authors (Jaynes 2003), as we shall see at the end of this chapter.

In the next section, when discussing statistical inference from a subjectivist standpoint, we shall look at another important concept introduced by de Finetti, exchangeability, which allows a clearer approach to inference.

De Finetti's axiomatization raises a number of problems, however. Let us examine a few of them.

The first problem was raised by de Finetti himself (1951). He failed to prove that there always exists a measure of probability that satisfies the following relationship for two events A and B:

$$p(A) \geq p(B) \text{ if and only if } A \succeq B.$$

In other words, if A is at least as probable as B, the weak-order relation between the events holds true. For an infinite set, de Finetti's axioms are not strong enough to ensure the existence of such a measure (Suppes and Zanotti 1975). For finite sets, Kraft et al. (1959) have shown that the relation may not be satisfied either. To this end, the authors built an example with five subsets that complies with the axioms; the addition of its probabilities yields an unacceptable result (0>0). Scott (1964) found that it was necessary, in this case, to impose an additional algebraic condition in order to fulfill the relationship. The condition concerns the characteristic function, which assigns a value of unity to elements of the set and zero to elements outside it. In such circumstances, however, we cannot formulate the condition using de Finetti's set-theory language (Suppes and Zanotti 1975), so the condition poses a tricky problem for the axiomatization procedure.

The second problem is due to de Finetti's belief that everyone must always have an opinion on events of whatever sort. Before presenting his axioms, he clearly stated:

> Let us consider a well-determined event, and let us suppose that we do not know in advance whether it will happen or not; our doubt as to its occurrence is open to comparison, and, consequently, to a gradation (de Finetti 1937).

In principle, it is difficult to assert this for all future events. If I am asked for my opinion on whether it will rain 10 years from now in my present location, I can express no opinion on the subject, and I shall not be able to compare this fact to the fact that the same event will occur in 20 years' time. It therefore seems beyond the possibilities of probability theory to assign a probability to any imaginable event. While we can take into account a far greater number of events than in objective probability theory, we cannot take them all into consideration.

We have also seen that de Finetti made extensive use of the coherence paradigm to establish his axioms but that he made virtually no use of the utility paradigm. Despite devoting considerable attention to it throughout his work (Muliere and Parmigiani 1993), he believed that the notions of probability and utility should be treated separately. For him, the notion of probability, cleansed of utility-related factors, belongs to a higher logical level than utility, and the development of probability calculus requires considerable elaboration unrelated to utility (de Finetti 1952).

Other authors (ex. Ramsey 1931; von Neumann and Morgenstern 1947; Savage 1954) believed, on the contrary, that utility was an integral part of the axiomatization of probability as decision-making theory. Here, we offer a more detailed presentation of Savage's axiomatization.

Savage set out to establish a theory of the behavior of a rational 'person' needing to take a decision. The 'person' could be defined as a unit differing from an 'individual' in the ordinary sense—for example, it could consist of a family, a company or a nation. Savage proceeds by examining a set S of elements $s, s', \ldots$, viewed as a complete description of the world, with subsets $A, B, C, \ldots$ constituting the various possible states of nature. The consequences of acts form another set F of elements $f, g, h, \ldots$. The acts producing these consequences are arbitrary functions from S to F, $\mathbf{f}, \mathbf{g}, \mathbf{h} \ldots$: we thus link the consequence $f(s)$ to the state s. The 'is not preferred to' relationship between acts is noted $\preceq$, implying that the 'is preferred to' relationship is written $\succ$. Savage offers seven axioms for subjective probability:

1. The relation $\preceq$ is a simple ordering.
2. For every f, g, and B, $f \preceq g$ given B or $g \preceq f$ given B.
3. If $f \preceq g$ for every, and B is not null; then $f \preceq f'$ given B, if and only if $g \preceq g'$.
4. For every A, B, $A \preceq B$ or $A \preceq B$.
5. There is at last one pair of consequences f, f' such that $f' \prec f$.
6. Suppose it is false that $g \preceq h$; then, for every f, there is a (finite) partition of S such that, if g' agrees with g and h' agrees with h except on an arbitrary element of the partition, g' and g' being equal to f there, then it will be false that $g' \preceq h$ or $g \preceq h'$.
7. If $f \preceq g(s)$ given B ($g(s) \preceq f$ given B) for every $s \in B$, then $s \in B$ given B ($g \preceq f$ given B).

In non-mathematical terms, the situation that Savage envisages is that of a 'person' who must take a decision, i.e., choose between different, mutually exclusive acts. Consequently, if the decision satisfies the axioms, it will generate a subjective-probability distribution and a utility function that will maximize its mean utility. The consequences of the person's decision will depend on the chosen act and the state of nature that will occur, whose occurrence is unrelated to the decision taken. The 'person' is supposed to know all possible acts, all possible states of nature, and the consequences of each act for each state of nature. The person's uncertainty stems from his or her ignorance of which state of nature will occur.

In these conditions, we can state the axioms in less abstract terms to show their implications more clearly. The first axiom indicates that the notion of preference between decisions must be transitive and that, given two decisions, one is at least as preferable as the other. This is to ensure coherence in decision-making. Savage considered replacing the simple-order relation by a partial-order relation in order to take feelings of indecision into account. However, he eventually rejected this solution, as it would have narrowed the scope of his axiomatics excessively. The second axiom extends the notion to the case where the field of definition of the decisions is restricted to any subset of states of nature B. We can say that if the person prefers act f to act g for each state of nature examined, (s)he will continue to prefer f to g when (s)he does not know which state will occur (this is the independence axiom,

2.2 Paradigm and Axiomatics of Subjective Probability

which expresses Savage's '*sure-thing principle*'). These first three axioms define the general conditions in which, according to Savage, decisions are taken.

The last four axioms spell out what he means by personal probability, which, of course, may vary from one person to another, as in de Finetti's system. The fourth axiom states that, given any two events A and B, one is at least as probable as the other. The fifth axiom excludes the trivial case where all consequences have equivalent utilities. The sixth axiom basically indicates that if event A is less probable than B, there exists a partition of S— the set of elements considered—such that the union of each of its elements with A is less probable than B. The seventh and final axiom extends the sure-thing principle to the case where an act may have an infinite number of consequences. Once we admit these axioms, we can replace the notion of qualitative subjective probability by the notion of quantitative probability, i.e., a measure of probability that displays its conventional properties given by the Kolmogorov's axioms.

To some extent, these axioms incorporate the notion of coherence introduced by de Finetti, adding the notion of utility, which arithmetizes the preference relationship between acts. Indeed, Savage recognizes that Daniel Bernoulli's general concept of utility (1738) matches the one he uses in his book (1954). An individual who complies with his axioms will maximize the utility, i.e., the subjective value, of the stakes.

Savage draws various consequences from his axioms. The first is that there exists a numerical probability that is associated with the states of nature and complies with the main rules of probability calculus, including the Bayesian rule. The second is that we can associate a numerical value with each consequence representing its utility for the person, which the person will seek to maximize. Lastly, the order of preference for the acts is that of their mathematical expectations, calculated from the probabilities and utilities defined earlier.

While promptly criticized, this axiomatization was used extensively, in particular by economists.

Savage had already presented his axiomatization of subjective probability at an international conference on risk in May 1952. Allais,[21] who was in attendance, had criticized what he called the postulates and axioms of the American school represented by Friedman, Marshack, Neumann-Morgenstern, Savage, and others. Allais gave a fuller version his criticism in an article in *Economica* (1953). We shall discuss only some of his criticisms here in order to illustrate the weaknesses of Savage's axiomatization.

Allais elaborated on various paradoxes that transgressed Savage's axioms. Let us examine one of them. Suppose we ask someone two questions:

1. *Do you prefer situation A to situation B?*

    Situation A: *Certainty of receiving 100 million.*
    Situation B $\begin{cases} 10 \text{ chances in } 100 \text{ of winning } 500 \text{ millions.} \\ 89 \text{ chances in } 100 \text{ of winning } 100 \text{ millions.} \\ 1 \text{ chance in } 100 \text{ of winning nothing.} \end{cases}$

---

[21] Allais won the Nobel Prize for Economics in 1988 for his contributions to market theory.

2. *Do you prefer situation C to situation D?*

$$\text{Situation } C \begin{cases} 11 \text{ chances in } 100 \text{ of winning } 100 \text{ millions.} \\ 89 \text{ chances of } 100 \text{ of winning nothing.} \end{cases}$$

$$\text{Situation } D \begin{cases} 10 \text{ chances in } 100 \text{ of winning } 500 \text{ millions.} \\ 90 \text{ chances in } 100 \text{ of winning nothing.} \end{cases}$$

Experience shows that most people will answer *A* to the first question and *D* to the second. Now according to Savage's axioms, people should seek to maximize the mathematical expectation of their gains. The gain is 100 million in situation *A* versus 139 million in situation *B*, which should lead to choice *B*; the gain is 11 million in situation *C* versus 50 million in situation *D*, so the choice here should be *D*. These results contradict Savage's axiom 2—more simply, the sure-thing principle, which states that all preferences should be independent of the outcomes of all possible options.

Many other examples have been described in which Savage's axioms are violated by observed human behavior (Kahneman and Tversky 1979; Grether and Plott 1979; Loomes and Sugden 1982). These examples challenge the very foundations of his axiomatization and, in some cases, the authors propose far more complex alternative theories. We shall not examine them here, for they provide only a partial solution to Allais's paradox (Weber 1998).

Allais's criticisms mostly concerned axioms 2 and 7, which formalize the independence principle. Suppes (1956, 1960, 1974) began by focusing on two classes of axioms. The first comprises the axioms that we may assume to be valid everywhere and at all times: he calls them *pure-rationality axioms*. In Savage's system, they consist of axioms 1–4 and axiom 7, i.e., those singled out for criticism by Allais. Suppes argues that these axioms seem to be ones with which a rational person would comply. The second class contains the axioms that postulate certain structural properties of the environment. Suppes calls them *structural axioms*. They consist of axioms 5 and 6, which Suppes proceeds to criticize.

Savage defended axiom 6 in particular, on the following grounds:

> Suppose, for example, that you yourself consider $B \prec C$, that is, that you would definitively rather stake a gain in your fortune on *C* than on *B*. Consider a partition of your own world into $2n$ events each of which corresponds to a particular sequence of *n* heads and tails, thrown by yourself, with a coin of your own choosing. It seems to me that you could easily choose such a coin and choose *n* sufficiently large so that you could continue to prefer to stake your gain on *C*, rather than on the union of *B* and any particular sequence of *n* heads and tails. (Savage 1954)

Suppes sees many objections to this argument. Specifically, he notes that:

> without radical changes in human thinking, it is simply not natural on the part of human beings to think of finite sequences of flips of a coin in evaluating likelihoods or probabilities, qualitative or quantitative, of significant events with which they are concerned. (Suppes 1974)

Savage's introduction of this idea is equivalent to a *deus ex machina* and undermines one of the main goals of his axiomatization: to extend the theory of rational

2.2 Paradigm and Axiomatics of Subjective Probability

behavior to fields of action where it is not natural to think in terms of random mechanisms. Suppes also criticized the use of acts that always produce the same consequences, whatever the state of nature, which Savage introduces to show that his axioms entail a maximization of the utility's expectation.

To test the axioms' validity in subjective probability, the work of psychologists (Preston and Baratta 1948; Kahneman and Tversky 1972; Tversky 1974) may be useful, as Luce or Suppes showed (Luce and Suppes 1965; Suppes 1974). As noted in Chap. 1, the axioms are not self-evident truths requiring no demonstration. Rather, they must be deduced from relationships that can be produced experimentally. For example, many trials show that, for events whose objective probabilities can be predicted (various types of games), subjects routinely overestimated low probabilities and underestimated high ones. There is also a major difficulty involved in using these subjective results, for most people modify their behavior in the face of risk when the sums wagered change. These criticisms partly converge with the objections by Allais.

One of Suppes's criticisms concerned the accuracy of the measures of probability used:

> to insist that we assign sharp probability values to all our beliefs is a mistake and a kind of Bayesian intellectual imperialism. (Suppes 1976)

In real life, the use of subjective probability is indeed an imprecise notion.

Shafer's article (1986) and comments provide a fuller view of the criticisms leveled at Savage, of which we have described the main ones here. Moreover many variations on these axioms have been developed since in order to answer some of these criticisms (Suppes 1956; Luce and Krantz 1971; Stigum 1972; Roberts 1974; Fishburn 1975; Narens 1976; Bernardo and Smith 1994). However, the main issue remains unchanged: to give the conditions which will permit a more precise definition of subjective probabilities checking with Kolmogorov axioms.

Simultaneously, this body of criticisms of Savage's axioms led Suppes (1974) to propose replacing the notion of a single probability for an event by that of upper and lower probabilities, already suggested by Cyril Smith (1961), Good (1962), and Dempster (1967, 1968).

This new axiomatization extends de Finetti's qualitative axioms and leads to an approximate measure of beliefs, which, however, is close to that of Savage. Let us take a closer look at the new proposal.

The basic idea is that there must be events whose probability can be estimated accurately and others whose probability is more approximate. A structure $\Omega = (\Omega, \mathfrak{J}, S, \succeq)$ is a rough measure of beliefs if and only if Ω is a non-empty set, $\mathfrak{J}$ and S are algebras of subsets of Ω, and $\succeq$ is a binary relationship on $\mathfrak{J}$. Intuitively, $\mathfrak{J}$ comprises all the events to which probabilities will be assigned, and S contains the standard events whose precise probability is known.

The first four axioms are identical to de Finetti's. Suppes (1974) adds three others:

5. S is a finite subset of $\mathfrak{J}$;
6. If $S \neq \emptyset$ then $S \succ \emptyset$;
7. *If $S \succeq T$ then there is a V in S such that $S \approx T \cup V$*

The fifth axiom simply states that standard events belong to the set of events whose probability we want to determine. The sixth ensures that any minimal element of S has a positive qualitative probability. The seventh lays down a simple solvability condition for standard events.

Suppes shows that in addition to a single measure of probability for standard events, we can use upper and lower probabilities to express the inaccuracy of the measure of arbitrary events. If $p_*(A)$ is the lower probability and $p^*(A)$ the upper probability of any event, both probabilities exhibit the following three properties:

I. $p_*(A) \geq 0$.

II. $p_*(\Omega) = p^*(\Omega) = 1$.

III. If $A \cap B = \emptyset$ then:

$$p_*(A) + p_*(B) \leq p_*(A \cup B) \leq p_*(A) + p^*(B) \leq p^*(A \cup B) \leq p^*(A) + p^*(B)$$

For standard events, naturally, we have $p(S) = p_*(S) = p^*(S)$. These lower and upper probabilities display several properties, including:

$$p^*(A) - p_*(A) \leq \frac{1}{n}$$

where n is the number of minimal elements in S. Moreover, if we define the relationship $* \succ$ for $\mathfrak{S}$ by: $A * \succ B$ if and only if there exists a set C in S such that $A \succeq C \succeq B$, we can then show that $* \succ$ defines a semi-order[22] for $\mathfrak{S}$, and that:

- if $A * \succ B$, then $p_*(A) \geq p^*(B)$,
- if $p_*(A) \geq p^*(B)$, then $A \succeq B$.

Suppes demonstrates these relationships (1974).

This axiomatization provides a measure that allows us to introduce a series of standard events whose probability is measured accurately. In turn, that probability will serve as an approximate measure for other events of interest.

But Shafer (1976) goes one step further by offering a reinterpretation of Dempster's work in a new theory, whose antecedents we can find in Hooper's credibility calculus (1699) and J. Bernoulli's combination rules (1713). The new theory is a generalization of subjective-probability theory:

> a reinterpretation that identifies his 'lower probabilities' as epistemic probabilities or degrees of belief, takes the rule for combining such degrees of belief as fundamental, and abandons the idea that they arise as lower bounds over classes of Bayesian probabilities.

This theory is often called the Dempster-Shafer theory. Let us see on what axioms it is based.

---

[22] The notion of semi-order rests on the idea that one alternative is preferable to another only if the utility of the first alternative exceeds the utility of the second by a certain constant threshold.

2.2 Paradigm and Axiomatics of Subjective Probability

Let Θ be a reference framework for a specific issue, which is a counterpart to the sample space. We postulate the existence of a numerical measure of the degree of belief $Bel(H)$ obtained from the information available on hypothesis $H \subseteq \Theta$. We lay down the following conditions, as axioms for the function Cr defined for a set:

1. $Bel(\emptyset) = 0$
2. $Bel(\Theta) = 1$
3. $Bel(H) \geq \sum \left\{ (-1)^{|I|+1} Bel(\bigcap_{i \in I} H_i) \right\}$ with $I \subseteq \{1, ..., n\}$ for any $n \geq 1$ and any set H, H_i being located in Θ with $H \supseteq H_i$.

Bel is called a belief function. From these axioms, we can construct the function *Pl*, called the plausibility of H: $Pl(H) = 1 - Bel(\bar{H})$, where $\bar{H}$ is the complementary set to H. We show that, unlike probability, the belief function is sub-additive, i.e.,:

$$Bel(H) + Bel(\bar{H}) \leq 1,$$

and plausibility is super-additive, i.e.,: $Pl(H) + Pl(\bar{H}) \geq 1$.

However, this axiomatization still relies on the notion of events not located in time. To remedy this problem, Shafer (2001) proposes a fuller theory that incorporates an event tree for events occurring over time. By introducing time, he generalizes Boolean algebra and sets up an event space where events can happen one after another. Shafer axiomatizes these event spaces (Shafer et al. 2000), which leads him to view subjective probability from a new angle. We shall not elaborate on this axiomatization, however, as it lies too far outside the scope of our book: it is no more related to classical logic, considered here, but to intuitionistic logic.

Belief theory has also been criticized for imposing an *ad hoc* combination rule—Dempster's orthogonal rule—without theoretical justification. And, while Shafer does note that Dempster's results can be used both for subjective probability and for belief functions, he does not clearly spell out the reasons.

In our opinion, the most general and most satisfactory axiomatics has been provided by Smets (1988, 1990), who distinguishes between two levels that were not separated in the subjective approach.

The first level is that of beliefs (*credal level*), a psychological level where beliefs are formed outside of any decision-making context. This concept of belief differs from the concept of subjective probability. Smets (1988) quantifies it by means of belief functions, defined without reference to the concept of probability. However, when we need to take a decision, the belief formed at the previous level induces a measure of subjective probability, which now lies at the decision level (*pignistic level*, from the Latin *pignus*, 'wager'). This measure of probability will serve to take decisions using subjective-probability theory.

As noted above, we shall not discuss in detail the axiomatization of belief functions, credibility, and the transition from the belief level to the decision level (and hence the probability level), as these issues would distract us from the subject of our book. We refer the interested reader to Smets's many articles on the axiomatization (Smets 1988, 1990, 1997, 1998). However, we offer a simplified version to show its considerable value.

We focus exclusively on Dempster's combination rule, $\oplus$, given earlier in Sect. 2.1, defined for a discernment space D, whose existence—as shown previously—requires axiomatization (Smets 1990):

1. $(Bel_1 \oplus Bel_2)(A)$ is a function only of A, Bel_1, and Bel_2.
2. The $\oplus$ rule is commutative.
3. The $\oplus$ rule is associative.
4. If $m_2(B) = 1$, then m_1 and m_2 must fulfill:
$$\left. \begin{array}{l} (m_1 \oplus m_2)(A) = \sum_{X \subset \bar{B}} m_1(A \cup X) \\ = 0 \quad otherwise \end{array} \right\} \forall A \subset B$$
5. The distribution must display internal symmetry (invariance through permutation).
6. For, $A \neq D$, $(m_1 \oplus m_2)(A)$ does not depend on $m_1(X)$ for all $X \subset \bar{A}$.
7. D contains at least 3 elements.
8. The distribution must satisfy a continuity hypothesis:
if $m_2(A) = 1 - \varepsilon$, $m_2(D) = \varepsilon$, $m_A(A) = 1$ then
$$\forall X, \lim_{\varepsilon \to 0}(m_1 \oplus m_2)(X) = (m_1 \oplus m_A)(X)$$
m_1 being a basic mass function.[23]

These axioms being satisfied, we can show that they imply the uniqueness of Dempster's combination rule (Smets 1990). Further axioms—not described in detail here—are needed to define more fully the belief functions and their transformation into probabilities for decision-making. These axioms use the three notions of the paradigm stated earlier: coherence, utility, and belief.

We have thus moved from an axiomatization of subjective probability to a more general axiomatization of belief functions, defined on incomplete data, which constitute a generalization of probability. We have simply noted the main stages here. For fuller details, we refer the interested reader to Fishburn's article (1986) on subjective-probability axioms, Shafer's book (1976) on the mathematical theory of evidence, the chapter of Suppes's book (2002a) on representations of

---

[23] Smets (1990) formulated these axioms slightly differently, calling the belief function bel:

1. *compositionality* axiom: $bel_{12}(A)$ is a function of A, bel_1 and bel_2 only.
2. *symmetry:* $bel_1 \oplus bel_2 = bel_2 \oplus bel_1$.
3. *associativity:* $(bel_1 \oplus bel_2) \oplus bel_3 = bel_1 \oplus (bel_2 \oplus bel_3)$.
4. *conditioning:* if bel_2 is such that $m_2(B) = 1$, then
$$m_{12}(A) = \sum_{C \to \bar{B}} m_1(A \cup C) \quad \text{for all } A \to B$$
$$= 0 \qquad otherwise$$
5. *internal symmetry:* the mass given by m_{12} to $A \in \Omega$ is independent of the masses given by m_1 (and m_2 to propositions $B \to \bar{A}$.
6. *auto functionality:* $\forall A \in \Omega$, $A \neq 1_\Omega$, $m_{12}(A)$ does not depend on $m_1(X)$ for all $X \to \bar{A}$.
7. *three elements:* there are at least three elementary propositions in D.
8. *continuity:* let $m_2(A) = 1 - \varepsilon$, $m_2(1_\Omega) = \varepsilon$. Let $m_A(A) = 1$. For any bel_1 defined on Ω, let $bel_{1A} = bel_1 \oplus bel_A$ then for all $X \in \Omega$, $\lim_{\varepsilon \to 0} m_{12}(X) = m_{1A}(X)$.

probability, and the articles by Smets, cited above, on the axiomatization of the transferable-belief model.

However, unlike objective probability, most of whose advocates have recognized Kolmogorov's axioms, the axiomatization of subjective probability has been more controversial and has often been fleshed out. It has not generated as solid a consensus, and some criticisms, such as those of Allais, remain relevant.

2.3 Subjectivist Statistical Inference

Recall that statistical inference may be viewed as a process for moving from data on a set of units to assertions about a new unit (Johnson 1932; Lindley and Novick 1981). In Sect. 1.3, we saw how hard it was to draw a correct inference in an objectivist framework and that, in fact, we could only respond to questions other than the simple problem posed above. In this section, we shall see that, when we take the subjective approach, the simple problem receives an answer that is naturally subjective but perfectly clear in its logic. The notion of exchangeable events introduced by de Finetti (1937) makes it easier to provide an answer.

Let us begin with the simple case of a game of heads or tails, using a coin of irregular appearance. The probability of obtaining heads on the $(n+1)$th toss, $p(E_{n+1})$, depends on the result A of the n previous tosses. As a rule, we shall need to assume that these probabilities depend on the order in which we obtained the result, whether we assume the influence of one toss on the next, or the existence of external conditions that vary with each toss. But, when we can assume that the probability does not depend on the order of trials, the results will be substantially simplified: the n trials will be said to be exchangeable if the joint probability distribution $p(E_1, \ldots, E_n)$ is invariant for all permutations of the n units. In this case, an additional unit E_{n+1} will be exchangeable for E if all $(n+1)$ units are exchangeable as well.

Let ω_r^n be the probability that in n tosses we shall obtain, in random order, r heads and $(n-r)$ tails: ω_r^n will be the sum of probabilities of the C_r^n distinct ways in which we can achieve this result. If the trials are exchangeable, then the probability that r outcomes will be positive and $(n-r)$ negative, in whatever order, will always be equal to:

$$p(A) = \frac{\omega_r^n}{C_n^r} \quad \text{and} \quad P(A \cap E_{n+1}) = \frac{\omega_{r+1}^{n+1}}{C_{n+1}^{r+1}}$$

where C_n^r is the number of combinations of r elements among n. This yields the following inference for the $(n+1)$th toss:

$$p(E_{n+1} \mid A) = \frac{(r+1)\omega_{r+1}^{n+1}}{(n+1)\omega_r^n}$$

which will be a function of r and n only. We can therefore state that:

> whatever the influence of the observation on the future prediction, it in no way requires us to *correct* the initial assessment of the probability p that was *invalidated* by the trial, by replacing it with another probability $p^*(E_{n+1})$ that is *consistent* with the trial and therefore probably *closer to the actual probability*; on the contrary, it is manifest only in the sense that, when the experiment gives us the result A of the first n trials, our opinion will be expressed not by the probability $p(E_{n+1})$, but by the probability $p(E_{n+1} \mid A)$, namely, the one that our initial opinion already assigned to the event E_{n+1} regarded as dependent upon the contingency A. This initial opinion, therefore, is in no way repudiated or corrected: it is not the function p that has been altered (replaced by another, p^*), but indeed the argument E_{n+1} that has been replaced by $E_{n+1} \mid A$, and it is precisely to abide by the initial opinion (as manifested in the choice of function p) and to remain consistent in our judgment that our predictions vary when a change occurs in known circumstances. (de Finetti 1937)

Thus, by adopting the subjective approach, we can find a perfectly logical solution to the inference problem. We evaluate the probability of each event in the series that has not yet occurred as being nearly identical to the frequency of earlier positive outcomes. Likewise, if we apply the insufficient-reason principle, we obtain:

$$\omega_n^0 = \omega_n^1 = \ldots = \omega_n^n = \frac{1}{n+1}.$$

Hence, in this case:

$$p(E_{n+1} \mid A) = \frac{r+1}{n+1} \times \frac{n+1}{n+2} = \frac{r+1}{n+2}.$$

This is a restatement of Laplace's succession rule, fiercely attacked by the objectivists.

We have taken the simple case of a heads-or-tails game, but we can generalize these results to the case of variables characterized not by a number (probability), as in our example, but by a function (distribution function, probability density, characteristic function, and so on). While theoretically self-evident, this general result is far from easy to demonstrate. De Finetti provided a rigorous presentation of it, which he summarized as follows:

> The laws of probability of a class of equivalent [*i.e., exchangeable*] random elements are the 'averages' of the laws of probability applying to cases of independence. (de Finetti 1937)

Similarly, we can define conditional exchangeability for two variables X and Y that are unknown to the researcher. In this case, a number of units n is said to be exchangeable for X, knowing that $Y = y$, if the joint conditional distribution, $p(X_1, \ldots, X_n \mid Y_i = y, \forall i)$, is invariant for whatever permutation of the units.

The objectivist statistical inference could not answer the question 'What is the probability that the unknown parameter lies in a given interval?' As shown in Chap. 1, it can only answer a far more complex question whose relevance was not self-evident. Subjectivist statistical inference, with the notion of exchangeability, enables us to answer the question directly—under certain assumptions, of course, but these can be clearly stated. It is the researcher who, given his or her subject and the available information, can say whether the events studied are exchangeable, conditionally exchangeable or non-exchangeable. Even in the latter instance, where events could be viewed as non-exchangeable, we may be led to modify the inferences in a very different way so as to take account of the circumstances in which they occur.

Let us take the case examined by de Finetti (1937), where the first n trials roughly yield a positive outcome and a negative outcome in alternation. Clearly:

> The most natural attitude will notably consist in predicting that the next toss will be more likely to produce a result that is contrary to the previous one.

It will be the researcher's task to draw the consequences of the non-exchangeability of earlier results in order to formulate in the most satisfactory manner the inference needed to predict the following outcome.

The comparison between the quotation from Fisher given in Sect. 1.3 and the concept of exchangeability led Lindley and Novick (1981) to note how the latter concept indeed enables us to properly define the notions of population and subpopulation, which Fisher left vague for practical applications. De Finetti defines a population on the basis of the exchangeability of units, and a sub-population on the basis of the conditional exchangeability of its units. This same notion makes it possible to identify the specific population to which an individual about whom we want to formulate an inference belongs. In the applications of this approach, we shall examine a specific example of the role of exchangeability in statistical inference.

2.4 Applications to Social Science

Regression analysis—used in such diverse disciplines as medicine, biology, demography, epidemiology, economic, and education science—may resort to Bayesian methods for estimating its models (Lee 1989; Gelman et al. 1995; Rouanet et al. 1998; Leonard and Hsu 1999; Gill 2008). Likewise, event-history analysis developed Bayesian methods (Ibrahim et al. 2001), although many frequentist approaches preceded it (Kalbfleisch and Prentice 1980; David Cox and Oakes 1984; Courgeau and Lelièvre 1989; Andersen et al. 1993). Multilevel analysis, as well, was to draw on Bayesian methods (Goldstein 2003). We shall not discuss all these analytical methods here. Instead, we shall describe one application to multilevel analysis—in particular, the subjectivist statistical inference that the methods allow with the aid of the concept of exchangeability. We have chosen our examples from the field of education science (Rouanet et al. 2002; Courgeau 2007b) and of testimonies, from Hooper (1699) to Dempster (1967).

The starting point is a paradox examined in detail by Edward Simpson (1951) but previously addressed by Cohen and Nagel (1934) and Chung (1942). We describe it here using a simple example: a comparison of test results for students in a given town (Courgeau 2007b), broken down by gender. Table 2.1 shows the total figures.

The pass rate in all of the town's high schools combined is 34.25% for girls versus 45.75% for boys, a gap of over 10 percentage points.

Table 2.1 Total results in the town's high schools

	Passes	Fails	Total
Boys	915	1,085	2,000
Girls	685	1,315	2,000

Table 2.2 Results for each high school

High school	Boys Passes	Fails	Total	Girls Passes	Fails	Total	Total
1	3	17	20	76	304	380	400
2	12	48	60	85	255	340	400
3	25	75	100	90	210	300	400
4	42	98	140	91	169	260	400
5	63	117	180	88	132	220	400
6	88	132	220	81	99	180	400
7	117	143	260	70	70	140	400
8	150	150	300	55	45	100	400
9	187	153	340	36	24	60	400
10	228	152	380	13	7	20	400
Total	915	1,085	2,000	685	1,315	2,000	4,000

However, as the town has ten high schools, we can also tabulate the number of boys and girls who pass or fail the exam at each high school (Table 2.2).

This time, the result is—on the face of it—surprising: the pass rate for girls is consistently five points higher than the boys' rate in every school. In the second school, for example, the pass rate is 25% for girls versus 20% for boys. Simpson's paradox stems from the comparison of the two results. In this case, the high school is referred to as a 'confusion factor' when we are studying the effect of gender on success in an exam. To resolve the apparent contradiction between these results, we therefore need to analyze the data more fully.

An initial solution is to perform separate logit analyses for each school, i, with gender as the only predictive variable:

$$p_i(R=1|S) = \left\{1 + \exp\left[-(a_{0i} + a_{1i}S)\right]\right\}^{-1}$$

In this case certain schools may have a very small student population and the standard deviation of the parameters a_{0i} and a_{1i} may be very large, making the results impossible to interpret.

For a more general result, valid for a larger high-school population (all French high schools, for example), we can suppose that the chosen schools will provide more general information on all French high schools. Just as we draw samples of individuals to deduce estimates for the entire population, so we can draw a sample of high schools to deduce information on all high schools. We can use either the objectivist approach to probability or the subjectivist approach.

2.4 Applications to Social Science

In the first case, we may regard the parameters a_{0i} and a_{1i} as random variables that we can write as:

$$a_{0i} = \alpha_0 + u_{0i} \text{ and } a_{1i} = \alpha_1 + u_{1i}$$

where u_{0i} and u_{1i} are zero-average random variables with the following variances:

$$\text{var}(u_{0i}) = \sigma_{u0}^2 \quad \text{var}(u_{1i}) = \sigma_{u1}^2 \quad \text{and} \quad \text{cov}(u_{0i}, u_{1i}) = \sigma_{u01}.$$

This gives us a multilevel logit model:

$$p(R=1|S) = \left\{1 + \exp\left[-\left(\alpha_0 + u_{0i} + \langle\alpha_1 + u_{1i}\rangle S\right)\right]\right\}^{-1}.$$

We know how to estimate the model's parameters and variance-covariance matrix, as well as the u_{0i} and u_{1i} values for each school (Goldstein 2003).

In the second case, we can estimate Bayesian models using MCMC (Markov Chain Monte Carlo) and Bootstrap methods, with Mlwin software (Goldstein 2003; Browne 1998). The MCMC method derives its name from the following notion: to produce acceptable approximations of integrals and other functions that depend on a distribution of interest, it is sufficient to generate a Markov chain whose limit distribution is the distribution of interest (Robert 2006). The Bootstrap method consists in building a series of samples similar to the observed sample and to use the series to estimate the distribution's parameters of interest. We have assumed here that a binomial model was well suited, as we can show in a non-Bayesian example. The assumption is easily confirmed here in the Bayesian example. We shall use the Bootstrap option only (for the results with the MCMC option, see Courgeau 2007b).

For the Bootstrap option, we have chosen to perform 500 draws in 10 sets: a test on 100 draws in 5 sets gave us bimodal estimates for the core density of certain parameters. We have used the parametric option, taking the least-squares estimators as our starting point. The non-parametric option yields unstable results, as the parameter values increase with the number of iterations: we have not reported the figures here.

Table 2.3 lists the resulting estimates of objectivist and subjectivist probabilities, with gender among the fixed parameters and high school among the random parameters.

In all models, we see that boys have a lower pass rate than girls, regardless of high school (36.87% versus 41.42%). By contrast, we find a significant random value at high-school level. It is similar for the objectivist model and for the Bootstrap estimate, and shows different results for each school. The girls' pass rate exceeds the boys' pass rate by approximately 5% in each school.

If we now include the percentage of boys, and estimate only a Bootstrap model with the non-Bayesian model, we obtain the results listed in Table 2.4.

As we can see, the two models yield nearly identical estimators, the random parameter at school level does not differ significantly from zero. Gender and the

Table 2.3 Results of objectivist and subjectivist multilevel models, without inclusion of percentage of boys (parameters and variances in parentheses)

Parameters	Objectivist method	Bootstrap (parametric)
Fixed:		
Intercept (α_0)	−0.538 (0.203)	−0.512 (0.191)
Gender (α_1)	0.191 (0.081)	0.197 (0.081)
Random:		
$\sigma^2_{u_0}$	0.384 (1.177)	0.439 (0.184)
$\sigma^2_{e_0}$	1	1

Table 2.4 Results of objectivist and subjectivist multilevel models, with inclusion of percentage of boys (parameters and variances in parentheses)

	Estimators (standard error)	
Parameters	Objectivist method	Bootstrap (parametric)
Fixed:		
Intercept	−1.641 (0.110)	−1.642 (0.108)
Gender (α_1)	0.221 (0.082)	0.223 (0.081)
Percentage of boys (α_2)	2.186 (0.147)	2.187 (0.145)
Random:		
$\sigma^2_{u_0}$	0	0
$\sigma^2_{e_0}$	1	1

percentage of boys in the school accurately explain the exam results, without the inclusion of the random parameter at school level. In this instance, where the populations observed are large, the objectivist and subjectivist results are thus virtually identical. As noted earlier, subjectivist inference will give clear results—with the aid of the notion of exchangeability—when we want to draw inferences on a new student's performance.

Before that, let us take a more detailed look at the results of the model including gender and school compared with the results of the model including gender and the percentage of boys in the school.

In the first, pure multilevel model, the underlying hypothesis is that pass rates are determined by the school and its teaching staff, for it includes no contextual variable enabling us to take account of students' specific characteristics. The random parameter at school level is large and substantiates the preceding argument.

The second, contextual multilevel model comprises a contextual variable specific to each school: the percentage of male students. Here, the percentage plays an essential role, single-handedly enabling us to distinguish between schools. By contrast, the random parameter for the school falls to zero. This shows that the differences are totally explained by the gender proportions and not by the teaching staff, as the previous analysis appeared to indicate.

2.4 Applications to Social Science

The model is, of course, somewhat artificial and yields this extreme result. In actual cases, the solution is less categorical, but it does lead to similar situations. One example concerns migration by Norwegian farmers compared with other occupations, which I examined in detail in an article (Courgeau 2002) and in a more general volume (Courgeau 2004b, 2007a). An initial, pure multilevel model enables us to distinguish migration behavior by region of origin, without bringing into play the percentage of farmers inhabiting each region; a second, contextual multilevel model allows us to qualify this result by showing the effect of the percentage. The effect sharply reduces the regional random parameter, without, however, offsetting it completely, as in the example analyzed here.

Let us now try to make predictions from these results. The notion of exchangeability will be essential for our purpose.

If we pick a new individual, can we predict his or her success with the aid of the parameters $\alpha_0, \alpha_1, \alpha_2, u_{0i}$, estimated from the population initially observed? This prediction will be a subjective probability one. Let us examine various cases, for we have seen that the hypotheses will change significantly depending on whether we perform a pure multilevel analysis or a contextual multilevel analysis.

We begin with the results of a pure multilevel analysis. As noted earlier, in different terms, the exchangeability test will concern the student's success in the exam, taking the school and gender as given, the effects of the two factors being independent here.

First, let us observe a new student of a given gender, attending a known school. For instance, in the case of a new male student taking the same type of exam in school no. 7, we can assume that the set of male students of school no. 7 is exchangeable as regards their chances of passing the exam, and that this new student is exchangeable with them as well. By factoring in the random parameter estimated for school no. 7, we can predict his success rate using the first model,[24] writing:

$$p(R = 1 | S = 0, L = 7) = 0.448$$

or 44.8%. The result for a girl in the same school would be 49.6% using the same model.

If we do not know the student's school, we can write:

$$p(R = 1 | S = 0) = p(R = 1 | S = 0, L = 1) p(L = 1 | S = 0) + \ldots$$
$$+ p(R = 1 | S = 0, L = 10) p(L = 10 | S = 0).$$

We know the probabilities of success for boys in each school thanks to the percentages observed, but we do not know the gender-specific probabilities for each school. For the pure multilevel model, we can assume that all these probabilities are

---

[24] We have taken the objectivist model here, as the population observed is large. However, as noted earlier, the results with the subjectivist model would be very similar.

identical and equal to one in ten, for the school effect seems independent of gender here. We thus obtain: $p(R=1|S=0) = 0.379$. A similar reasoning applied to girls yields $p(R=1|S=1) = 0.421$, in other words, a higher probability of success for a new girl than for a new boy. These results show that Table 2.1, which we should use for an objectivist inference, is not recommended, as it yields an opposite result owing to the non-exchangeability of the observations.

Let us turn to the results obtained with a contextual multilevel analysis. The exchangeability test now concerns the student's success and the school effect, measured by its percentage of boys, given the new student's gender.

We return to the first case in the previous analysis, i.e., a new male student taking the same kind of exam in school no. 7. Absent information on the percentage of boys taking the test, we can assume the percentage is unchanged from the previous example, and the second model gives us:

$$p(R=1|S=0, L=7) = 0.445$$

For girls attending the same school, the figure is 0.500, a result very close to the one given by the first model.

However, if the percentage of boys in the school has changed for this exam—a fact that could be known in advance—the second model will supply a different estimate of the pass rate. For instance, if the percentage of boys falls from 65% to 55%, the expected pass rate will drop to 39.2%, for the relevant characteristic is now the percentage of boys in the school and not the school itself, whereas the first model will still indicate 44.8%. In this case, the predictions generated by the two models may diverge sharply.

If we now disregard the school, we can continue to write the same formula as before. But the hypothesis posited for the first model does not apply to the second, for we see that it is the percentage of boys in the school that introduces the differences in exam outcome. Absent new information on the percentages of boys taking this new exam, the most relevant percentages will be those observed in the previous test. This yield the following probabilities of success:

$p(R=1|S=0) = 0.4575$ for a new boy and
$p(R=1|S=1) = 0.3425$ for a new girl.

In other words, this result corresponds to that of the combined schools and contradicts the result supplied by the first model. If we know the new percentages of boys in each school for the new test, we must apply them to determine a new boy's probability of success. If the percentages of boys are equal to those of girls in all schools for the new test, then the result will be close to that obtained with the first model. Indeed, using the results from Table 2.4, we find:

$p(R=1|S=0) = 0.377$ for boys as against 0.379 in the first model, and
$p(R=1|S=1) = 0.419$ for girls as against 0.421 in the first model as well. On the other hand, each possible distribution will now yield a different result.

2.4 Applications to Social Science

This example shows how the use of the concept of exchangeability allows statistical inference from the observation of a sample. However, the link between the initial sample and the new individual for whom we perform the inference must be formulated with great care if we want to obtain an accurate prediction. The link requires underlying hypotheses for the data, which we must combine with the data in order to set the precise conditions for exchangeability and arrive at the final inference. For instance, with the first pure multilevel model, we assumed that exchangeability concerned the result determined by school and gender, whereas in the second contextual multilevel model, exchangeability concerned the result and the school conditional only upon the student's gender. The latter analysis was confined to two sub-populations: boys and girls. A student's enrollment in a particular school is taken into account only through the percentage of boys in the school, not through the quality of the teaching staff.

Our other example of the application of subjective probability concerns the combination of testimonies. It is relevant to such areas as jurisprudence, artificial intelligence, and signal processing (particularly speech, images, and information). We assume that $p_1 = 0.8$ and $p_2 = 0.6$ are the degrees of confidence attributed to two witnesses of a given event—the equivalent to the degree of belief in the Dempster-Shafer theory. It is easy, of course, to generalize this example to any number of testimonies.

We begin with Hooper's processing of testimonies (1699), already described in theoretical terms at the start of this chapter. We need only generalize his formulas to the situation where degrees of confidence vary from one witness to another. Thus, the degree of confidence that he attributes to two testimonies when they are simultaneous (assuming that the two witnesses concur, but that each is assigned a different degree of confidence) is:

$$p = p_1 + p_2(1 - p_1) = 0.92.$$

As we can see, this confidence exceeds the degree assigned to the two testimonies considered separately, and it tends toward unity when the number of testimonies tends toward infinity. This rule is effectively justified if the two witnesses are not conniving with each other. If they are, the rule is less relevant. By contrast, when the two testimonies are given in succession, the degree of confidence falls to:

$$p = p_1 p_2 = 0.48.$$

In this case it is lower than the confidence assigned to the two testimonies considered separately, and it tends toward zero when the number of testimonies tends toward infinity.

Naturally, these rules are based on very simple hypotheses, which we need to make more specific in order to obtain robust results.

Jacob Bernoulli (1713) pursued this path and set more detailed conditions for what he designated more broadly as *arguments*—which could consist of testimonies, but also of signs or circumstances relating to the event studied.

He examines two types of argument: (1) what he calls *pure arguments*, which support a proposition in some cases and provide no support in others; (2) what he calls *mixed arguments*, which prove a proposition in some cases and prove the contrary in others. He gives the example of two testimonies in a murder case. The first comes from people who witnessed the crime at a distance and only saw that the perpetrator was wearing a black tunic. Now Gracchus and three other persons near the victim were wearing black tunics. This can therefore be viewed as a mixed argument against Gracchus. The second argument is that Gracchus was pale when being interrogated. This is a pure argument, for Gracchus may be pale for a reason other than the murder, yet he may be the perpetrator.

When two pure arguments are involved, Bernoulli shows that the confidence we can assign to them is:

$$p = 1 - (1-p_1)(1-p_2) = 0.92.$$

This situation is indeed identical to Hooper's two instantaneous witnesses. In the case of two mixed events, Bernoulli shows that the confidence we can assign to them is:

$$p = \frac{p_1 p_2}{p_1 p_2 + (1-p_1)(1-p_2)} = 0.857.$$

The confidence value here is lower than in the previous example. When we have a pure argument and a mixed argument, if the pure argument proves the case, then the mixed argument does not come into play; when the pure argument proves nothing, the mixed argument comes into play:

$$p = p_1 + (1-p_1)p_2 = 0.92.$$

We are back to the situation with two pure arguments.

Lambert (1764) fleshed out this reasoning by distinguishing between (1) cases where we believe the witness, (2) cases where we do not believe the witness, but can say nothing more, and (3) cases where we believe the opposite of what the witness states. We shall not discuss his arguments here (see detailed presentation in Bloch 1996), but we shall give the results obtained in the example studied (Table 2.5).

We see that Lambert replicates Bernoulli's formula when both arguments are pure or mixed, and the two witnesses agree. By contrast, Lambert's formula seems more satisfactory than Bernoulli's one when one argument is pure and the other mixed, and the formula is different when the first or the second argument is pure.

Laplace (1812) introduces four probabilities instead of two: P_1 and P_2, which he calls the witnesses' veracity (the probability that they are not trying to deceive), and the probabilities r_1 and r_2 that the witnesses are not mistaken. When the witnesses are not mistaken and agree, Laplace finds the same probabilities as Bernoulli and Lambert in the mixed case; when the witnesses are mistaken and disagree, Laplace finds the same probability as Lambert, again in the mixed case.

2.4 Applications to Social Science

Table 2.5 Lambert's combination rules for $p_1 = 0.8$ and $p_1 = 0.6$

	Witnesses agree	Value	Witnesses disagree	Value
2 pure	$1-(1-p_1)(1-p_2)$	0.920	$\dfrac{p_1(1-p_2)}{1-p_1 p_2}$	0.615
2 mixed	$\dfrac{p_1 p_2}{p_1 p_2 + (1-p_1)(1-p_2)}$	0.857	$\dfrac{p_1(1-p_2)}{p_1 + p_2 - 2 p_1 p_2}$	0.421
1st pure, 2nd mixed	$\dfrac{p_2}{1-p_1(1-p_2)}$	0.882	$\dfrac{1-p_2}{1-p_1 p_2}$	0.769
1st mixed, 2nd pure	$\dfrac{p_1}{1-p_2(1-p_1)}$	0.909	$\dfrac{p_1(1-p_2)}{1-p_1 p_2}$	0.516

If Laplace simultaneously introduces the probabilities that the witnesses are not mistaken, he obtains more complex results. Consequently, when the witnesses agree, the confidence that we can assign to their testimony becomes:

$$p = \frac{p_1 p_2 r_1 r_2 + p_1 r_1 (1-r_2) + p_2 r_2 (1-r_1)}{p_1 p_2 r_1 r_2 + r_1 (1-r_2) + r_2 (1-r_1) + (1-p_1)(1-p_2) r_1 r_2}.$$

We see that when $r_1 = r_2 = 1$, i.e., when witnesses have observed the phenomenon accurately, we reproduce the result obtained by Bernoulli and Lambert in the mixed case, i.e., $p = 0.857$. By contrast, the results differ according to their chances of being wrong. Thus, when the two witnesses have one chance in two of being wrong, $r_1 = r_2 = \dfrac{1}{2}$, we obtain:

$$p = \frac{p_1 + p_2 + p_1 p_2}{2 + p_1 p_2 + (1-p_1)(1-p_2)} = 0.734.$$

This indeed reduces the probability that the phenomenon occurred.

Lastly, let us examine the results obtained by applying the Dempster-Shafer theory—which introduces belief functions and plausibility—to the same example.

For the first witness, by definition, the belief function is P_1 and his or her plausibility is $P_1 + P_2 = 1$. The weightings applied to this person satisfy the following equations:

$$Bel_1(A) = p_1 = m_1(A)$$
$$Pl_1(A) = 1 = m_1(A) + m_1(A \cup \bar{A}).$$

Hence:

$$m_1(A) = p_1 = 0.8$$
$$m_1(\bar{A}) = 0$$
$$m_1(A \cup \bar{A}) = 1 - p_1 = 0.2$$

We obtain similar equations for the second witness and, if the witnesses agree, Dempster's combination rule gives:

$$(m_1 \oplus m_2)(A) = m_1(A)m_2(A) + m_1(A)m_2(A \cup \bar{A}) + m_2(A)m_1(A \cup \bar{A})$$
$$= p_1 + p_2 - p_1 p_2 = 0.92$$
$$(m_1 \oplus m_2)(A \cup \bar{A}) = m_1(A \cup \bar{A})m_2(A \cup \bar{A}) = (1 - p_1)(1 - p_2) = 0.08$$

This yields the belief function and the plausibility for both witnesses: The belief function effectively matches the result obtained for pure arguments by Hooper, Bernoulli, and Lambert.

If the witnesses disagree, we obtain the following values for the second witness:

$$Bel_2(\bar{A}) = p_2$$
$$Pl_2(\bar{A}) = 1$$

and the mass functions become:

$$m_2(A) = 0$$
$$m_2(\bar{A}) = p_2$$
$$m_2(A \cup \bar{A}) = 1 - p_2.$$

In this case, Dempster's combination rule will yield:

$$(m_1 \oplus m_2)(A) = \frac{m_1(A)m_2(A) + m_1(A)m_2(A \cup \bar{A}) + m_2(A)m_1(A \cup \bar{A})}{m_1(A)m_2(\bar{A}) + m_1(\bar{A})m_2(A)}$$
$$= \frac{p_1(1 - p_2)}{1 - p_1 p_2} = 0.615$$

and

$$(m_1 \oplus m_2)(A \cup \bar{A}) = \frac{m_1(A \cup \bar{A})m_2(A \cup \bar{A})}{m_1(A)m_2(\bar{A}) + m_1(\bar{A})m_2(A)} = \frac{(1 - p_1)(1 - p_2)}{1 - p_1 p_2} = 0.154,$$

with the following values for the belief function and plausibility:

$$Cr(A) = \frac{p_1(1 - p_2)}{1 - p_1 p_2} = 0.615$$

$$Pl(A) = \frac{1 - p_2}{1 - p_1 p_2} = 0.769$$

The value of the belief function is the one found by Lambert for two pure arguments when the witnesses disagree; the plausibility value is the one found by Lambert for one pure argument and one mixed argument when the witnesses disagree.

While each of the methods that we have just examined uses a different model, the results show a great similarity between the seventeenth- and eighteenth-century models and the ones recently developed by Dempster, Shafer, Smets, and others.

In conclusion, all the examples given here[25] show that subjective probability can be used successfully in all social sciences and that it allows inferences whose significance is far clearer than those obtained with objective probability. However, when the number of individuals observed becomes large, the two probability estimates converge to identity. Depending on the issue studied, the importance of that number can vary: we have seen, for example, that the comparison between two sex ratios at birth required a very substantial number of individuals.

Nevertheless, this approach has raised and continues to raise problems that we have already mentioned and shall now discuss.

2.5 Problems Posed by the Subjective Approach

The paradigms and axioms on which subjective probability is based have been the target of many criticisms, of which we have already presented a selection. We shall now try to review the main objections and problems, without aiming for exhaustiveness.

First, we can challenge the assumption that a behavior defined by the axioms of subjective probability is a rational behavior. These axioms describe decisions taken in a situation of uncertainty by a theoretical individual, the rationality of whose behavior may not necessarily apply to real persons. We shall examine shortly the criticisms aimed at the notion of the utility of a risk and particularly at the independence axiom. Beforehand, let us look at the more general paradigm of behavioral coherence. To be fully corroborated, the paradigm should be based on the observation of the behavior of real individuals or at least should be verified *ex post* by tests of these behaviors. But subjectivists reject the results of such trials outright, arguing that some of the people tested do not display that consistency and are not rational. But the subjectivists are in a false situation, for they cannot extract from the experiment what they mean by coherence. Without observing actual behavior, they can hardly build an acceptable system of axioms. We shall see later that the testing of the independence axiom—what Savage calls the sure-thing principle—through psychological experiments leads to its rejection.

However, a number of psychologists have conducted tests to substantiate the notion of coherence. The results have shown that the probabilities assigned by people to events are influenced by the way in which the events are described and do not

---

[25] Apart from Allais's criticisms discussed above, we could, of course, have quoted many examples in economics making extensive use of subjective probability. Some of these examples are reported in the next section on issues raised by subjective analysis.

satisfy the coherence condition. Several studies (Frischhoff et al. 1978; Ayton 1997; Mandel 2005) have shown that the probability attributed to a disjunction described in a condensed manner in a sentence, $p(X)$, tends to be lower than when the disjunction is spelled out, $p(X_1 \cup X_2 ... \cup X_n)$; the latter probability, in turn, tends to be smaller when it is spelled out in even greater detail, $p(X_1) + p(X_2) ... + p(X_3)$. This leads psychologists to conclude that 'probability judgments are attached not to events but to descriptions of events' (Tversky and Koeler 1994). Psychologists are continuing their research in an effort to understand why the processes underlying human judgments systematically diverge from the coherence criteria on which subjective probability rests (Mandel 2008).

It is interesting to see the reaction of probabilists—here, Lindley discussing the article by the psychologist Tversky (1974), reviewed jointly with another article by Suppes:

> Why do you spend your time studying how people make decisions when we know how they *should* make decisions? Would it not be better to devote your energies to teaching them the principles of maximum expected utility?

To which the philosopher Suppes replied:

> it is the obligation of the psychologist qua scientist to teach the truth, not to preach it. I am all in favour of teaching sound methods of decision-making, but the psychological task of understanding the subtle and complex problems of how real choices are made remains still largely unexplored territory.

We would add that, despite major advances by psychologists in the understanding of these issues since 1974, many questions remain unanswered. In any event, the issue of an axiomatization without experimental foundations is still on the table.

A second line of criticism has been directed against the notion of transitivity, which we find in de Finetti's third axiom and Savage's first axiom. The notion postulates the existence of a complete order of preference for acts, and we have seen that Savage had tried in vain to replace it with a partial order of preference. But here as well, we know that real acts are not always transitive: we may prefer act B to act A and act C to act B, without necessarily preferring act C to act A. That is what happens if people react in discontinuous fashion when sensitivity thresholds are crossed. However, we may assume that when people are incapable of making choices transitively, or cannot make up their minds as to the right choice, then their feelings of uncertainty cannot be represented by subjective probabilities. This does, however, limit the scope of application of subjective probabilities.

Thirdly, we have already described the scathing criticisms leveled against Savage's independence axiom, particularly by Allais (1953). Recall that these attacks focused on the hypothesis that if an agent is indifferent to the choice between two simple lotteries L_1 and L_2, then the same agent is indifferent to the choice between the combination of L_1 and another simple lottery L_3, with a probability p, and the combination of L_2 and L_3, displaying the same probability p. When the winnings offered by L_3 increase, L_1 and L_2 become consolation prizes, and the agent will modify his or her preferences between the two lotteries in order to minimize risk in the event that (s)he would not win the bigger prize offered by L_3. Allais

2.5 Problems Posed by the Subjective Approach

decided to verify the validity of this independence axiom experimentally back in 1953, and his results refuted the axiom. This protocol was reproduced several times, with lower winnings and real choices where the subjects received the winnings associated with the chosen lottery (Morrison 1967; Slovic and Tversky 1974). The results show 30–60% incompatibility with the independence axiom. The more thorough study by McCrimmon and Larson (1979) concluded that the frequencies of violations of the axiom depended on the value of possible winnings and on probabilities. Again, experiments have shown that one cannot arbitrarily posit an axiom unsupported by experimental verification. It should be noted that many other violations of the theory of expected utility have been recognized: for a detailed description, see Starmer (2000).

We cannot dwell here on all the debates that have taken place between economists on this topic, for we would digress from our subject. However, these debates have too often addressed the psychological foundations of subjective probability for us to ignore them altogether.

Two psychologists—Kahneman[26] and Tversky (1979) with their 'prospect theory'—were the first to loosen the over-restrictive conditions of the independence axiom. Their studies generalized the results obtained by Allais and led to the use of probability weighting to allow for inconsistencies with respect to Savage's theory. The weighting describes the way in which people evaluate their loss and gain prospects asymmetrically. The authors generalized their theory (Tversky and Kahneman 1992) by proposing a cumulative function for the transformation of probabilities rather than weightings. This approach remains highly experimental and hardly provides a theoretical justification of the functions chosen or of the reasons why low probabilities are overestimated and high probabilities underestimated.

Also in 1982, Loomes and Sugden, as well as Bell followed by Fishburn, published an approach now commonly known as 'regret theory', which attempts to incorporate the regret or pleasure felt by people when receiving a sum x instead of a sum y that they would have received by choosing another alternative. The theory proposed by these authors therefore allows for such feelings when persons must make decisions amid uncertainty. However, it does not provide an explanation for what is known as the 'framing effect'. Depending on how a question is framed, Tversky and Kahneman (1981) have shown that people may choose one solution in one case and the opposite solution in the other, even though the content of the questions itself remains identical. Regret theory, therefore, fails to rebut the criticisms against the theory of expected utility.

Machina (1982), for his part, tries to develop a modified theory of expected utility by eliminating the independence axiom, which Allais had attacked. But Machina's theory, as well, does not avoid some other violations of the original expected-utility theory (Starmer 1992).

To sum up, the multiple variants of expected-utility theory, while immune to certain criticisms based on test results, consistently succumb to other criticisms

---

[26] Kahneman won the Nobel Prize for Economics in 2002 for this theory.

based on other tests. As a recent article on the subject (Bethany Weber 2007) concluded:

> Ultimately it may be that no single theory can provide an adequate explanation of all the phenomena related to the Allais paradox.

In light of these findings, the solution that consists in stating a theory's axioms first and trying to verify them later is doomed to fail without a deeper examination of the rationale behind the axioms.

Subjective-probability axioms also pose a major problem that did not exist with objective probability. Knowledge [for objective probability] 'is reduced to a pure recording by the subject of data that are already fully organized, independently of the subject, in an external world (physical or ideational)' (Piaget 1967). In these circumstances, the estimated probability allows us to measure an ideal probability whose existence and value are independent of experiment. The measure will be all the more accurate as the observations are numerous and repeated in identical fashion. By contrast, for subjective probability, 'the subject is actively involved in knowledge' (Piaget 1967).

Accordingly, many authors have shown that people's willingness to behave in compliance with subjective-probability axioms is proportional to the degree of their prior awareness of this mode of reasoning. We therefore seem to be in the presence of a circle from which we cannot escape without a deeper study of

> the formation of normative facts pertaining to chance, i.e., the genesis of standards taken as an object of empirical investigation (Matalon 1967).

Savage (1954), for example, believed that his axiomatics was normative, as it supplied

> a set of criteria by which to detect, with sufficient trouble, any inconsistencies there may be among our beliefs and to derive them from the beliefs we already hold such new ones as consistency demands.

Allais's criticisms and the changes that have been proposed by a number of economists and psychologists for over 50 years clearly show that Savage was mistaken. Even more forcefully, Shafer (1985) demonstrated the need to give up this pretense and recognize that when we perform a Bayesian estimate, in the subjectivist sense, we are merely constructing an argument by analogy with a similar problem relating to games of chance, but we have no proof of the argument's validity. Shafer believes that a conditional probability can be valid not in all cases—as the subjectivists claim—but only when a clear protocol has been established. The protocol must specify all the elements that should be observed and determine a partition or at least a system of partitions for the set of events concerned.

Freund's example (1965) of the two-aces puzzle clearly illustrates this. Let us describe it briefly. Suppose we have only four cards on a table: the ace and two of spades, and the ace and two of hearts. I pick two cards at random. There are six possible pairs: ace and two of spades, ace and two of hearts; ace of hearts and two of spades; ace of spades and two of hearts; two of hearts and two of spades. The probability of each pair is: $\frac{1}{6}$. Someone with no other information than the fact that I

2.5 Problems Posed by the Subjective Approach

have picked two cards can determine the probability of event A: I have both aces; the probability of event B_1: I have at least one ace; and the probability of event B_2: I have the ace of spades. All these probabilities are subjective:

$$p(B_1) = \frac{5}{6} \quad p(B_2) = \frac{1}{2} \quad p(A) = p(A \cap B_1) = p(A \cap B_2) = \frac{1}{6}.$$

Now I tell the person: I have one ace. With this information, (s)he can use the theorem of compound probabilities to determine the probability of A conditional upon B_1, i.e., the fact that I have said that I have an ace:

$$p(A|B_1) = \frac{p(A \cap B_1)}{p(B_1)} = \frac{1/6}{5/6} = \frac{1}{5}.$$

This information therefore allows the person to increase the probability that I have two aces, which seems perfectly normal.

I give the person new information by announcing that I have the ace of spades. To introduce the new information, (s)he will again make the probability conditional upon the event B_2:

$$p(A|B_1 \cap B_2) = p(A|B_2) = \frac{p(A \cap B_2)}{p(B_2)} = \frac{1/6}{1/2} = \frac{1}{3},$$

since my holding the ace of spades necessarily implies that I have an ace. This reasoning apparently allows the person to improve his or her estimate of the probability that I have two aces. This claim seems paradoxical in that it does not seem that knowing that I have the ace of spades instead of one ace should have any effect on the probability that I have two aces.

In fact, in this case, no protocol has been established to determine what information I would give the. Consequently, the application of the theorem of compound probabilities is not legitimate, as Shafer (1985) has shown. Shafer proposes supplementing Savage's axioms with the hypothesis that we have a protocol for all new information. But, if so, the axiomatics of subjective probability can no longer be regarded as normative, and Savage's wish is therefore unfulfilled.

Even more importantly, de Finetti and Savage built their axiomatization not on fair wagers—as earlier authors had since Pascal—but on people's personal degrees of belief. As de Finetti stated very clearly (1937):

> given a complete class of incompatible events E_1, E_2, ... , E_n, all assessments of probability that assign random non-negative values to p_1, p_2, ... , p_n that sum to unity are acceptable assessments: each of these assessments reflects a consistent opinion, an intrinsically legitimate opinion, and each individual is free to adopt, among these opinions, the one that he prefers, or, to put it more accurately, the one that he *feels*.

De Finetti and Savage thus freed the estimation of a probability from the fair-wager criterion, but at the same time they incorporated a degree of personal belief for each individual who must take a decision. This introduces all the earlier-mentioned difficulties encountered by psychologists and economists. Jaynes (2003), for instance, attacks this approach with the following arguments:

> If probabilities are thought to be defined basically in terms of betting preferences, then for assigning probabilities one's attention is focused on how to elicit the personal probabilities of different people. In our view, that is a worthy endeavour, but one that belongs to the field of psychology rather than probability theory [.]

He also notes that he feels uncomfortable with the idea of defining an approach in terms of profits.

Far more forcefully, Seidenfeld et al. (1990) have shown that these degrees of belief or personal probabilities may fail to provide fair wagers for certain events characterized by 'state-dependent utilities'. This runs counter to the assumption made by Ramsey, de Finetti, and Savage in defining subjective probability—namely, that people's beliefs were consistent with the sums they bet (the 'Dutch Book argument'). We thus need to reconsider the subjectivists' attempt to dispense with the notion of fair wager through their axiomatics.

Our criticisms of the subjective approach advocated by Ramsey, de Finetti, and Savage are also largely applicable to the more complex theories of Shafer and Smets. They too examine the consistent choices made by a rational person. Smets (1998) notes:

> They (*Ramsey*, 1931; *Savage*, 1954; *DeGroot*, 1970) have shown that if decisions must be 'coherent', the uncertainty over the possible outcomes must be represented by a probability function. This result is accepted here except that *such probability functions quantify the uncertainty only when a decision is really involved.* ... We also accept that this probability function is induced from the beliefs entertained at the credal level.

But of course these beliefs are situated at another level, where coherence is no longer called for.

As noted earlier, we have not been able to explore all the issues raised by the subjectivist approach, which would have taken us too far. For readers seeking further information, we particularly recommend the following philosophical studies on the subject: Kyburg 1978; Kaplan 1996; Gillies 2000; Eriksson and Hájek 2007; Hájek 2008a, b.

However, the issues raised in this chapter show that another epistemic approach may address some of them and dispense with the reliance on psychology, which the subjective approach cannot avoid. In particular, a logical epistemic approach has been developed for as long as the subjective epistemic approach, and it may answer some of these questions. We shall therefore devote the next chapter to examining it.

Chapter 3
The Epistemic Approach: Logicist Interpretation

At the start of Chap. 2, we noted that Leibniz (written around 1703 but only published in 1765), after reviewing the progress of studies by Pascal, Huygens, and de Witt on wagers, had advanced the notion of a new logic of probability. In particular, Leibniz showed that the basis on which such a discipline had been built consisted in adopting an arithmetical mean between several equally plausible propositions:

> This is the axiom: *aequalibus aequalia*, like hypotheses must receive like consideration. But when the hypotheses are unlike, we must compare them with one another.

At that point, we may ask: how can we compare them? Leibniz goes one step further:

> I have said more than once that a new sort of logic would be required, which would address degrees of probability, since Aristotle did nothing less than that, and merely put some order into certain popular rules distributed according to commonplace thoughts, which can be of use in some occasion where the purpose is to enlarge upon discourse and give it some countenance, without troubling to give ourselves the necessary scales for weighing appearances and forming a solid judgment thereupon.

But Leibniz hardly provided the systematic treatment of knowledge that we might have expected of him—a treatment that would be needed to determine the degrees of probability associated with that knowledge.

The efforts of eighteenth-century and early-nineteenth-century authors partly satisfied his wishes. As noted in Sect. 2.1, Nicolas Bernoulli, Bayes, Laplace, and others succeeded in establishing a theory of epistemic probability by using the fair-wager hypothesis, a fuller version of the axiom cited at the start of that section. For example, Laplace's principle (1812) stated that:

> The probability of a future event is the sum of the products of the probability of each cause, derived from the observed event, times the probability that, with this cause obtaining, the future event will occur.

As shown in Sect. 2.1, this principle allowed Laplace to define the probabilities of many events. However, he was obliged to add the following condition, still known today as the *principle of insufficient reason*, in the case where certain probabilities are unknown:

> When the probability of a simple event is unknown, we can also assume that its values range anywhere from zero to unity.

But Laplace had no principle for determining these *prior* probabilities in cases where the information available does not enable us to assume their uniform distribution over the interval [0, 1].

As noted earlier, there was no distinction between subjective probability and logical probability in this period; the only form discussed was epistemic probability. Indeed, Laplace used his theory to address both astronomical problems, for which logical probability may be suitable, and problems in political and moral science (testimonies, court rulings, etc.), for which subjective probability would seem better appropriate. Divergences between the two approaches did not truly emerge until the early twentieth century.

For a start, Poincaré (1912) described the definition of a probability as a sort of *petitio principii* (begging the question):

> how can one recognize that all cases are equally probable? A mathematical definition is not possible here; therefore, in each application of the conventions, we must state that we regard such and such a case as equally probable.

Here, the conventions will elude the mathematician's mind, but they may be either subjective, as the subjectivists advocated, or logical, as we shall see now.

3.1 Logical Probability

The representatives of this current share with the subjectivists the notion that probability expresses a degree of belief. The entire approach presented in Chap. 2 remained identical for subjective probability and logical probability until the early twentieth century. However, contrary to the principles that guided the subjectivists from Ramsey onward, the logicist approach views the degree of belief not as a personal opinion but as a logical relationship between propositions, valid for all.

First, let us see how the approach developed starting in the early twentieth century. Rather than attempting an exhaustive survey, we shall focus on the most prominent and active advocates.

One of the first representatives of the logicist approach was the economist John Maynard Keynes, who turned the 1909 dissertation that won him a Fellowship at King's College (Cambridge) into a book on probability (1921).

In the Chap. 1 of this book, one page after the Leibniz epitaph on the need for a new logic of probability, he clearly stated his program:

> The Theory of Probability is logical, therefore, because it is concerned with the degree of belief which is *rational* to entertain in given conditions and not merely with the actual beliefs of particular individuals which may or may not be rational.

3.1 Logical Probability

Keynes was restating the goal of developing a logical theory of probability without reference to individual beliefs, which—as seen in Sect. 2.5 —are highly problematic. He also announced his intention

> to discuss the truth and the probability of *propositions* instead of the occurrence and the probability of *events*.

On this count as well, he parted company with the subjectivists, who treated the event as a significant fact (de Finetti 1937), and the objectivists, whose reasoning centers on sets of events rather than on propositions. By contrast, Keynes concurred with the subjectivists in recognizing the principle of inverse probability—or, more simply, the Bayesian principle—as essential to his approach, whereas the objectivists hardly ever applied it.

While resorting to the notion of 'single probability' for certain propositions, Keynes suggested the use of an interval when a single measure is both theoretically and practically impossible. He noted that:

> Many probabilities, which are incapable of numerical measurement, can be placed nevertheless *between* numerical limits. And by taking particular non-numerical probabilities as standards a great number of comparisons or approximate measurements become possible.

The term 'non-numerical' here denotes the inability to assign a single measure to a probability. Keynes elaborated on these measures in some chapters of his book. The notion was taken up in subjective probability by Koopman (1940, 1941) and later by Dempster (1967) and Shafer (1976). In contrast, logicists hardly ever applied it.

Lastly, Keynes introduced the notion of the 'weight' of an argument independent of its probability, which 'measures the *sum* of favourable and unfavourable evidence, [*while*] the probability measures the *difference*'. In a sense, this resembles the notion of entropy, which Richard Cox (1961) later applied to probability theory. Keynes states:

> As the relevant evidence at our disposal increases, the magnitude of the probability of the argument may either decrease or increase, according as the new knowledge strengthens the unfavourable or the favourable evidence; but *something* seems to have increased in either case,—we have a more substantial basis on which to rest our conclusion. I express this by saying that an accession of new evidence increases the *weight* of an argument. New evidence will sometimes decrease the probability of an argument, but it will always increase its 'weight'. (Keynes 1921)

The description of this weight and the attributes assigned to it—which he discusses in an entire chapter devoted to the subject—are strongly reminiscent of, but not identical to, the notion of entropy, to which we shall return. The latter notion was taken up by the economist Ellsberg (1961, 2001) in his critique of Savage's axioms.

Unfortunately, Keynes never spells out exactly what he means by 'degree of rational belief'. He indicates that some persons may have greater logical intuition than others, and that the perception of certain probability relationships may be beyond the capabilities of some of us, but he provides no evidence for the existence of these logical entities (Mattheu Wilson 2007).

Keynes also rejects Laplace's indifference principle, which states that mutually exclusive propositions concerning the same data should be assigned equal probabilities, if nothing allows us to choose between them. It would be too long to rehearse his arguments here, but let us quote his conclusion:

> This account may seem rather confusing; but it is not easy to give a lucid account of such a confusing doctrine.

Later, we shall see that Jeffreys rebutted this position by showing that it totally prevents the establishment of a theory of logical probability.

Keynes's book was criticized by many authors such as Fisher (1922a), Ramsey (1926), and Jeffreys (1939), who respectively addressed probability from objectivist, subjectivist, and logicist standpoints. Jeffreys (1939) voices a definitive—and, in our view, wholly justified—opinion on the author:

> This book is full of interesting historical data and contains many important critical remarks. It is not very successful on the constructive side, since an unwillingness to generalise the axioms has prevented Keynes from obtaining many important results.

Keynes actually seems to have abandoned a number of his ideas on probability, particularly after Ramsey's critique (Bateman 1987; Davis 2003; Gillies 2003), although some scholars argue otherwise (Gerrard 2003; O'Donnel 2003). Keynes virtually stopped writing about probability after these attacks. On his own admission (Keynes 1971):

> probability is concerned not with objective relations between propositions but (in some sense) with degrees of belief ... [, and] the basis of our degrees of belief—or the a priori probabilities as they used to be called[—]is part of our human outfit, perhaps given us merely by natural selection, analogous to our perceptions and our memories rather than formal logic. So far I yield to Ramsey—I think he is right.

Later in the text, however, he adds:

> in attempting to distinguish 'rational' degrees of belief from belief in general he [*Ramsey*] was not yet, I think, quite successful. It is not getting to the bottom of the principle of induction merely to say that it is a useful mental habit.

Clearly, while Keynes believed that his probability theory faced theoretical difficulties, he was dissatisfied with Ramsey's inability to distinguish between personal opinion and rational belief (Mattheu Wilson 2007).

In the event, it was Jeffreys—geophysicist, astrophysicist, and statistician—who effectively launched logical probability. He first set out his ideas on scientific inference with Wrinch (1919, 1921, 1923), then expanded them in a book (1931). Next, he established his conceptual framework of logical epistemic probability (1932, 1933b, 1934, 1937), which he presented in a seminal work on probability (1939).

In their very first article, Wrinch and Jeffreys (1919) described probability as an extension of classical logic to cases where the premises do not allow us to draw conclusions with absolute certainty. This notion was later elaborated in detail by Richard Cox (1946, 1961). The focus was no longer—as in de Finetti's work—on events whose occurrence a person may put in doubt, but on propositions about which this new logic should enable us to reason. Wrinch and Jeffreys thus took up

3.1 Logical Probability

the fundamental notion that Leibniz had pledged to develop and that Keynes had already advocated.

The article also defended the indifference principle, which Keynes rejected. The authors did so by introducing the notion of a base of data combinations that can be ranked:

> If in one combination the proposition is more probable relative to the data than in another, the number corresponding to the first is greater than that corresponding to the second.

The obvious corollary to this notion is Laplace's principle, which assigns equal probabilities to propositions about data that give us no reason to expect one proposition to be more valid than the other. As we shall see later, however, this principle is not applicable to any estimation problem whatsoever.

Wrinch and Jeffreys also distance themselves from the objectivist theory of probability, and reject the adoption of objectivist probability as the limit of a ratio when observations tend to infinity:

> The existence of a probability on this theory requires that a limit shall exist to which a certain ratio tends in the long run; and one is led to ask what the evidence is for the existence of such a limit.

Jeffreys never regarded that limit as mathematically viable.

In their following, two-part article, Wrinch and Jeffreys (1921, 1923) showed that, while the number of possible quantitative laws for a given phenomenon is infinite, this set is countable and the laws can be ranked by decreasing order of simplicity. The laws' probabilities form the terms of a convergent series of sum 1. We can thus choose the simplest law, in keeping with a simplicity principle, and this choice is based on a reasonable degree of belief.

In his singly-authored book, Jeffreys (1931) reconsidered the problem of scientific inference and shifted the issue of the validity of scientific methods toward the notion of probability:

> When we make a scientific generalization we do not assert the generalization or its consequences with certainty; we assert that they have a high degree of probability on the knowledge available to us at that time, but that this probability may be modified by additional knowledge.

This relativism led him to a notion of probability defined as a relationship between a proposition and a data set, which makes it possible to draw statistical inferences from past observations to predict future outcomes.

Jeffreys also discusses the application Laplace's principle to any given estimation problem. He shows that the prior distribution of the unknown standard deviation,[1] σ, of a normal distribution cannot be uniform, under pain of generating an inconsistency with the fact that we know nothing about the standard deviation, except that it

---

[1] What Jeffreys used was not the standard deviation but the quantity $h = \dfrac{1}{\sigma\sqrt{2}}$. This does not alter the reasoning significantly.

is positive. If we take a random finite value α for σ, Jeffreys shows that the probability of $\sigma < \alpha$ is always null, which is inconsistent with the fact that we know nothing about it σ. The distribution should in fact be proportional to the inverse of the standard deviation, which is based on the fact that this prior is invariant for the change in parameter $\sigma' = 1/\sigma$ as well as for any power of σ'. By contrast, for the random variable itself, Jeffrey adopts the standard distribution, which is a uniform prior. We shall see later how to complete his imperfect demonstration to show that this distribution is perfectly non-informative, for he omits to recognize that other transformations that preserve the positivity of σ lack the invariance property.

On the same subject, a bitter argument erupted in the *Proceedings of the Royal Society* in the early 1930s between Jeffreys (1932, 1933a, 1934), who advocated a logicist approach to probability, and Fisher (1933, 1934), then a pure objectivist.[2] Interestingly, it was in the same period that Kolmogorov proposed his axiomatization of objective probability. Jeffreys (1933a) attacked the argument that objective probability was the limit of a frequency:

> That a mathematician of Dr. Fisher's ability should commit himself to the statement that the ratio of 2 infinite numbers has an exact value can only be regarded as astonishing.

In rebuttal, Fisher (1934) totally rejected any definition of subjective probability:

> Jeffreys' definition of probability is subjective and psychological: 'We introduce the idea that the idea of a relation between one proposition p and another proposition q, expressing the *degree* of knowledge concerning p provided by q'. In this it resembles the more expressive phrase used by Keynes, 'the degree of rational belief'. Obviously no mathematical theory can really be based on such verbal statement.

This exchange clearly shows the gulf between the objectivist and logicist approaches, which prevented the two protagonists from speaking a common language. Jeffreys (1933a, b) took advantage of this exchange to engage in a searching examination of what a theory of probability should be:

> It seems to be necessary therefore for me to begin at an earlier stage and explain why a theory of probability is necessary, and what is the scope of such a theory.

In these articles, he began a project that he extended in his 1939 book, namely, to establish a theory of logical probability or, rather, of logical probabilist inference. In the following section, we look at the basic principles and the axiomatics that he proposed. Here, suffice it to note that the book is regarded as the founding text of logical probability, even though, as discussed later, it still avoided many basic issues concerning that type of probability (Robert et al. 2009). In particular, like Laplace, Jeffreys failed to present his principles as the necessary consequences of the analysis of incomplete information. As a result, for more than 30 years, he was subjected to the same attacks as Laplace had been.

---

[2] Fisher later moved toward a different definition of probability that, as noted earlier (Section 1.1), was regarded as lacking clarity and was barely used after him.

3.1 Logical Probability

In his wake, the physicist and probabilist Richard Cox (1946, 1961), rather than quibbling endlessly on whether it is legitimate or not to use logical or objectivist probability, asked the crucial question: is it possible to construct a consistent set of mathematical rules to establish a reasoning that is no longer deductive but plausible?

He showed that, if we try to represent the degrees of plausibility of a proposition by real numbers, then we can state the consistency conditions in the form of functional equations, whose general solutions can be expressed. At this point, Cox was reasoning only on propositions, without yet bringing the concept of probability into play. His propositions conform to the rules of Boolean algebra, already used by Jeffreys. Cox introduces a function of these arguments here. He shows that, among all the functions that can theoretically meet these conditions, there is one with particularly simple properties, which he calls 'probability' and which complies with all the rules of probability calculus. He thus proves that all inference methods in which we represent the degrees of plausibility by real numbers are necessarily equivalent to Laplace's method, otherwise they are inconsistent. We shall examine his axioms in greater detail in the axiomatics section below.

In his book (1961), Cox also draws a major parallel between probability and 'entropy', on which, as we shall see, Jaynes was to elaborate (1963, 1979). The term entropy was introduced in thermodynamics by Clausius in 1865, in the wake of Carnot's paper (1824). This is not the place for a detailed account of the history and evolution of the concept in physics—a field where it was elaborated most notably by Maxwell (1859), Boltzmann (1871), Gibbs (1902), and Shannon (1948). We merely want to show its link to probability theory.

The notion of entropy complements that of logical probability. It can supply a measure of the information provided by a full set of propositions on a subject. To define entropy, let us consider the hypothesis, h, and the comprehensive set of not equally probable but mutually exclusive propositions, $a_1, a_2, \ldots a_m$, that partly result from the hypothesis. Entropy H is accordingly defined as a numerical measure whose value is equal to:

$$H(a_1, a_2, \ldots, a_m | h) = -\sum_i p(a_i | h) \ln p(a_i | h),$$

where $p(.)$ is a probability measure. This formula is deduced from Shannon's (1948) with the addition of the hypothesis h. As $p(a_i | h) \leq 1$, we see that the measure can never be negative, and if one of the propositions is certain, all the others are impossible and the system's entropy is null. We can also show that the entropy of a set of mutually exclusive propositions reaches maximum when the propositions are equally probable. It will then equal $\ln(m)$, where m is their number.

We can generalize entropy to a system of propositions that form a comprehensive set. Each proposition in the set implying a proposition in the system belongs to the system.

As noted earlier, Cox made full use of Boolean algebra—already employed by Jeffreys, who, oddly, does not mention Boole. Cox notes that the algebra is perfectly suited to a theory of logical probability. We shall expand on his axiomatics in the next section. For the moment, let us provide a more general picture of Cox's algebra of propositions. The purpose is:

> to investigate the fundamental laws of those operations of the mind by which reasoning is performed; to give expression to them in the symbolical language of Calculus, and upon this foundation to establish the science of Logic and construct its method. (Boole 1854)

Thus a given proposition may have a fixed meaning, such as the sentence 'Socrates is a man,' or a meaning that may vary according to the context of the utterance, such as the sentence 'I agree with what all previous speakers have said.' We can show that these rules are roughly the same as those of classical algebra, where a quantity may be regarded as a constant or a variable.

We can spell out the three basic operations of this algebra of propositions: negation of a proposition, conjunction of two propositions, and disjunction of two propositions. The combination of these operations suffices to represent all the propositions that we can deduce from the two initial ones. However, some results of these operations differ substantially from those observed in classical algebra. For instance, the conjunction of a proposition with itself yields that same proposition, whereas in classical algebra it yields the square of the number considered.

As in classical algebra, we can define a proposition as a function of one or more propositions. But we can readily see, as the previous example shows, that the functions generated by the propositions of Boolean algebra display far less variety than those of classical algebra.

Cox's direct successor was Jaynes, who already cited Cox in a 1957 report on plausible reasoning that was not published until many years later (1981). Jaynes carried on Cox's work all his life; a posthumous book by Jaynes, *Probability theory: the logic of science*, was published in 2003. Let us examine Jaynes's contribution to logical probability.

Like Polya (1954) before him, Jaynes showed how our brain performs not only deductive reasoning, but also plausible reasoning. Deductive reasoning was formalized by Aristotle in the fourth century B.C.E. In volume V of the *Organon*, *Topics*, he lists strong syllogisms of the two following types:

$$\frac{A \text{ implies } B}{A \text{ is true}}$$
$$B \text{ is true}$$

or:

$$\frac{A \text{ implies } B}{B \text{ is false}}$$
$$A \text{ is false}$$

3.1 Logical Probability

In most cases, however, we cannot apply these deductive reasonings, but must make do with plausible reasonings of the following third and fourth type:

$$\frac{A \text{ implies } B}{B \text{ is true}}$$
$$\overline{A \text{ becomes more plausible}}$$

or:

$$\frac{\text{If } A \text{ is true, then } B \text{ becomes more plausible } B \text{ is true}}{A \text{ becomes more plausible}}$$

It is easy to see that these statements no longer enable us to assert the conclusion, but they are essential for correct reasoning. Logical probability should allow such reasoning.

The next problem is to how to calculate these plausibilities, so that from the initial information and from observations we can obtain a single value for each proposition whose logical probability we wish to estimate. To avoid all controversy, Jaynes introduced a robot and gave it rules that it would apply blindly to compute the probabilities, by comparing its reasoning to ours when confronted with the same problem. The robot must therefore be given unambiguous information of a simple logical type. Of course the truth or falsehood of this proposition, which has two possible values, is generally not known, since the purpose of the search is to determine its logical probability.

The paradigm and axioms proposed by Jaynes were thus very similar to those of Richard Cox, as we shall see in the following section. Consistently with the notations of Keynes (1921) and Cox (1961), Jaynes uses the symbol $A|B$ to represent the conditional plausibility that A is true, given that B is true. Using two rules—'product'[3] and 'sum'—he is able to define a continuous function, ranging between 0 and 1, that matches the conditional plausibility with a conditional probability $P(A|B)$.

The qualitative properties of his theory make it possible to put the syllogisms shown above into a clear quantitative form. Let us see in greater detail how this is done.

For the first two strong syllogisms, let C be their major premise:

$$C \equiv A \Rightarrow B.$$

These syllogisms will therefore obey the 'product' rule, which defines its probabilities[4]:

$$P(A \cap B|C) = P(A|C)P(B|A \cap C) \text{ and } P(A|\overline{B} \cap C) = P(\overline{B}|C)P(A|\overline{B} \cap C).$$

---

[3] Accordingly, the 'product' rule is written:

$P(A \cap B|C) = P(A|C)P(B|A \cap C) = P(B|C)P(A|B \cap C)$

[4] In order to differentiate sets from propositions, some authors use the terms $\cup$ or $\cap$ for the disjunction or the conjunction of sets, and the terms $\vee$ or $\wedge$ for the conjunction and disjunction of propositions. As the same rules hold for sets and propositions we will use here the same terms, but we inform the reader that we are now speaking about propositions.

As these syllogisms tell us that $P(A \cap B|C) = P(A|C)$ and $P(A \cap \overline{B}|C) = 0$, then, in terms of logical probability, the syllogisms reduce to:

$$P(B|A \cap C) = 1 \text{ and } P(A|\overline{B} \cap C) = 0.$$

This shows us that Aristotle's deductive logic is the limit form of logical probability, when the robot becomes increasingly certain of its conclusions.

For the two plausible reasonings that follow, we can again apply the 'product' rule:

$$P(A|B \cap C)P(B|C) = P(B|A \cap C)P(A|C) \text{ and}$$

$$P(B|\overline{A} \cap C)P(\overline{A}|C) = P(\overline{A}|B \cap C)P(B|C).$$

In the first case, we know that $P(B|A \cap C) = 1$, and since $P(B|C) \leq 1$, then:

$$P(A|B \cap C) \geq P(A|C),$$

as the plausible reasoning predicted.

In the second case, we know that $P(\overline{A}|B \cap C) \leq P(A|\overline{C})$, hence:

$$P(B|\overline{A} \cap C) \leq P(B|C),$$

as the plausible reasoning predicted.

We also see that Jaynes's approach to probability is fundamentally different from that of the objectivists, who measure it from observed frequencies:

> The probabilities assigned to individual measurements are not measurable frequencies; they are only a means of describing a *state of knowledge*; just the original sense in which Laplace and Jeffreys interpreted a probability distribution. (Jaynes 1979)

The only difference with Laplace's approach is that the principle of insufficient reason can now be generalized with the aid of the notion of entropy. Let us take a more detailed look at how Jaynes achieved this generalization.

As noted earlier, Laplace proposed a prior uniform distribution of probability when he lacked information on the phenomenon studied, but he hardly knew how to proceed when he had fuller information available. Later, Jeffreys exposed the danger of estimating the prior probability of a standard deviation as being uniformly distributed, but he realized that his demonstration was incomplete. Jaynes showed the need for a more clearly formulated method that his robot could use to estimate these prior probabilities correctly.

For a discrete distribution, these requirements were met by Shannon's entropy,[5] a notion that Richard Cox applied to probability. Shannon's entropy made it possible

---

[5] It is preferable to distinguish this notion using the term Shannon's entropy, which concerns information, from classical entropy, which concerns a thermodynamic system.

3.1 Logical Probability

to quantify the preconditions for a reasonable measure of the degree of uncertainty, by writing more simply the entropy as:

$$H(p_1, p_2, \ldots, p_n) = -\sum_{i=1}^{n} p_i \log(p_i).$$

for the n values of this distribution $(p_1, p_2, \ldots, p_n)$, to which we add as many constraints as we have prior information items available. The distribution that maximizes this equation, under the specified constraints, will give a complete description of all the prior information at our disposal.

A schematic but simple example will illustrate this method. Let us imagine that nineteenth-century archeologists had recorded the ages at death in an ancient cemetery, but that their report gives only the mean age at death (36 years) of the population, whose cemetery was later destroyed. Let us try, from this information, to estimate the distribution of the population into three broad age groups (30 years or less, 31–60, and 61–90), which we can deduce from the prior information.

For simplicity's sake, we note the three age groups (1, 2, 3), so the mean age measured is equal to 1.2, and the sum of probabilities is equal to 1. We can then resolve the system by means of Lagrange's multipliers:

$$\partial\left[H - \lambda \sum_{i=1}^{3} i p_i - \mu \sum_{i=1}^{3} p_i\right] = \sum_{i=1}^{3}\left[\frac{\partial H}{\partial p_i} - \lambda i - \mu\right] \partial p_i = 0,$$

hence the solutions:

$$p_i = \exp(-\mu - 1 - \lambda i).$$

By defining the function:

$$f(\lambda) = \sum_{i=1}^{3} \exp(-\lambda i),$$

we see that Lagrange's multipliers can be written:

$$\mu + 1 = \log f(\lambda),$$

$$1.2 = -\frac{\partial \log f(\lambda)}{\partial \lambda}$$

This leads to the following age distribution:

$$p_1 = 0.8263$$
$$p_2 = 0.1474$$
$$p_3 = 0.0263$$

We see how far removed this distribution is from the equal-probabilities distribution. In fact, we can compute it for any mean-age value, and verify that the estimates do lie within the limits [0, 1] of the probabilities. To improve the estimate, we can, of course, introduce other information drawn from the observation of other populations of the same type.

For a continuous distribution, the problem is more complex, but can be partly solved by the entropy maximization—as Jaynes showed (2003)—provided that we can compute a correct measure for the problem posed.

We can grasp the notion by means of Bertrand's paradox (1889). This problem—which, at first glance, seems simple—consists in determining the probability that a chord of a given circumference is larger than the side of the equilateral triangle inscribed in the circumference. Depending on the method used to address the problem, which we need not describe in detail here, the results differ. In the first case, we find a probability of $\frac{1}{3}$; in the second case, it is equal to $\frac{1}{2}$. This contradiction stems from the underlying hypotheses framed for each estimation, concerning how to measure the probability that a point lies inside a given area. In the case of a continuous distribution, this measure is also—to within one constant factor—the prior distribution describing total ignorance of the variable examined. Continuous group theory enables us to solve most of these problems. Let us see how, in a specific case.

Take a sample $\{x_1, x_2, ..., x_n\}$, obtained from a two-parameter distribution $P(x|\mu,\sigma) = \Phi(x,\mu,\sigma)dx$, for which we want to estimate the mean μ and the standard deviation σ. We can introduce a prior distribution of the probability of μ and σ:

$$P(\mu,\sigma|I)d\mu\,d\sigma = f(\mu,\sigma)d\mu\,d\sigma.$$

But, as noted above, there are many ways to measure this probability and we do not know which function f to use. Let us suppose, however, that we change variables as follows:

$$\mu' = \mu + b$$
$$\sigma' = a\sigma$$
$$x' - \mu' = a(x - \mu)$$

where $0 < a < \infty$ and $-\infty < b < \infty$, which constitutes a transformation group. With these new variables, the prior distribution becomes:

$$P(x'|\mu',\sigma') = \Psi(x',\mu',\sigma'),$$

giving us:

$$\Psi(x',\mu',\sigma') = a^{-1}\Phi(x,\mu,\sigma)$$

and the *prior* distribution g:

$$g(\mu',\sigma') = a^{-1}f(\mu,\sigma).$$

3.1 Logical Probability

We can then show (Jaynes 2003) that the invariance condition under the preceding transformation group leads to the solution:

$$\Phi(x,\mu,\sigma) = \frac{1}{\sigma} h\left(\frac{x-\mu}{\sigma}\right),$$

where h is an arbitrary function. In this case we see that under the term 'complete ignorance' we actually put a 'state of knowledge' such that, under the transformations examined here, this state of knowledge will not be modified. Let us now see what we can determine about the prior distribution.

Let us consider a sample $\{x'_1, x'_2, \ldots, x'_n\}$, and try to estimate its mean μ' and standard deviation σ'. If we are in a state of 'complete ignorance' as defined earlier, then the problem is equivalent to the previous one. The functions f and g must therefore be identical whatever the values of a and b. Consequently, f must satisfy the functional equation:

$$f(\mu,\sigma) = af(\mu + b, a\sigma)$$

whose general solution is:

$$f(\mu,\sigma) = \frac{C}{\sigma},$$

where C is a constant term. This gives us a now complete demonstration of Jeffreys's rule, described above in this Sect. 3.1.

But this rule applies only to the preceding transformation group. Had we chosen another group, another rule would have applied. We must always, therefore, specify the transformation group on which we are working. However, in the example studied here, the transformation group used seems to be the most logical choice, as we can change the mean through translation, and modify the standard deviation through multiplication by a constant.

In our view, it was the physicist and computer scientist Knuth (2003a, b, 2005, 2007, 2008) who succeeded—using lattice theory—in showing that the theory of logical probability and information theory were dual with respect to each other. He thus combined probability and entropy into a more general theory: we will discuss it in more details on Sect. 3.2 and Conclusion of Part I.

We have deliberately refrained from discussing the work of the philosopher Rudolf Carnap on probability (1950). First, unlike all the authors cited here, Carnap considers two different concepts of probability: (1) what he calls 'probability$_1$' or 'inductive probability' denotes a logical probability that people use constantly, so there can be no doubt as to its meaning; (2) what he calls probability$_2$' is objective probability. A great many authors have criticized this dualist approach (Suppes 2002a). Second, Carnap's more general philosophical program to reduce the theoretical language of science to a language of observations (Carnap 1928, 1933) is now recognized by most philosophers as a failure (Pierre Jacob 1980; Sarkar 1996). Mainly for these reasons, we shall not address his approach to logical probability here.

In sum, the logicist approach to probability enjoyed a powerful expansion in the late eighteenth and early nineteenth centuries with Laplace. It then declined to the point of total extinction in the second half of the nineteenth century and the early twentieth century, but has since been revitalized, winning a growing number of advocates.

3.2 Paradigm and Axiomatics of Logical Probability

As in the previous chapters, we begin by defining the underlying paradigm of this approach before examining the main axiomatizations that have been derived from it over time.

3.2.1 The Paradigm

We shall try to identify here the main ideas that inform the logicist approach. Let us begin by specifying which aspects of its predecessors it rejected.

The notion of frequency had been crucial for objective probability and had prevailed for over a century: the probability of an event was regarded as equivalent to its frequency. This notion was stripped of its importance in logical probability:

> The essence of the present theory is that no probability, direct, prior, or posterior is simply a frequency. (Jeffreys 1939)

Jeffreys recognized that a numerical estimate is often the same as a frequency but, even so, this does not imply that probability and frequency are identical concepts. Likewise, Jaynes attacked the objectivists for rejecting the notion of probability defended by Bernoulli, Bayes, and Laplace as describing a state of knowledge and for replacing it with the notion of frequency. Examining Shannon's results, he observed:

> The probabilities assigned to individual messages are not measurable frequencies; they are only a means of describing a *state of knowledge*; just the original sense in which Laplace and Jeffreys interpreted a probability distribution. (Jaynes 1979)

This overall rejection of the notion of frequency by the logical-probability theorists shows the extent of their hostility to von Mises's axiomatization. However, as we shall see, their axiomatization converges in many respects with that of Kolmogorov, replacing the latter's notion of 'set' with that of 'propositions' in the Boolean sense.

Similarly, they rejected any definition of probability in terms of infinite sets of possible observations, as noted earlier for Jeffreys. He is very clear about this:

> In fact, no 'objective' definition of probability in terms of actual or possible observations, or possible properties of the world, is admissible. (Jeffreys 1939)

3.2 Paradigm and Axiomatics of Logical Probability

At the other end, the logicists rejected the subjective notion of personal probability. Let us take a more detailed look at the reasons for this rejection.

After showing how the principle of inverse probability describes a process of learning from experience, Jeffreys states:

> Differences between individual assessments that do not agree with the results of the theory will be part of the subject-matter of psychology. Their existence can be admitted without reducing the importance of a unique standard of reference. (Jeffreys 1939)

Polya (1954) went further by engaging in a very interesting discussion on the subject. He took up the third plausible reasoning outlined in Sect. 3.1, where we know that A implies B and that B is true. In this situation, two people may agree that A becomes more probable, but may disagree on the strength of this evidence. The subjectivists recognize this by allowing the probability to take all values from zero to unity, as the person wishes, subject only to compliance with consistency rules. The logicists, for their part, do not accept that this probability can be individual. They conclude that two people agree that A becomes more probable, but set aside the strength of its evidence. Their plausible reasoning is thus unilateral, leaving open a wide margin of disagreement between different individuals.

Jaynes also clearly stated his disagreement with the subjectivist approach:

> Subjective Bayesians face an awkward ambiguity at the beginning of a problem, when one assigns prior probabilities. If these represent merely opinions, then they are basically arbitrary and undefined; it seems that only private introspection could assign them, and different people will make different assignments. (Jaynes 2003)

In his view, prior probabilities should be determined solely by logical analysis, not introspection.

We can now distinguish more clearly the fundamental differences between the three approaches to probability, and we can more easily define the paradigm of the logicist approach.

This approach examines propositions, not events (as in the objectivist approach), and it submits them to logical analysis in order to extract all the objective information that they contain. The information is therefore independent of the individual, contrary to the subjective-probability approach. The paradigm of logical probability is predicated upon the notion of consistency. Let us take a more detailed look at its implications.

First, if a probability can be obtained through several different reasonings, then all possible reasonings should yield the same result. Second, to determine a probability, we need to factor in all the objective information available on this proposition, without neglecting any of that information. We can therefore state that the notion of consistency effectively characterizes the logical relationship that should exist between a proposition and the information available on it, whether this consists of observed data or a more complex form of information.

As we can see, this notion of consistency is different from that of coherence, used in subjective probability. Coherence encompasses not only objective information but also the subjective information that a person may possess on the phenomenon studied. For the logicists, a rule that exhibits the property of coherence, but not that of consistency, cannot be accepted as a rule of logical inference, for it may lead to different results for different individuals, depending on their psychological conditions.

We also saw that the theory of logical probability needs to introduce the notion of Shannon's entropy in order to produce a more correct formulation of the prior information at our disposal. This notion allows us, by maximizing its expression, to formalize the incomplete information often available on a phenomenon of interest. Maximization will therefore generalize the principle of insufficient reason, also called the indifference principle. On this score, logical probability stands in opposition to subjective probability: for the latter, prior probabilities represent totally subjective degrees of belief, which can therefore take on any value provided that the coherence principle is met. For this reason, we believe it is useful to incorporate entropy into the paradigm of logical probability.

3.2.2 The Axioms

Before examining the various axiomatics proposed for logical probability, we must define what we mean by *proposition* and review the axiomatics of the algebra established by Boole (1854) for the logic of propositions.

We cannot engage in a detailed discussion of the significance of the notion of proposition. However, we should state the specific meaning that we assign to it here. We shall define a proposition as a syntactical construct for which it makes sense to speak of truth. Accordingly, we shall not regard optative statements (expressing a wish), imperative statements, and questions as propositions. We also adopt the framework of classical logic, which does not consider propositions of indeterminate status: a proposition will be either true or false. The propositional calculus becomes an algebraic structure called *Boolean algebra*. Let us see in greater detail how to define it.

We start from a set of propositions $P = \{A, B, ...\}$, defined in the sense that we have given them. We introduce two laws of internal composition: the logical product or conjunction of two propositions A and B is noted $A \cap B$; the logical sum or disjunction of two propositions A and B is noted $A \cup B$; the negation of a proposition A is noted $\overline{A}$. Two specific elements must be included in set P: 'the proposition is true', noted I, 'and the proposition is false', noted Φ; these elements are the truth values of any proposition. In this algebra, the sign = does not denote equal values, but equal truth. In writing complex propositions, we can use parentheses to indicate the order in which the propositions must be combined.

The algebra is accordingly defined by the following axioms, which apply both to the two laws of internal composition and to all elements of P:

1. Idempotence: for all A, $A \cup A = A$ and $A \cap A = A$.
2. Commutativity: for all A and B, $A \cup B = B \cup A$ and $A \cap B = B \cap A$.

3.2 Paradigm and Axiomatics of Logical Probability

3. Associativity: for all A, B and C, $A \cup (B \cup C) = (A \cup B) \cup C = A \cup B \cup C$ and $A \cap (B \cap C) = (A \cap B) \cap C = A \cap B \cap C$.
4. Distributivity of one law relative to the other: for all A, B, C, $(A \cup B) \cap C = (A \cap C) \cup (B \cap C)$ and $(A \cap B) \cup C = (A \cup C) \cap (B \cup C)$.
5. Absorption: for all A and B, $A \cap (A \cup B) = A$ and $A \cup (A \cap B) = A$.
6. Bounds: for all A: $I \cup A = I$ and $\Phi \cap A = \Phi$.
7. Complementarity: for all A, $A \cup \overline{A} = I$ and $A \cap \overline{A} = \Phi$.

As we can see, this algebra meets the duality principle, where we can consistently permutate the symbols $\cup$ and $\cap$ conjointly with the symbols I and Φ. If a result is true in one Boolean algebra it is true in its dual.

The implication $A \Rightarrow B$ is merely a simplified expression of $A = A \cap B$. In this case, the two strong syllogisms stated in Sect. 3.1 can be represented simultaneously as follows: if $B = I$, and is therefore true, then the formula shows that $A = I$, therefore A is also true; if $B = \Phi$, and is therefore false, then the formula shows that $A = \Phi$, therefore A is also false. By contrast, this formulation can no longer say anything about the two plausible propositions given after the previous syllogisms. We need to extend Boolean deductive logic in order to take account of these cases, whose examination is essential here.

We shall now present some axiomatics of logical probability. However, we shall set aside the axioms put forward by Keynes (1921). Unlike his highly original ideas on probability, they are not of great interest, as they merely formalize a classical view of probability (Suppes 2002a).

Let us move directly to the far more original approach developed by Jeffreys. His basic idea is to represent the probability of a proposition A, for which we have information I, by $P(A|I)$. He does not immediately offer a measure of this probability, but tries to determine which axioms must satisfy the probabilities (Jeffreys 1939):

1. Given I, A is either more, equally, or less probable than B, and no two of these alternatives can be true.
2. If I, A, B, C are four propositions, and, given I, A is more probable than B and B is more probable than C, then, given I, A is more probable than C.
3. All propositions deducible from a proposition I have the same probability on data I; all propositions inconsistent with I have the same probability on data I.
4. If, given I, A and A' cannot both be true, and if, given I, B and B' cannot be both true, and if, given I, A and B are equally probable and A' and B' are equally probable, then, given I, $A \cup A'$ and $B \cup B'$ are equally probable.
5. The set of possible probabilities on given data, ordered in terms of the relation 'more probable than', can be put into one-one correspondence with a set of real numbers in increasing order.
6. If $I \cap A$ entails B, then $P(A \cap B|I) = P(A|I)$.[6]

The first axiom entails the comparability of probabilities and resembles de Finetti's first axiom—with, however, the additional condition on available informa-

---

[6] For the sake of consistency with the rest of our discussion in this volume, we have designated the propositions in different terms from those used by Jeffreys.

tion. The second axiom entails the transitivity of the probability relation and resembles de Finetti's third axiom. The third axiom states that the extreme degrees of probability are certainty and impossibility. It ensures consistency between deductive logic and the logic of the probable. It is similar to de Finetti's second axiom. The fourth axiom states that the equality of probabilities is preserved in the logical sum of propositions: we can qualify it as the additivity axiom. De Finetti formulated his fourth axiom in more general terms, since he considers inequalities as well as equalities, but, again, the two axioms resemble each other. The fifth axiom states that we can establish a correspondence between the set of probabilities and a set of real numbers in increasing order. The sixth axiom is an extension of the third axiom: 'it is impossible, given I, that either A or $A \cap B$ should be true without the other'.

Most interestingly, although Jeffreys (1939) completely ignored de Finetti's axioms (1931a, b, 1937), his first four axioms closely resemble the first four axioms of the Italian subjectivist. Naturally, they differ by being subordinated to prior information, which subjectivists did not consider in their axioms. Conversely, in 1938 (published in English in 1985), de Finetti reviewed Jeffreys's book on scientific inference (1931). He was therefore aware of Jeffreys but proposed his axioms before Jeffreys.

Jeffreys showed that the principle of inverse probability—first enunciated by Bayes—follows very easily from this axiomatics, as do many other logical-probability results. This set of axioms, as we shall see later, also allowed him to build a theory of statistical inference. He named it 'induction' in contrast to 'deduction', which he regarded as inapplicable to non-mathematical sciences.

Immediately on publication, however, his axiomatics and many of his positions were met with considerable skepticism by most critics, who included some of the best probabilists and statisticians of his day (Neyman 1940; Irwin 1941; Wilks 1941). Irwin (1941), for instance, stated unequivocally that

> one must reject its axioms. Jeffreys' first axiom is 'Given p, q is either more or less probable than r, or both are equally probable; and no two of these alternatives are true'. Now, if probability is a degree of rational belief, this axiom might legitimately be rejected. Its acceptance is a matter of opinion. It by no means follows that degrees of rational belief can be ordered in a linear series.

Clearly, Irwin failed to grasp Jeffreys's argument and attributed to him the subjectivist stance that individuals should be left free to choose the probability they prefer, as de Finetti argued. Irwin concluded with a very clear rejection:

> From a scientific point of view it is doubtful that there will be many scholars thoroughly familiar with the system of statistical inference initiated by R.A. Fisher and extended by J. Neyman, E.S. Pearson, A. Wald and others who will abandon this system in favour of the one proposed by Jeffreys in which inverse probability plays the central role.

Interestingly, Wald, mentioned here, was an ardent advocate of von Mises's objectivist approach: he supplied the general strategies for decision-making in the presence of uncertainty (1950). In so doing, he merely revived the rules given by Bayes and Laplace in the eighteenth century. Wald openly recognized this, calling

3.2 Paradigm and Axiomatics of Logical Probability

his rules 'Bayes strategies.' Some objectivist authors have been honest enough to recognize the principles of the logicist school, when necessary.

The response from philosophers and mathematicians was just as negative as that of the probabilists (Nagel 1940; Braithwaite 1941). Let us simply quote Nagel (1940), echoing criticism of the first axiom:

> But if there is no common agreement on the degree of probability a proposition possesses, of what value is the elaborate calculus?

This is another example of the confusion between subjective probability and logical probability. The quotations above are merely a small fraction of the criticisms directed against Jeffreys.

The critics' tone did not change until the second edition of the work, in 1961, at a time when the Bayesian approach was making a comeback. For instance, Lindley (1962), one of the first to revive the Bayesian approach, wrote:

> This is probably the most original and important book on statistics that had appeared in the last 40 years. The only serious competitor is Fisher's 'Statistical analysis for research workers'. The distinction between the two is that Fisher is usually right for the wrong reason, whereas Jeffreys gets the reasoning broadly correct, as well as the answers.

There is no other way to describe this than as an outspoken tribute to Jeffreys's work.

Since Lindley, a great many authors have recognized the importance of Jeffreys's book and cite it as the seminal text for this approach. The article by Robert et al. (2009), while noting the book's weaknesses, states that one can reasonably claim it as the main reference in the field. On the axiomatics, the authors write that these

> paragraphs derive standard mathematical logic axioms that directly follow from a formal [modern] definition of a probability distribution, with the provision that this probability is always conditional on the same data.

We shall not go as far as these authors, for—as they themselves noted for the book as a whole—Jeffreys's axiomatics requires further adjustments. Let us see, specifically, whether it meets the conditions of the paradigm outlined earlier.

The axiomatics does indeed satisfy the first condition for the consistency of the theory:

> For any assessment of the prior probability the principle of inverse probability will give a unique posterior probability. (Jeffreys 1939)

The result obtained is therefore unique. But, for the prior distribution, it is harder not to set aside a part of the total information available in order to estimate it. Jeffreys gives some principles—particularly for non-informative prior elements such as the estimation of a standard deviation—but he notes:

> These principles sometimes indicate a unique choice, but in many problems some latitude is permissible, so far as we know at present. (Jeffreys 1939)

Thus the two conditions for the theory's consistency are not always met by the prior distribution. Lastly, for these prior distributions, Jeffreys could not use the maximization provided by Shannon's entropy, which was not known at the time when he wrote his book. His axiomatics, therefore, does not cover the entire field defined by our paradigm.

Let us now see how Richard Cox (1946, 1961) axiomatizes these probabilities. When using the rules of Boolean algebra, he argued, only two axioms are useful:

1. The probability of an inference on given evidence determines the probability of its contradictory on the same evidence.
2. The probability on given evidence that both of two inferences are true is determined by their separate probabilities, one on the given evidence, the other on this evidence with the additional assumption that the first inference is true. (Cox 1961)

These axioms do not use the term probability in its conventional meaning, for Cox does not define this probability until later, as we shall see. The term 'likeliness', proposed by Shafer (2004), seems to us to be the most satisfactory term, if we want to avoid all confusion with the terms likelihood (used by Cox but problematic today owing to its use by Fisher 1956), credibility[7] and plausibility,[8] already used with another meaning, as seen in Chap. 2. We shall therefore use likeliness in the subsequent discussion.

Cox introduces the notation $A|I$ for the likeliness of inference A, on the basis of hypothesis I. He then states that the second axiom implies that the likeliness of $A \cap B|I$ is a function of the likeliness $B|I$ and of $A|B \cap I$:

$$(A \cap B)|A = F(A|B \cap H, B|H).$$

But, because of the rules of Boolean algebra, as:

$$A \cap B \cap C|I = (A \cap B) \cap C|I = A \cap (B \cap C)|I,$$

the function F must satisfy the functional equation:

$$F[F(x,y),z] = F[x,F(y,z)],$$

for $x = A|B \cap C \cap I$, $y = B|C \cap I$, and $z = C|I$.[9] Under certain conditions that the function F must meet—here, it must be twice differentiable with a continuous second derivative —the result is a function of a single variable P satisfying:

$$P(A \cap B|I) = CP(A|I)P(B|A \cap I),$$

---

[7] Authors using the term with this meaning include Paris (1994), Halpern (1999a, b), and Colyvan (2004, 2008).
[8] Authors using the term in this sense include Jaynes (2003) and Arnborg and Sjödin (2001).
[9] This equation was solved by Abel (1826).

3.2 Paradigm and Axiomatics of Logical Probability

where the constant C can be taken as equal to unity. As this function is arbitrary, we can take $P(u) = u$, which entails the following relationship, also known as multiplicative rule:

$$A \cap B | I = (A|I)(B|A \cap I).$$

Similarly, Cox shows that the negation of a proposition A must satisfy the equation $\bar{A}|I = f(A|I)$, so that the function f satisfies the functional equation $f[f(x)] = x$. Cox's axioms therefore entail the following additive rule:

$$A|I + \bar{A}|I = 1$$

and the generalized additive rule:

$$A \cup B | I = A|I + B|I - A \cap B|I.$$

Using the first two rules alone, Cox was able to demonstrate what is known as Cox's theorem, which states that the only measures of likeliness satisfying the rules are the measures that are isomorphic with the measure of logical probabilities. This theorem and its axioms suffice to demonstrate the validity of the entire analysis of logical probability—Bayes's theorem included—except the procedure for assigning prior probabilities. Cox thus appears to have fully met his goal, namely:

> [to] try to show that ... it is possible to derive the rules of probability from two quite primitive notions which are independent of the notion of ensemble and which ... appeal rather immediately to common sense. (Cox 1946)

However, some authors have pointed out weaknesses in his reasoning. In particular, they have shown that Cox omitted various axioms and hypotheses that consequently remained implicit, and their non-fulfillment can actually invalidate his approach altogether. In fact, Cox's approach relied less on common sense than he claimed.

Paris (1994), for instance, showed the need to add five hypotheses to prove Cox's rules. Some are not problematic, but others are more inconvenient. In particular, the possible values of the probability thus defined must form a dense set to ensure that Cox's measure is isomorphic with the measure of probability.

Halpern (1999a), using an explicit counter-example, shows that Cox's rules are not always met in a finite set, even under hypotheses more heroic than Cox's own. Moreover, in infinite sets, these rules are inadequate. But in a later article (1999b), Halpern shows that two conditions are necessary for the validity of Cox's rules, even in finite sets. First, the previously mentioned function F should be associative—a condition that Cox stated in his 1946 article but did not in his 1961 book. Second, the function f should also meet the condition:

$$yf\left(\frac{x}{y}\right) = f(x)f\left[\frac{f(y)}{f(x)}\right].$$

Halpern notes, however, that it is hard (at least for himself) to see this condition as grounded in common sense. More simply, Shafer (2004) suggests adding the following two axioms, which are implicit in Cox's text, on the existence of the two functions, with the notations used here:

3. The likeliness $\overline{A}|I$ is determined in some way by the likeliness $A|I$:
   $\overline{A}|I = f(A|I)$, where f is some function of one variable.
4. The likeliness $A \cap B|I$ is determined in some way by the two likeliness $A|I$ and $B|A \cap I$: $A \cap B|I = F(A|B \cap I, B|I)$, where F is some function of two variables.

Regarding #4, Cox (1946) observed that: 'Written in symbolic form, this assumption may not appear very axiomatic'. He thus apparently recognized that this was indeed an axiom. However, in his book (1961), he no longer described it as an axiom.

Van Horn (2003) set out in detail the arguments needed to obtain complete proof of Cox's theorem. First, he clearly made the likeliness of a proposition conditional upon the information available. Second, he showed the conditions for ensuring the accountability of the likeliness in the propositional calculus.

Van Horn proposed the following axioms, which provide a substantial response to the criticisms discussed above:

1. The likeliness $A|I$, is a single real number. There exists a real number T such that $(A|I) \leq T$ for every I and A.
2. Likeliness statements are compatible with the propositional calculus (Boolean algebra).
3. There exists a nonincreasing function f such that $(\overline{A}|I) = f(A|I)$, for all A and consistent I. Define $F = f(T)$.
4. There exists a nonempty set of real numbers P_0 with the following two properties:

   - P_0 is a dense subset of (F, T). That is, for every pair of real numbers a, b such that $F \leq a < b \leq T$, there exists some $c \in P_0$ such that $a < c < b$.
   - For every $y_1, y_2, y_3 \in P_0$ there exists some consistent I with a basis of at least three atomic propositions—call them A_1, A_2 and A_3—such that $(A_1|I) = y_1$, $(A_2|A_1,I) = y_2$ and $(A_3|A_2,A_1,I) = y_3$.

5. There exists a continuous function $F: [F,T]^2 \to [F,T]$, strictly increasing in both arguments on $[F,T]^2$, such that $(A \cap B|I) = F[(A|B,I),(B|I)]$ for any A, B and consistent I.

We have slightly modified some notations. In particular, we have replaced the term 'plausibility' with 'likeliness' and have not reproduced the Boolean algebra axioms.

Van Horn took up Cox's initial axioms, giving a clearer formulation of their conditions. To these, Van Horn added the conditions that were often implicit in Cox's own presentations—an absence regretted by his critics. Van Horn then revisited Cox's demonstration, adding some of the previous hypotheses, and obtained the multiplicative and additive rules. He thus arrived at a theorem that supplies the basic conditions for probability calculus:

> There exists a continuous, strictly increasing function P such that, for every A, B, and consistent I,

3.2 Paradigm and Axiomatics of Logical Probability

1. $P(A|I) = 0$ iff A is known to be false given the information I.
2. $P(A|I) = 1$ iff A is known to be true given the information I.
3. $0 \leq P(A|I) \leq 1$.
4. $P(A \cap B|I) = P(A|I)P(B|A,I)$.
5. $P(\overline{A}|I) = 1 - P(A|I)$ if I is consistent.

We can see again that the main Kolmogorov's axioms are also valid for logical probabilities, with the difference that these probabilities apply to propositions and are information-dependent (Jaynes 2003). Van Horn also gives them here in theorem form, as he has deduced them from more fundamental axioms.

Cox (1961), as well, introduced Shannon's entropy, devoting an entire chapter to it. He showed that the notion makes it possible to measure the quantity of information contained in a set of inferences that we want to assess in probabilistic terms. Unfortunately, Cox does not use Shannon's entropy to estimate prior probabilities in the presence of incomplete information.

Jaynes took up most of Cox's analysis, additionally seeking general principles in order to determine prior probabilities from qualitative information on the phenomena studied.

As shown earlier, Jaynes's main contribution was to apply the notions of entropy and transformation group for the purpose of estimating prior probabilities under various conditions of information availability. We have already given specific examples. Let us now examine the more general solution.

In the discrete case, let us assume that a variable, x, can take n values, $(x_1, x_2, \ldots, x_n)$, corresponding to n different propositions, and that there are m different functions of x, yielding the constraints:

$$f_k(x)$$

where $1 \leq k \leq m < n$, and we have at our disposal a set of values, F_k, which must satisfy the expectations of these functions. In this case, therefore, we must find the values of the n parameters p_i that comply with the system of k equations:

$$F_k = \sum_{i=1}^{n} p_i f_k(x_i).$$

These values are the ones possessing maximum entropy, subject to all the constraints. Incorporating Lagrange multipliers, λ_j, this configuration is written:

$$\partial \left[H - (\lambda_0 - 1) \sum_{i=1}^{n} p_i - \sum_{j=1}^{m} \lambda_j \sum_{i=1}^{n} p_i f_j(x_i) \right] = \sum_{i=1}^{n} \left[\frac{\partial H}{\partial p_i} - (\lambda_0 - 1) - \sum_{j=1}^{m} \lambda_j f_j(x_i) \right] \partial p_i = 0$$

The solution of this equation system yields the solutions:

$$p_i = \exp\left[-\lambda_0 - \sum_{j=1}^{m} \lambda_j f_j(x_i)\right],$$

with the condition that the probabilities must sum to unity. If we now define the function:

$$Z(\lambda_1, \lambda_2, \ldots, \lambda_m) = \sum_{i=1}^{n} \exp\left[-\sum_{j=1}^{m} \lambda_j f_j(x_i)\right],$$

we can show that the values of F_k are equal to:

$$F_k = -\frac{\partial \log Z(\lambda_1, \lambda_2, \ldots, \lambda_m)}{\partial \lambda_k},$$

which allows us to estimate all the P_i values.

For continuous distributions, the solution is more complex and requires a prior distribution describing our complete ignorance of the variable x. Earlier, we gave the solution for estimating the standard deviation of such a distribution, using a transformation group that involves a change of scale and a translation. We refer the interested reader to the chapter of Jaynes's book, *Ignorance priors and transformation groups* (2003), which shows the complexity of a general solution to this problem.

However, not all Bayesians join in this quest for a unique prior (see, in particular, the issues discussed by Seidenfeld (1987) and Robert (2006), and their criticisms of the entropy-maximization approach). Even Jeffreys—who initially believed in the existence of a single logically correct prior—later altered his position. In the first edition of *Scientific inference* (1931), he wrote:

> Logical demonstration is right or wrong as a matter of the logic itself, and is not a matter for personal judgement. We say the same about probability. On a given set of data *p* we say that a proposition *q* has in relation to these data one and only one probability. If any person assigns a different probability, he is simply wrong, and for the same reason as we assign in the case of logical judgements.

A similar passage occurs in the first edition of *Theory of probability* (1939). But in an article published many years later, he noted:

> It may still turn out that there are many equally good methods [...] if this happens there need be no great difficulty. Once the alternatives are stated a decision can be made by international agreement, just as it had been in the choice of units of measurement and many other standards of reference. (Jeffreys 1955)

In the second edition of *Scientific inference* (1957), Jeffreys omitted the passage quoted above. He actually proposed another method (1946) than those used in

3.2 Paradigm and Axiomatics of Logical Probability

Theory of probability (1939) to estimate the prior by means of the information matrix, introduced by Fisher (1925a):

$$I(\theta)_{ij} = E\left(-\frac{\partial^2 l}{\partial \theta_i \partial \theta_j}\right),$$

where *l* is likelihood, and θ the parameters to estimate. The rule consists in taking the *prior* values:

$$\pi_\theta(\theta) \propto det\left[I(\theta)\right]^{1/2}$$

where the symbol $det[.]$ denotes the determinant of the information matrix. This estimate possesses the invariance property. Many other prior estimates have been suggested, which we cannot describe in detail here. For a fuller presentation and discussion, see Kass and Wasserman (1996).

Admittedly, entropy maximization is not the only method that can be used to estimate priors. Yet we believe the concept is essential for understanding entropy's deeper connections to logical probability. Let us begin by looking at the axiomatization of entropy.

As we said in Sect. 3.1, Shannon (1948) set the conditions that a measure must fulfill in order to characterize the information contained in a discrete distribution of probabilities $p_1, p_2, \ldots, p_n$. The measure $H(p_1, p_2, \ldots, p_n)$ needs to meet the following conditions:

1. *H* should be continuous in the p_i.
2. If all the p_i are equal, $p_i = \frac{1}{n}$, then *H* should be a monotonic increasing function of *n*. With equally likely events there is more choice, or uncertainty, when there are more possible events.
3. If a choice be broken down into two successive choices, the original *H* should be the weighted sum of the individual values of *H*.

Shannon then showed that the only function *H* satisfying these three axioms is of the form:

$$H = -K \sum_{i=1}^{n} p_i \log p_i.$$

If we set the constant $K = 1$, we effectively obtain Shannon's entropy. Other authors (Shore and Johnson 1980; Skilling 1988; Caticha 2004) have naturally discussed these axioms and suggested more precise ones, in particular with a view to making them applicable to more complex conditions and continuous probability distributions. We shall not describe them here, for the subject of this book is not entropy: our focus will be, instead, on the links between probability and entropy, which we have not yet fully spelled out. Our discussion is then confined to the three simple axioms given by Shannon for discrete probabilities.

As noted earlier, Richard Cox (1946, 1961) showed that probability is the only logically consistent measure of the relative degree of implication between propositions.

He did so by generalizing the Boolean implication and defining probability as the degree below which an assertion B is implied by another assertion A. We can thus write:

$$P(B|A) \equiv (A \rightarrow B),$$

where the right-hand member represents the value of the degree of implication of proposition B by proposition A. Cox was able to show that probability was the only measure (to within one constant) allowing an estimation of that value. Cox (1979) pursued this path of inquiry by defining a *question* as the exhaustive set of propositions that answer the question. He showed that, in the same way as one had considered the relations between propositions under the effect of implication, one can consider the relations of one question with another question. One can then define the relevance of a question with respect to another as the degree to which a question a provides information on another question b. As a result, we can write:

$$H(a|b) \equiv (a \rightarrow b),$$

where the right-hand member represents the value of the relevance of question a to question b. We shall see below that this measure is performed by entropy, hence the notation $H(.)$, which we have already used for entropy.

While generalized Boolean algebra supplies a suitable mathematical structure for propositions, hence for logical probability, it is not fully appropriate for questions, and thus for the concept of relevance. True, we saw Sect. 3.2 that we can define the complementary of an assertion as the negation of that assertion, such as $A \cup \overline{A} = 1$ and $A \cap \overline{A} = \Phi$. But we shall now demonstrate that this complementarity does not obtain for questions (Knuth 2002).

To do so, it is useful to examine in greater detail how we can define Boolean algebra from more general notions of partly ordered sets, lattices, and distributive lattices, which eventually lead to Boolean algebra. In particular, lattice theory—in which the order relationship is fundamental—will provide a better understanding of this process (Birkhoff 1935; Barbut 1968; Barbut and Monjardet 1970). Let us see more specifically how the theory developed.

Lattice theory works on sets partly ordered by the order relation $\subseteq$,[10] which satisfies the following axioms for all elements a, b, c:

1. Reflexivity: for all a, $a \subseteq a$.
2. Antisymmetry: if $a \subseteq b$ and if $b \subseteq a$, then $a = b$.
3. Transitivity: if $a \subseteq b$ and if $b \subseteq c$, then $a \subseteq c$.

---

[10] This relation is more general, as it applies not only to sets but also to aggregates in which objects can be individualized even though the total definition of the objects is not possible. In such cases, the relation is written $\leq$.

3.2 Paradigm and Axiomatics of Logical Probability

This order relation is partial, as there are elements that cannot be placed in a relation either in one direction or another. This relation possesses the duality property, which means that the reciprocal of a partial order is itself a partial order. In consequence, the relation $\supseteq$ is also a partial order. Each pair of elements admits of an upper bound and a lower bound, as does the complete set itself.

From these partly ordered sets, we can define a *lattice* that will satisfy the first two axioms of Boolean algebra stated earlier: commutativity and associativity. Moreover, the relation $a \subseteq b$ can be expressed by the two consistency relations:

$$a \cup b = a \text{ and } a \cap b = b.$$

We can therefore define such a lattice either as an algebra provided with the two internal distributions noted $\cap$ and $\cup$, or as a set provided with an order relation $\subseteq$

When we introduce the third axiom of Boolean algebra (distributivity), we obtain a distributive lattice. And when we introduce the complementarity axiom, we end up with a Boolean lattice.

Richard Cox (1946, 1961) showed that the propositions formed a Boolean lattice and that the two internal composition laws comply with Boolean algebra. If the number of elements of the most detailed partition of the propositions is n, the total number of partitions will be 2^n. By contrast, Knuth (2002, 2003a, b) has shown that the questions form a distributive lattice composed of the set of proposition subsets. The number of subsets follows a monotonic Boolean function,[11] which increases far more rapidly than the number of distinct propositions: for example, if for 4 propositions, of which two are join-irreducible, we have 5 questions, then for 8 propositions there will be 19 questions; for 16 propositions, 167 questions; for 32 propositions, 7,580 questions; for 64 propositions, 7,828,353 questions, and so on. But this distributive lattice, even reduced to the questions that can be answered by a proposition, is not Boolean: Knuth (2002) has shown that—unlike with the propositions—some questions have no complement. However, the distributive lattices share the associativity and commutativity properties of Boolean lattices. This allows a calculation similar to that of the propositions with a multiplicative rule and an additive rule, as well as the equivalent of a Bayesian theorem. The symmetry between the two calculations, in this case, is remarkable.

We can go even further by showing that, as probability is the only valid measure of the Boolean space of the propositions (to within one constant) (Cox 1946, 1961), entropy will be the only valid measure (to within two constants) of the question space (Aczèl et al. 1974; Knuth 2009), which forms a lattice. We can thus normalize this measure between zero and unity, as well as the probability.

This approach also justifies the entropy maximization advocated by Jaynes, for it involves the use of available information on the question space to assign priors in the proposition space (Caticha 2004). We will develop in the Conclusion of Part I, its use in order to show a cumulativity in probability theories.

---

[11] For more details, see the website http://research.att.com/~njas/sequences/.

3.3 Logicist Statistical Inference

The subjectivists, such as Ramsey and de Finetti, emphasized the individual in order to introduce probability and statistical inference. By replacing the fair price of a wager—suggested by Pascal and the eighteenth-century probabilists—with the personal price that an individual was willing to pay for a wager, they introduced a purely individual psychological behavior. In contrast, the logicists were to leave no room for the individual in probability theory, as Jaynes (2003) stated very clearly:

> When we apply probability theory as the normative extension of logic, our concern is not with the personal probabilities that different people might happen to have, but with the probabilities that they 'ought to' have, in view of their information [...]

For Jaynes, the probabilist must concentrate on the incomplete information available on a phenomenon in order to draw an inference, i.e., to derive a forecast of a future phenomenon from a similar past phenomenon.

Likewise, we saw in Sect. 1.3 how difficult it was, if not impossible, to draw a correct inference under an objectivist approach. We must now examine, therefore, the logicist approach to inference.

In fact, the main goal of the logicist approach is to supply a method that will allow us to draw inferences from observation data. These inferences must be consistent with one another and usable for predicting the outcome of future experiments in the same field. The term 'predict' is important here, for it in no way implies a mathematical form of deduction.

As seen earlier, the objectivists reject the notion of a probability of a hypothesis; they therefore refuse induction and accept only deduction as valid.[12] By contrast, the logicists clearly accept both: they effectively recognize the role of deduction in mathematics,[13] but reject the notion that any scientific method can somehow be reduced to deductive logic (Jeffreys 1931, 1939). They go well beyond, by emphasizing that inference of future behavior from past observations is not deductive but, on the contrary, fully inductive:

> A common argument for induction is that induction has always worked in the past and therefore may be expected to hold in the future. It had been objected that this is itself an inductive argument and cannot be used in support of induction. What is hardly ever mentioned is that induction has often failed in the past and that progress in science is very largely the consequence of direct attention to instances where the inductive method has led to incorrect predictions. (Jeffreys 1931)

Jeffreys later showed that this process not only suffers no contradiction but constitutes the only possible method for scientific progress. It is this method of successive

---

[12] Induction—a term coined by Francis Bacon (1620)—'consists in discovering the principles of a system through the study of its properties, by means of observation and experiment' (Franck 2002). It has been used by all the pioneers of modern science, including Galileo, Descartes, and Newton.

[13] The term *mathematical induction*—which implies that mathematicians use induction—is misleading, for the method is actually a form of deduction whose conclusion contains no more information than was present in latent form in its premises.

approximations that underpins the strength of the approaches used not only in logical probability but also in most natural and social sciences.

At the same time, unlike the subjectivists who rely solely on coherence as defined by de Finetti, the logicists emphasize the consistency property spelled out by Cox. The latter is the only means of setting rules for logical inference:

> Yet it is consistency—not merely coherence—that is essential here, and we find that, when our rules have been made to satisfy the consistency requirements, then they have automatically (and trivially) the property of coherence. (Jaynes 2003)

This notion of consistency, already advanced by Jeffreys (1931), has for Jaynes a triple significance:

- If we can reach a conclusion in several ways, then each reasoning must lead to the same result.
- We must take into account all the information available regarding a question, without overlooking any of it.
- We must always represent equivalent states of knowledge by equivalent plausibility distributions. In other words, if we have the same information set on two questions, we must assign the same plausibility to both sets.

Jaynes (2003) actually introduced a robot to avoid all subjective judgment in assessing consistency: in this case, each individual, with access to the same information, must arrive at the same estimate of the probability of a given event. In contrast, the coherence condition merely implies the absences of basic contradiction in an individual's subjective assessment of probabilities: under this condition, each person can assign a different probability to a given event (de Finetti 1937).

The consistency condition is the most significant difference between subjectivist statistical inference and logicist statistical inference. On the other hand, the subjective and logical approaches to probability are in total accord on the notion of exchangeability and the theorem demonstrated by de Finetti, which we discussed in the section on subjectivist statistical inference in Sect. 2.3. Hence logicist statistical inference also enables us to answer the question on which objectivist inference stumbled: what is the probability that an unknown parameter lies in a given interval? In this connection, it is interesting to note that, in order for his subjectivism to be able to predict the outcome of coin tosses (heads or tails), de Finetti needed to set certain symmetry principles for logical probability, enabling certain probabilities to be identical to others (de Finetti 1964). He did not explain, however, why a symmetry argument may be acceptable in certain cases and not more generally in all cases (Franklin 2001), as the logicists admit.

We can conclude that, in the logical approach to probability, statistical inference and probability theory form an inseparable whole.

3.4 Application to Social Science

From his paper *La probabilité des causes par les événements* (*The probability of causes by events*) (1774) to his book *La théorie analytique des probabilités* (*The analytical theory of probability*) (1812), Laplace extensively examined the application of inverse probability to various fields such as population sciences, astronomy, and human testimony. We begin with the application to the sex ratio at birth (Laplace 1778), while recalling that Laplace addressed many other demographic phenomena including mortality, nuptiality, and fertility in different population groups.

Laplace begins with an initial estimate based on the observation of $m+n$ births, of which m are boys and n are girls.[14] He seeks the probability p that $r+s$ future births will include r boys and s girls. Assuming that all values of the prior probability x that a future birth will be a boy are equally probable, he shows that:

$$p = \frac{(r+s)!}{r!\,s!} \times \frac{\int_{x=0}^{1} x^{m+r}(1-x)^{n+s}\,dx}{\int_{x=0}^{1} x^{m}(1-x)^{n}\,dx}.$$

After calculation of the integrals and simplifications owing to the fact that m and n are large numbers, the equation becomes:

$$P = \frac{(r+s)!}{r!\,s!} \times \frac{m^r n^s}{(m+n)^{r+s}}.$$

The value p is identical to the one that we would obtain under the assumption that the likelihoods of male and female births stand in the ratio $\frac{m}{n}$. Laplace therefore concludes that these likelihoods also stand in the same ratio.

Next, Laplace seeks the probability p' that the birth of an another boy in addition to the $m+n$ births observed lies between $\frac{m}{m+n}-\theta$ and $\frac{m}{m+n}+\theta$. Positing $m=\frac{1}{\alpha}$ and $n=\frac{\mu}{\alpha}$, and using a calculation procedure of the same kind as the previous one although even more complex, Laplace obtains the following approximate value for the probability (ignoring the $m^{-\frac{5}{2}}$-order quantities), which—when the number of observations is large—becomes very small:

$$p' = 1 - \frac{\sqrt{\alpha\mu}}{\sqrt{2\pi}\,(1+\mu)^{\frac{3}{2}}\theta}\left\{1 - \alpha\frac{12\mu^2 + (1+\mu)^2(1+\mu+\mu^2)\theta^2}{12\mu(1+\mu)^3\theta^2}\right\}$$

$$\left\{\frac{[1-(1+\mu)\theta]^{m+1}\left[1+\frac{1+\mu}{\mu}\theta\right]^{n+1}}{+[1+(1+\mu)\theta]^{m+1}\left[1-\frac{1+\mu}{\mu}\theta\right]^{n+1}}\right\}$$

---

[14] We have changed Laplace's notations to preserve consistency with our notation system.

3.4 Application to Social Science

Examining the values of all the terms of the equation above, he therefore concludes that the value of p' will differ all the less from certainty or unity as m and n will be larger numbers.

As an example, Laplace takes the births that occurred in Paris between 1745 and 1770. In those 26 years, 251,527 boys and 241,945 girls were born, which translates into a masculinity proportion of 50.971%. Using a formula similar to the previous equation, Laplace computes the probability of a male birth being equal to or less than one-half at 1.1521×10^{-42}. He draws this conclusion:

> As it is exceedingly small, we can assert, with the same certainty as any other moral truth, that the difference observed in Paris between births of boys and those of girls is due to a greater likelihood for births of boys (Laplace 1781).

Thanks to the preceding equation, we can also determine the probability that the chances of a male birth lie within the limits 0.50971 ± 0.001. That probability is roughly 0.99984, i.e., very close to unity. For this estimation, we have used the series development given by Laplace (1781) for the final terms of the equation:

$$\left[1 - (1+\mu)\theta\right]^n \left[1 + \frac{1+\mu}{\mu}\theta\right]^n = e^{-\frac{(1+\mu)^3 \theta^2}{2\mu} \frac{1}{\alpha} - \frac{(\mu-1)(1+\mu)^4 \theta^3}{3\mu^2} \frac{1}{\alpha} \cdots}$$

replacing θ by $-\theta$ in the other term.

For London, data equivalent to the Parisian numbers show 737,629 male births versus 698,958 female births in the period 1664–1758. The resulting masculinity proportion is 51.346%, which is even higher than the Paris value. This leads Laplace to wonder whether the higher proportion is evidence of a higher probability.

Let u be the probability of a male birth in Paris, m the number of male births and n that of female births in Paris, $u - x$ the probability of a male birth in London, m' the number of male births and n' that of female births in London. The probability of this double event is thus:

$$K u^m (1-u)^n (u-x)^{m'} (1-u+x)^{n'},$$

where K is a constant coefficient. In consequence, the probability that a male birth will be less likely in London than in Paris is:

$$p = \frac{\int_{x=0}^{x=1} \int_{u=0}^{u=x} u^m (1-u)^n (u-x)^{m'} (1-u+x)^{n'} \, du \, dx}{\int_{x=0}^{x=1} \int_{u=0}^{u=1} u^m (1-u)^n (u-x)^{m'} (1-u+x)^{n'} \, du \, dx}$$

After a series development of this quantity, and taking the first three terms of the series, Laplace obtains the approximate value of p:

$$p = \frac{1}{410458}.$$

He can therefore conclude:

> the odds are over four hundred thousand to one that boys are more easily born in London than in Paris; we can thus regard as a very probable thing that there exists in the former town one more cause than in the second that facilitates the birth of boys, and that depends either on climate, or on food and customs. (Laplace 1778)

He even seeks to identify one of the causes for Paris:

> parents in the surrounding countryside, having found it advantageous to keep their male children with them, had sent them to the Hospice des Enfants-Trouvés de Paris [Paris Foundling Home] in a smaller proportion than that of the sex ratio at birth. (Laplace 1812)

This marked the start of a fuller demographic analysis that used various characteristics of the two cities in order to attempt to explain the differences.

Curiously, later demographers showed scarce interest in pursuing such a refined analysis of the phenomena that they were studying. In this connection, we should bear in mind that the introduction of population censuses sidelined some earlier concerns, particularly by supplying exhaustive populations at risk. This avoided the use of civil-registration data for population estimates (multiplier method). Moreover, as pointed out in Chap. 1, the variances of the computed rates were so small as to become of negligible value for analytical purposes.

The second example that we shall examine is the application of logical probability to judicial affairs, which will take us from Condorcet (1785) to Laplace (1812) and Poisson (1837), with a brief detour via Quetelet's non-Bayesian approach (1835), to finish with nowadays Bayesian jurisprudence (Vignaux and Robertson 1996).

Condorcet (1785) published a 500-page memoir on the application of probability to court rulings. Although he did not use the term 'logical probability'—which none of his contemporaries did either—we can easily see that it is indeed the subject of his book. Condorcet was already familiar with Bayes (Condorcet 1778) and, in particular, with Laplace's memoir on probability (1778), which elaborated on these concepts. The third section of Condorcet's memoir (1785) opens with a very explicit statement of the need to begin by establishing, in a general way,

> the principles according to which one can determine the probability of a future or unknown event, not from the knowledge of the number of possible combinations that produce the event, or the opposite event, but only from the knowledge of the order of known or past events of the same kind.

There is no clearer definition of this approach, which takes the observation of past events as the starting point to determine the probability of a future event through logical reasoning.

Condorcet applied the approach to legal and social science. He stated the purpose of his memoir in the *Discours préliminaire* [introduction]:

> Reason, with a modicum of reflection, will make one feel the need to compose a Tribunal in such a manner as to make it nearly impossible for a single innocent to be sentenced, even over a long lapse of time; but it will not indicate the bounds that can be set on that probability, nor how to obtain it, without multiplying the number of Judges beyond the limits that one can hardly exceed.

3.4 Application to Social Science

From the outset, Condorcet thus saw the need to protect the individual from a wrong decision by the jury, but reason would not suffice for the purpose. Probability calculus was indispensable for measuring the desired goal with precision. Poisson (1837) pointed out the originality of Condorcet's use of probability calculus:

> but, as regards the probability of rulings, it is fair to say that Condorcet deserves credit for the ingenious idea of making the solution depend on the Bayesian principle, by considering successively the defendant's guilt and innocence, as an unknown cause of the ruling handed down, which accordingly becomes an observed fact, from which we must deduce the probability of that cause.

Let us see how Condorcet proceeds to attain his objective.

Condorcet begins by defining a general model for a jury's decision-making process, which takes various characteristics of the jury into account. The model depends on the jury's size, $2q+1$, the majority of votes required to make a decision, $1, 2, \ldots, 2q'+1$, and a qualitative characteristic of the jury, which he will then try to estimate, namely, the probability that the opinion of one of the voters will be consistent with the truth, v, and therefore simultaneously the probability that it will be contrary to the truth, e, with $v+e=1$.

Condorcet initially assumes that the characteristics are known, in order to express the probability that there will be at least one more vote for the truth, expressed by V^q:

$$V^q = v^{2q+1} + \binom{2q+1}{1}v^{2q}e + \binom{2q+1}{2}v^{2q-1}e^2 \ldots + \binom{2q+1}{q}v^{q+1}e^q,$$

where, more generally, $\binom{n}{m}$ designates the binomial coefficient of m among n.

Condorcet shows that this equation may be written more simply as:

$$V^q = v + (v-e)\left[ve + \binom{3}{1}v^2e^2 + \binom{5}{2}v^3e^3 + \binom{7}{3}v^4e^4 \ldots + \binom{2q-1}{q-1}v^qe^q\right].$$

Thus, when $v > e$, we can—by increasing the number of voters—obtain as high a probability as we want that the decision will be consistent with the truth, for the value of V^q constitutes the first terms of the series expansion of the quantity:

$$v + (v-e)\left[-\frac{1}{2} + \frac{1}{2}(1-4ve)^{-\frac{1}{2}}\right],$$

equal to 1 when we use the relation $v = 1-e$:

$$1 - e + (1-2e)\left[-\frac{1}{2} + \frac{1}{2}(1-2e)^{-1}\right] = 1.$$

Condorcet easily generalizes the expression to the case where the majority will be of $2q'+1$ votes:

$$V^q = 1 + \binom{2q'-1}{0} e^{2q'}(-e) + \binom{2q'+1}{1} e^{2q'+1} v \left(\frac{v}{2q'+1} - e \right)$$
$$+ \binom{2q'+3}{2} e^{2q'+2} v^2 \left(\frac{2v}{2q'+2} - e \right) \ldots + \binom{2q-1}{q-q'} e^{q+q'} v^{q-q'} \left(\frac{q-q'}{q+q'} v - e \right)$$

We can see that when $q = q'$ —i.e., in a unanimous vote, the second term being negative and the following terms being null—a defendant who is not guilty may nevertheless be convicted, as $V^q = 1 - e^{2q+1}$ in this case. When $q > q'$, we shall always need to remove the second term, and then—as long as the factor in parentheses is negative—the subsequent terms. This will initially diminish V^q, which will later increase. For instance, assuming that a seven-vote majority is required, and that the probability of error is 1/3, V^q will be lowest when the number of voters is 19.

Next, Condorcet needs to determine the probability, v, of a correct verdict. He proposes two approaches, but we shall examine only the second. The first consists in forming a Tribunal of truly enlightened men to examine the evidence: it is of little relevance to our discussion, and Condorcet himself preferred the second—which requires the use of logical Bayesian methods.

The second approach consists in assuming that the probability of a fair decision by each juror ranges between one-half and unity, i.e., that it will conform to truth rather than error. Otherwise, it would be absurd to hand down a verdict by majority vote in order to approach the truth. In this case, therefore, Condorcet clearly does not take all possible values for these priors, but he estimates them as greater than ½.

To this end, he must now determine the prior that each judge will individually make the right decision, v. Condorcet proposes three alternatives:

> 1. that in each decision the votes of all Voters have a constant probability; 2. that in each decision and for each Voter, the probability varies; 3. that we admit both hypotheses together, by multiplying the probability resulting from each by the probability that this hypothesis will occur.

We shall set aside the first hypothesis—which Condorcet does not find natural to accept on its own—and explore the second in greater detail. We shall then report only the outcome obtained under the third hypothesis, which results from the combination of the other two and has Condorcet's preference.

Under the second hypothesis, the problem is to determine the probability of obtaining a new vote consistent with the truth, knowing that in the past the vote was

3.4 Application to Social Science

m times consistent and n times inconsistent with the truth. If we know, in addition, that this prior lies between ½ and 1, then we can write it as equal to[15]:

$$\frac{\left[\int_{\frac{1}{2}}^{1} x\, dx\right]^{m+1} \left[\int_{\frac{1}{2}}^{1}(1-x)\, dx\right]^{n}}{\int_{0}^{1} x\, dx \left[\int_{\frac{1}{2}}^{1} x\, dx\right]^{m} \left[\int_{\frac{1}{2}}^{1}(1-x)\, dx\right]^{n}} = \frac{\left(\frac{3}{8}\right)^{m+1}\left(\frac{1}{8}\right)^{n}}{\frac{1}{2}\left(\frac{3}{8}\right)^{m}\left(\frac{1}{8}\right)^{n}} = \frac{3}{4}.$$

As a result, this probability, v, will be independent of m and n, and we need only replace v by this value and e by ¼ in the equations above to obtain the values of V^q.

By contrast, under the third hypothesis, Condorcet shows that the probability can be written as:

$$\frac{\int_{\frac{1}{2}}^{1}\left[x^{m+1}(1-x)^{n} + x^{n+1}(1-x)^{m}\right] dx + \frac{3^{m+1} + 3^{n+1}}{4^{m+n+1}}}{\int_{0}^{1} x^{m}(1-x)^{n}\, dx + \frac{3^{m} + 3^{n}}{4^{m+n}}}.$$

In this case the probability v will depend on m and n, and the equation giving V^q will be more complex.

We shall not pursue Condorcet's reasoning further, for our aim was to show his use of logical probability. Let us simply note that one of the consequences of Condorcet's arguments is his condemnation of the death penalty:

> Justice would demand no less strongly that a defendant should not be condemned as long as the crime was not proven, and there can be no injustice in acquitting a defendant whenever the probability of his crime, however high, does not reach the limit that we have found to be the starting point for genuine certainty.

However, the principle of probability calculus ensures that this certainty can never be obtained.

Condorcet's memoir met with little approval from his contemporaries. La Harpe (1799) viewed the intrusion of probability in legal science as one of the misuses of late-eighteenth-century philosophy. Destutt de Tracy, in the second edition of his work (1801, 1804–1818), launched an argumentative attack on Condorcet's

---

[15] Condorcet uses a different notation for his integrals than the standard form used today. For example, he writes $\dfrac{\frac{1}{2}\text{—}m}{\int x\partial x}$ the expression now written $\left[\int_{\frac{1}{2}}^{1} x\, dx\right]^{m}$. Here, we use the modern notations.

probability plan, while calling Condorcet himself 'a superior man who shall always be regretted'. Destutt de Tracy had this to say on probability:

> Under this collective and common noun are wrongly gathered a multitude of sciences or portions of sciences, all of them different, alien to one another, and impossible to bring together without creating total confusion. Indeed, what is commonly known as the science of probability comprises two distinct parts, namely, on the one hand, the search for an evaluation of the data, and, on the other hand, the calculation or combinations of those data.

He continued by listing a number of things, far more numerous than generally believed, that are not amenable to probability-based treatment:

> the degrees of capability of men, the degrees of the energy and power of their passions, their prejudices, their habits [...]

He therefore did not totally reject the usefulness of probability, but placed limits on their use. His text contains many of the arguments later wielded by the objectivists, particularly Venn (1866).

Most jurists merely ignored the intrusion of probability calculus in their discipline. Bentham (1823), for instance, rejected the false certainty implied by the mathematical treatment of legal and social matters.

Despite these objections, Laplace (1816) resumed these investigations more than 20 years later, in order to assess the French legal system that, in 1808, allowed juries to return verdicts by seven votes to five. He noted from the outset that:

> To convict a defendant, the judge must not await mathematical evidence, which is unattainable in moral matters. But when the probability of the crime is such that citizens would have more to fear from the attacks that could arise from its impunity than from court errors, it is in the interest of society to demand the defendant's conviction.

Unlike Condorcet, who required certain proof—never attainable—to convict a presumed criminal, Laplace sought to avert social danger by demanding not a certainty but a degree of probability, a: he 'assumes that the judge who convicts a defendant thereby asserts that the probability of his crime is at least a'. On the other hand, he adopted Condorcet's hypothesis by assuming that the probability that each judge will make the right decision, which he calls x, is equal to or greater than ½, and that it varies uniformly over the entire interval. Lastly, he supposed that the tribunal consists of $p+q$ judges, of whom p find the accused guilty and q absolve him.

In this case, the probability of a correct verdict will be:

$$a = \frac{x^p (1-x)^q}{x^p (1-x)^q + (1-x)^p x^q}.$$

We now need to multiply this probability by the probability of the value of x, for the observed event:

$$\frac{\left[x^p (1-x)^q + (1-x)^p x^q \right] dx}{\int_0^1 x^p (1-x)^q \, dx},$$

3.4 Application to Social Science

which, by multiplying this probability by a, gives the probability of a correct verdict relative to a value of x:

$$\frac{x^p (1-x)^q}{\int_0^1 x^p (1-x)^q \, dx}.$$

Summing this quantity from ½ to 1, we find that the probability of a correct verdict for all possible values of x is:

$$\frac{\int_{1/2}^1 x^p (1-x)^q \, dx}{\int_0^1 x^p (1-x)^q \, dx},$$

and that the probability of the potential error in the verdict will be:

$$\frac{\int_0^{1/2} x^p (1-x)^q \, dx}{\int_0^1 x^p (1-x)^q \, dx}.$$

The latter value is easy to calculate:

$$\frac{1}{2^{p+q+1}} \left\{ \begin{array}{l} 1 + \dfrac{p+q+1}{1!} + \dfrac{(p+q+1)(p+q)}{2!} + \dfrac{(p+q+1)(p+q)(p+q-1)}{3!} + \cdots \\ + \dfrac{(p+q+1)(p+q)(p+q-1)\cdots(p+2)}{q!} \end{array} \right\}.$$

If we demand unanimity, the quantity, i.e., the probability of error, simplifies to $\dfrac{1}{2^{p+1}}$.

In the French legal system of the time, the jury was composed of 12 members and the court comprised five judges. When a defendant was declared guilty by eight votes to four, he had to be convicted. Applying the equation above, the probability of an error by the jury was $\dfrac{1093}{8192}$, or ap.proximately 13%, which is very far from negligible. In special courts, where five votes out of eight sufficed to convict, the probability of the potential error was $\dfrac{125}{512}$, or nearly one-quarter: we can understand Laplace's conclusion that 'the magnitude of this fraction is frightening'. In England, by contrast, when the jury comprised 12 members, a unanimous vote was required for a conviction. The jury's potential error fell to $\dfrac{1}{8192}$, i.e., less than one-thousandth of the French value. The difference between the two countries was glaring. Moreover, in France, when a defendant was found guilty by a simply majority of seven jurors to five, entailing a risk of error of $\dfrac{2380}{8192}$ or 29%, the five judges had to cast votes as well: when only three of the five judges joined the minority of the jury, the defendant was convicted.[16] In this case, and assuming the judges' opinion followed the same distribution as that of the jurors, the probability of an erroneous verdict was $\frac{106762}{262144}$ or 40.73%. This system, which claimed to protect citizens, therefore exposed them to an even greater risk of conviction. Laplace did not perform that calculation[17]—for he assumed that judges were more qualified to make a reliable judgment than jurors—but it led him to criticize the *Code d'instruction criminelle* harshly on this issue in his treatise.

Laplace also realized the complexity of the problems posed by the calculation of the probability of court rulings:

> For this, one needs to know the probability of the crime below which a defendant cannot be convicted, without the citizens having to fear miscarriages of justice more than the attacks that could result from the impunity of an absolved culprit. One must then determine the probability of the crime resulting from the court's decision and set the majority so that these probabilities are equal.

Laplace was forced to admit that those probabilities are impossible to obtain and that—given our ignorance of these two elements of the calculus—we can only solve the problem of the probability of error in the court's decision, as we have just done.

As with Condorcet, many authors rejected Laplace's work on the probability of court rulings. For example, it was condemned by Pope Pius VII and poorly received by the university (Barbin and Marec 1987).

Let us now examine how Quetelet (1835) and Poisson (1837) both used the same source: the *Comptes généraux de l'administration de la justice criminelle* (French criminal justice statistics), one using an objectivist approach, the other a logicist approach.

Initially, Quetelet had proposed using Laplace's method, particularly in demography (1827). But he changed his mind when establishing his theory of the *average man* (1835). The theory lies too far outside our subject for us to elaborate on it here. However, we should note that it was based on the analysis of exhaustive data, such as census data but also administrative sources. For instance, in the same work, Quetelet analyzed French data on convictions from 1825 to 1830.[18] He began by observing a slight decline in the proportion of convictions of defendants without testing its significance as Laplace would have done. He then tried to identify the characteristics that could alter these proportions, such as the type of

---

[16] Article 351 of the Criminal Procedure Code (Code d'Instruction Criminelle) stipulated as follows 'If, notwithstanding, the defendant is found guilty only by a simple majority [of jurors], the judges shall deliberate in private on this same matter; and if the opinion of the minority of jurors is adopted by the majority of judges, so that by adding the number of votes, the total exceeds that of the majority of jurors and the minority of judges, then the finding in favor of the defendant shall prevail'. In the case examined here, we have nine 'guilty' votes (seven jurors and two judges) and only eight 'not guilty' votes (five jurors and three judges): the defendant will therefore be convicted.

[17] It was performed by Gergonne (1818–1819).

[18] Quetelet's figures for 1825 differ from Poisson's, for the latter had been corrected by the Justice Ministry in 1827 (Stigler 1986).

Table 3.1 Data from French criminal justice administration on personal crimes, 1825–1831

Year	Number of individuals charged with personal crimes	Number of convictions	Probability of conviction
1825	1,897	882	0.4649
1826	1,907	967	0.5071
1827	1,911	948	0.4961
1828	1,844	871	0.4723
1829	1,791	834	0.4657
1830	1,666	766	0.4598
1831	2,046	743	0.3631

offense (property crime, personal crime), sex and age of defendant, literacy level, and educational attainment. His chosen approach for processing the data was therefore already fully objectivist, and thus lies outside the scope of this chapter.

Poisson's approach (1835, 1836b, 1837) to the same data—to which he added those of the period 1831–1833[19]—was quite different: he applied the logical probability method to measure the effects of various jury characteristics on the final verdicts. In the opening of the very first chapter of his book (Poisson 1837), he clearly stated that 'the probability of an event is our reason for believing that it will occur'. Specifically, his main goal was to test his model of jury decision-making rather than to estimate, as Quetelet had done, the effects of different characteristics on the jury's verdict. He used the annual figures provided by the ministry, and not only the computed rates as Quetelet had done.

The Table 3.1 gives the personal-crime data for the period 1825–1831 used by Poisson.

From these data, Poisson examined in greater detail the hypotheses formulated by Laplace for his analysis of verdicts. In particular, Poisson found it hard to justify *a priori* the hypothesis that the probability of a juror reaching the correct decision should range between ½ and unity. As a result, the equation expressing the probability of errors by jurors never factors in their degree of knowledge of the case submitted to them. It also assumes that, prior to the jury's decision, there was no presumption of the defendant's guilt. Poisson found this hypothesis, as well, unacceptable.

Poisson sought to avoid any hypotheses, relying solely on the laws of probability calculus. He decided to use factual data—here, the *Comptes généraux de l'administration de la justice criminelle* (Table 3.1)—to estimate two quantities that depended on the conditions in which the verdicts were reached:

> One expresses the probability that a juror picked at random from the list available to a criminal court will not cast the wrong vote; the other is the probability, before the opening of proceedings, that the defendant is guilty.

---

[19] Poisson corrected the 1825 data to harmonize the definitions for the entire period, for, contrary to the data concerning other years, they included persons not present at the verdict (note in the report for 1827). Poisson also stated that he had to process separately the data for the years after 1830, because new legislation changed the majority required for a verdict from 7 jurors out of 12 to 8 out of 12.

Although Poisson did not want his probabilistic reasoning to rest on any hypothesis, we can see that he did set up an underlying model, whose parameters he proceeded to estimate.

For the estimation, he began by assuming that the parameters are known. He noted u the probability that a juror would not cast the wrong vote (the probability could vary from one juror to another) and k the probability that the defendant was guilty or, more accurately, susceptible to conviction—for, as Poisson clearly stated, we can never obtain mathematical proof of a defendant's guilt.

Poisson started with the simplest case, involving a single juror. It can easily be shown that the probability of a defendant's conviction, γ, is written:

$$\gamma = ku + (1-k)(1-u), \tag{3.1}$$

for the event occurs either when (1) the defendant is guilty and the juror does not cast the wrong vote (an event with a probability of ku) or (2) when the defendant is not guilty and the juror casts the wrong vote (an event with a probability of $(1-k)(1-u)$). After the juror's decision, let us assume that the defendant is guilty, with a probability of p. Using the Bayesian rule, we can write:

$$p = \frac{ku}{ku + (1-k)(1-u)}, \tag{3.2}$$

for the event observed here is the defendant's conviction, whose probability is ku.

In the general case, when each juror is equally likely to cast the wrong vote, Poisson showed that, if we consider the probability γ_i that the defendant will be convicted by $n-i$ votes to i, Eq. 3.1 becomes:

$$\gamma_i = \binom{i}{n}\left[ku^{n-i}(1-u)^i + (1-k)u^i(1-u)^{n-i}\right], \tag{3.3}$$

and Eq. 3.2, which gives the probability of the defendant's guilt, p_i, can be written:

$$p_i = \frac{ku^{n-i}(1-u)^i}{ku^{n-i}(1-u)^i + (1-k)(1-u)^{n-i}u^i}, \tag{3.4}$$

We can readily verify that this probability of a correct verdict depends only on the majority $m = n - 2i$ of votes cast in favor of it, not on the total number of jurors n.

These equations can easily be generalized to the case where we only know that the defendant was convicted by a majority of at least m votes, i.e., that the number of votes could have ranged from m to $m+2$, ... , up to $m+2i$—that is, a unanimous decision. That was the procedure applied in France between 1825 and 1830, when convictions required a majority of at least seven votes to five, and between 1831 and

3.4 Application to Social Science

1833, when the minimum required majority was eight votes to four. The probability that the defendant will be found guilty by at least $n-i$ votes will therefore be:

$$c_i = kU_i + (1-k)V_i, \tag{3.5}$$

where:

$$U_i = u^n + \binom{1}{n}u^{n-1}(1-u) + \binom{2}{n}u^{n-2}(1-u)^2 + \cdots + \binom{i}{n}u^{n-i}(1-u)^i,$$

$$V_i = (1-u)^n + \binom{1}{n}(1-u)^{n-1}u + \binom{2}{n}(1-u)^{n-2}u^2 + \cdots + \binom{i}{n}(1-u)^{n-i}u^i.$$

Similarly, the probability of the defendant's guilt will become:

$$P_i = \frac{kU_i}{kU_i + (1-k)V_i}. \tag{3.6}$$

We must now examine the actual case where the probability of a juror's casting the wrong vote is unknown. We must therefore introduce an unknown function $\varphi(u)du$, which represents the probability of a value of that possibility, u, such that $\int_0^1 \varphi(u)du = 1$. In this case, the probability λ_i that the chances of not making the wrong decision will range between the given limits l and l', when the defendant is convicted by $n-i$ votes to i, is:

$$\lambda_i = \frac{k\int_l^{l'} u^{n-i}(1-u)^i \varphi(u)du + (1-k)\int_l^{l'} u^i(1-u)^{n-i}\varphi(u)du}{k\int_0^1 u^{n-i}(1-u)^i \varphi(u)du + (1-k)\int_0^1 u^i(1-u)^{n-i}\varphi(u)du}. \tag{3.7}$$

Poisson showed that, in some cases, notably when l and l' are symmetrical with respect to ½, this probability becomes independent of the defendant's guilt, k. He also calculated the probability of guilt, ζ_i, under the same conditions:

$$\zeta_i = \frac{k\int_0^1 u^{n-i}(1-u)^i \varphi(u)du}{k\int_0^1 u^{n-i}(1-u)^i \varphi(u)du + (1-k)\int_0^1 u^i(1-u)^{n-i}\varphi(u)du}. \tag{3.8}$$

Unlike the situation where u is assumed to be known, ζ_i will not depend exclusively, like P_i, on the majority m, but also on the total number of jurors and on the probability distribution of the chances of not casting the wrong vote.

As earlier, when the accused is convicted by a majority of at least $n-2i$ votes, the equations, for λ_i, become:

$$Y_i = \frac{k\int_l^{l'} U_i \varphi(u)du + (1-k)\int_l^{l'} V_i \varphi(u)du}{k\int_0^1 U_i \varphi(u)du + (1-k)\int_0^1 V_i \varphi(u)du}, \tag{3.9}$$

and for ζ_i:

$$Z_i = \frac{k\int_0^1 U_i \varphi(u)du}{k\int_0^1 U_i \varphi(u)du + \int_0^1 V_i \varphi(u)du}. \quad (3.10)$$

Poisson calculated approximate values of the integrals contained in these equations, which will depend only on the constant k and the function $\varphi(u)$.

We can now compare these equations to Laplace's results. Laplace assumed that $\varphi(u)$ is zero for $u < \frac{1}{2}$, and is constant for $u \geq \frac{1}{2}$. The preceding equation accordingly becomes:

$$\zeta_i = \frac{k\int_{\frac{1}{2}}^1 u^{n-i}(1-u)^i du}{k\int_{\frac{1}{2}}^1 u^{n-i}(1-u)^i du + (1-k)\int_0^{\frac{1}{2}} u^{n-i}(1-u)^i du}. \quad (3.11)$$

Moreover, Laplace did not consider the probability k of guilt before the verdict. In other words, guilt is neither more nor less probable than non-guilt, i.e. $k = \frac{1}{2}$. Under this new condition, we have:

$$\zeta_i = \frac{\int_{\frac{1}{2}}^1 u^{n-i}(1-u)^i du}{\int_0^1 u^{n-i}(1-u)^i du}. \quad (3.12)$$

We effectively obtain the probability that the accused will be not guilty, $(1-\zeta_i)$, by performing the integrations. The value will be the same as the one given earlier, substituting n for $p+q$ and i for q. However, Laplace's hypotheses are no doubt too simple to be plausible. This led Poisson to state:

> If, therefore, only one verdict had been reached by the jurors picked from this list, the preceding equations would have no useful application; the same would still be true if a modest number of verdicts had been reached; but we know that, on the contrary, very large numbers of convictions, in known proportions, have been handed down by juries picked successively at random from the same general list[.]

Poisson used this argument as the basis for applying Eqs. 3.3, 3.4, 3.5, and 3.6, which would allow an estimation of the unknown constants k and u from observations, as well as of the probability of a correct verdict.

By aggregating the years 1825–1830, when verdicts were reached by seven votes to five, Poisson obtained an estimate, $\hat{c}_5$, of the probability of conviction by at least seven votes, c_5, equal to 0.4782. There was also a near-certain probability (P=0.9953) that the unknown probability and its estimate would differ by no more than 0.0135. Likewise, for 1831, when convictions were reached by eight votes to four, Poisson obtained a far lower estimate: $\hat{c}_4 = 0.3631$. Subtracting this estimate from the previous one, he found a difference of 0.1151, indicating a strong effect of the change in legislation, which reduced the probability of conviction.

3.4 Application to Social Science

Poisson could then estimate the two parameters of his model. First, he observed that the expression c_i did not change when k and u are replaced by $(1-k)$ and $(1-u)$. The problem therefore had two solutions: a pair of values greater than ½, and a pair smaller than ½. He assumed that, in normal times, the average probability of guilt must exceed that of innocence. However, he noted that in more troubled times, such as the French Revolution of 1789, the legal innocence of defendants could be more probable than their guilt. He thus arrived at an estimate of the probability k of a defendant's guilt of 0.5354, and an estimate of the probability u that a juror would not cast the wrong vote of 0.6786 (recall that this probability is assumed to be identical for each juror here). It should be borne in mind that these estimates concern personal crimes, and that both parameters remained constant throughout the 7-year period studied. For property crimes, the probabilities were different: Poisson found $k' = 0.6744$ for a defendant's guilt and $u' = 0.7771$ for a juror's error. Poisson presented these results more vividly:

> Before a verdict was handed down, someone who would not have known the identity of the jurors, or even the place where the case would be tried, could have wagered, at that time, slightly over two to one, that each juror would not cast the wrong vote if the crime were of the first type [personal crime] and nearly seven to two, for a crime of the second kind [property crime]. We use the crude expression *wager so much to so much* here in order to convey in a more vivid manner the significance that we should attach to the values of u and u', and despite the fact that our hypothetical wager is a fallacy, since we would never know who won.

The final step was to show the probability that a verdict was correct, once it had been reached. For 1831, Poisson showed that $P_4 = 0.9811$, in other words, of the 743 persons convicted that year, five should not have been. This is a far cry from Laplace's estimates of the chances of error in criminal verdicts—results that Laplace had described as terrifying.

Like earlier studies, Poisson's work on the probability of verdicts attracted little notice from jurists but sparked lively debates among scientists. For instance, the discussions that followed the reading of his papers on the law of large numbers (1836a) and on probability calculus (1836b) pitted the members of the Academy of Science hostile to the application of probability to moral matters—Poinsot and Dupin—against those who were in favor, including Navier and, of course, Poisson, all alumni of the École Polytechnique:

> M[onsieur] Poinsot regards probability calculus in moral matters, such as court verdicts and assembly votes, as an erroneous application of mathematical science: he believes that one can draw no consequence from it that could serve to improve human decisions (comment made in the discussion of Poisson's paper, 1836a).

He reiterates his doubts in the discussion of the other paper from Poisson (1836b). Similarly, Dupin believed that probability could not apply to human affairs, which are too complex to be reduced to sufficiently simple hypotheses amenable to mathematical treatment. Conversely:

> M[onsieur] Navier believes that the facts of all kinds on which our observations may dwell, and even the political or judicial facts that involve human passions and interests, equally depend on determined and subsisting laws, based on human nature. This principle having

been accepted, we shall necessarily conclude that the attentive and regular observation of facts can shed light on events to come, by highlighting the effects of the laws at work, and can lead to the determination of results that we may accept with a certain degree of confidence, the main purpose of probability calculus being to provide a measure for them. [comment made in the discussion of Poisson's paper, 1836a].

These quotations clearly show a clash between a probabilist conception of the world and a positivist conception, as defended by Comte in his *Cours de philosophie positive* (1830–1842). For instance, in the 27th lecture, he wrote:

It is the basic notion of assessed probability that I find directly irrational and even sophistic: I view it as essentially unfit to guide our conduct in any instance, or at most in games of chance. It would routinely lead us in practice to reject, as numerically implausible, events that will occur nevertheless.

Many authors have analyzed Comte's rejection of probability (Coumet 2003; Brian 2006). Brian ties it to the 'three states' law, which charts a path from the theological state defended by Leibniz, Süssmilch, and others, to the metaphysical state, of which Laplace is a privileged representative, and lastly to the positive state championed by Comte.

Even without going as far as Comte's rejection, Poisson's study was nevertheless one of the last nineteenth-century manifestations of Laplace's logical probability, which disappeared to give way to objective probability. Only after the mid-nineteenth-century did the Laplacian approach enjoy a revival, with growing applications to judicial issues.

Nearly a century and a half later, Gelfand and Solomon (1973) took up Poisson's model to show its applicability to a decision by the United States Supreme Court concerning 12-member juries. Lindley (1977) examined the problem of determining the probability that two items of evidence found on the crime scene and the suspect come from the same source.

With the development of new criminal-investigation methods (such as fingerprint analysis and techniques based on DNA collected from blood samples), many researchers have pursued these methods by using logical probability to better assess a defendant's guilt or innocence (Robertson and Vignaux 1991, 1993, 1995; Vignaux and Robertson 1996; Dawid and Mortera 1996). We shall not describe these many applications in detail here but simply outline the spirit in which they were conceived (Jaynes 2003), taking a fictitious example.

Let us suppose that a crime has been committed in the Paris area but that, at the outset, we know nothing about the circumstances. All we know is that the capital has ten million inhabitants. This information item alone gives us the odds that a random inhabitant of the Paris area, x, is guilty:

$$O(guilty \mid x) = \frac{1}{10^7}$$

which gives the plausibility of the person's guilt measured in decibels:

$$e(guilty \mid x) = 10 \log_{10} O(guilty \mid x) = -70 \; db.$$

3.5 Problems Posed by the Logicist Approach

Let us now suppose that we know that the criminal did not act gratuitously, but had a motive. We can then recalculate the plausibility of guilt:

$$e(guilty \mid motive) = e(guilty \mid x) + 10\log_{10}\left[\frac{P(motive \mid guilty)}{P(motive \mid not\ guilty)}\right]$$

$$\cong -70 - \log_{10} P(motive \mid not\ guilty)$$

for the probability of the event $P(motive \mid guilty) \cong 1$, as it is highly improbable that the crime was unmotivated. Let us now suppose that the investigation has determined the number of persons with a motive for the crime. If the number is N, then:

$$P(motive \mid not\ guity) = \frac{N-1}{10^7 - 1} \cong 10^{-7}(N-1),$$

and the plausibility of guilt becomes:

$$e(guilty \mid motive) \cong -10\log_{10}(N-1).$$

As we can see, the Paris-area population has disappeared from the equation, which now refers only to the number of motivated persons—a number that may actually be very low. Any new information can thus be added, so that the measure of plausibility will become sufficient to allow the defendant's conviction or, on the contrary, will be insufficient for that purpose, even if each information item taken separately is totally inadequate to prove the defendant's guilt.

To conclude, we can see that logical probability—like subjective probability—can be used successfully in social science. The basic difference between the two approaches is that a logical probability cannot be obtained from any random prior, provided that it lies between zero and unity; rather, it must justify this prior using arguments based on the information about the phenomenon studied available before the experiment.

3.5 Problems Posed by the Logicist Approach

Like the other approaches, the logicist approach has been subject to various criticisms, which we shall now examine.

First, the logicists regard a degree of probability not as a personal feeling (as in the subjectivist conception), but as a logical relation valid for all. The limit probabilities of zero and one represent logical impossibility and necessity. One of the criticisms directed at logical probability is that logical impossibility for the logician is not compatible with the zero probability of certain events, which may nevertheless occur. The logicists make the following claim:

> This linkage of probability and logical necessity, while reflecting a common usage in many cases, is not consistent with the modern conception. To say that an event has zero

probability in no way means that it is logically impossible but only that, at the limit, the ratio of the number of occurrences of the event to the total number of trials converges toward zero. (Matalon 1967)

For these authors, therefore, logic and probability do not exist on the same level of analysis.

Jeffreys (1939) had shown that his third axiom introduced a new logic of which deductive logic, actually used by Matalon, was only a part. Under this new logic, Jeffreys was able to demonstrate the following theorem:

If p is consistent with the general rules, and p entails $\bar{q}$, then $P(q \mid p) = 0$.

But the theorem's inverse:

If $P(q \mid p) = 0$, then p entails $\bar{q}$,

is false according to his proposed convention 3, a rule generally adopted:

If p entails q, then $P(q \mid p) = 1$.

For example, a continuous random variable can take all values between zero and unity. The probability that its value will be exactly ½ is zero, although the value ½ is not impossible for the variable.

We may therefore conclude that this criticism is easily answered, and was actually aimed at Carnap's approach, not examined here.

We also mentioned some of the discussions concerning Cox's axiomatization (1946, 1961)—which at the outset had not been sufficiently elaborated, particularly by Halpern (1999a, b)—and the solutions offered by van Horn (2003), which seem satisfactory to us. However, these issues have generated a broader discussion (Snow 1998; Skilling 1998; Arnborg and Sjödin 2000, 2001; Arnborg 2006; Shafer 2004; Colyvan 2004, 2008), notably on the uniqueness of Cox's axiomatics for probability calculus and, more generally, for a logic of uncertain reasoning. We have partially described the forms of logic of uncertain reasoning in our chapter on subjective probability (Dubois and Prade 1988; Smets 1997; Shafer and Vovk 2001), and shown that these belief functions defined on incomplete data generalized probability by leading to weaker structures that use the theory of *fuzzy sets* (Zadeh 1965, 1978). We are no longer dealing here with probability *per se* but with a theory of beliefs, which we shall not discuss in detail, as noted in Sect. 2.2.

We have also pointed out the difficulty in finding a unique prior in some cases, and cited Jeffreys's doubts about obtaining a general solution to this problem. Arnborg (2006), for instance, observes:

A lot of effort went into the idea of finding a canonical and unique prior, an idea that seems to have failed except for finite problems with some kind of symmetry, where a natural generalization of Bernoulli's indifference principle has become accepted. The problem is that no proposed priors are invariant under arbitrary rescaling of numerical quantities or non-uniform coarsening or refinement of the current frame of discernment.

It has been shown that priors complying with the principles of sufficient reason or maximum entropy are not invariant under rescaling (Robert 2006). For example, if the masculinity proportion, p, obeys a uniform distribution in the interval [0, 1], then the sex ratio, $\pi = \frac{p}{1-p}$, will obey a prior of density $\frac{1}{(1+\pi)^2}$. However, we did

3.5 Problems Posed by the Logicist Approach

note that the use of well-defined transformation groups allowed the introduction of invariance with respect to these groups (Harr 1933). Robert (2006) writes:

> The corresponding measures can accordingly be viewed as non-informative priors derived from the invariance structure.

As a change of scale or location shows us a problem from a different angle, we can state that we do have some knowledge of the problem:

> 'complete ignorance' of a location and a scale parameter is a state of knowledge such that *a change of scale and shift in location does not change that state of knowledge*. (Jaynes 2003)

Consequently, invariance gives us an alternative method to construct these non-informative priors. Current research in this area holds the promise of significant advances on the subject.

We also know that the way in which a problem is presented influences the ease with which we can solve it. To take this issue one step further, one frequent criticism of logical probability is even more basic: to what extent may priors, which we assign by means of entropy maximization, depend on the problem's presentation?

Let us take an example from social science: the probability of being an internal migrant. We assume that we have no information on whether a person is a migrant or non-migrant. What prior should we assign to the proposition 'is a migrant' represented by m? As we have no other information, the person is equally likely to be a migrant or non-migrant. Using symmetry arguments, we can say that the prior of this proposition is ½. Entropy maximization yields the same conclusion when our language $\mathcal{L}$ possesses only the proposition m. But let us now suppose that we have a new language $\mathcal{L}'$ that enables us to distinguish three categories of internal migrants, and that we now have propositions for each category: 'is an intra-municipal migrant,' *mic*, 'is a migrant between municipalities of the same *département* [French administrative division],' *mid*, and 'is a migrant between *départements*,' *med*. Proposition m can thus now be expressed as $mic \cup mid \cup med$. In this case, entropy maximization gives us the prior of $mic \cup mid \cup med$, ⅞, which differs from the previous value. We can therefore conclude, at first glance, that entropy maximization yields a prior that depends on the language used.

As early as 1961, Salmon defined a linguistic-invariance criterion, but was dissatisfied with the proposed method. More recently, Halpern and Koller (1995; article completed in 2004), Paris and Vencovská (1997), and Williamson (2009) have examined this language dependency in greater detail. Paris and Vencovská (1997) suggest the following interpretation. The entropy-maximization method has been incorrectly applied to such cases. In fact, when an agent states that proposition m becomes a new proposition $mic \cup mid \cup med$, we must take into consideration not the language $\mathcal{L}'$ but the language $\mathcal{L}'' = \{m, mic, mid, med\}$, which allows us to spell out the relation $m \leftrightarrow mic \cup mid \cup med$. In this new language, entropy maximization effectively yields a *prior* of ½, equal to the first one, for $mic \cup mid \cup med$. Here, language dependency no longer obtains.

Unfortunately, however, this solution works only in cases where an individual refines his or her language. If, instead, two persons use languages $\mathcal{L}$ and $\mathcal{L}'$

respectively, then they will assign two different priors to what we know—but they do not know—to be the same proposition. Here, the probabilities generated by entropy maximization will depend on the language used.

This dependency, however, is not confirmed solely by entropy maximization: it also applies to subjective probability. The dependency property is thus inherent in all epistemic probabilities, and cannot serve as an argument against logical probability alone: it is a general property of epistemic probabilities.

We shall not elaborate on the other criticisms directed against logical probability (Pearl 1988; Hunter 1989): for a detailed presentation, see Williamson (2005, 2009) and Paris and Vencovská (1997). Most often, these criticisms are fully compatible with the logical-probability approach and can be incorporated into it.

In conclusion, logical probability has been the target of a fair number of attacks, but its advocates have succeeded in countering a great number of them.

Conclusion of Part I

After the detailed examination of the main approaches to probability, the statistical inference that they imply, and their application to social science, we can now try to provide more detailed and better-informed answers to questions to which we have so far offered only partial answers: What is probability? What type of inference does it allow, particularly for social science? Can the three approaches be combined to any extent?

A Unique Notion or a Multi-faceted Notion?

It will be useful to begin with a historical summary of the emergence of the three approaches described separately in the preceding chapters. This will give us a clearer idea of how the main approaches to probability have developed over time.

From the earliest reasonings on events related to chance (Pascal and his correspondence with Fermat 1654, 1922) until the first half of the nineteenth century, when Cournot (1843) began to distinguish between objective probability and subjective probability, the notions of belief, frequency, and, more generally, of logic of the probable were in fact perceived as forming an inseparable whole.

In the last four chapters of their treatise on logic, Arnauld and Nicole (1662) clearly distinguish the truths that concern the essence of things from those that concern the belief in events:

> the ones that concern only the nature of things and their immutable essence, independently of their existence; and the others, which concern existing things, and especially human and contingent events, which may or may not occur when we are dealing with the future, and which may not have occurred when we are dealing with the past.

We see that mathematical logic will apply to the first sort of truths, for 'we must conclude that a thing is false, if it is false in a single case'. We can recognize the principle of deductive reasoning formalized by Aristotle in the fourth century B.C. (see Chap. 3). But to the second sort of truths, another form of logic must apply:

> One must therefore posit as a certain and indubitable maxim in this respect, that the mere possibility of an event is not a sufficient reason for me to believe it; and that I may have reason to believe it, even though I deem it not impossible that the contrary may have occurred. Accordingly, of two events, I may have reason to believe the one and not believe the other, even though I believe both are possible.

This is a clear manifestation of the principle of plausible reasoning, which Leibniz later sought to introduce (1765), but was eventually developed in the twentieth century by Polya (1954) and Jaynes (1981), as we have described it in Chap. 3. Arnauld and Nicole add the following to this mode of reasoning:

> To judge the truth of an event, and to persuade myself into a resolution to believe or not to believe it, it must not be considered nakedly, and in itself, like a proposition in geometry; but all the circumstances that accompany it, both internal and external, are to be weighed with the same consideration. I call internal circumstances those that belong to the fact itself; and external, those that relate to the persons whose testimonies induce us to believe it.

Once again, therefore, they pit geometrical reasoning—more generally, mathematical reasoning—against probabilistic reasoning and introduce the circumstances attending the latter. Thus, for an event that may have occurred several times in the past, we need to consider its frequency, which is one of its internal circumstances. Regarding thunder, for example, they note:

> Out of two million people, it is very much if there is one who dies in that manner, and we may even say that there is hardly any violent death that happens more rarely.

In other cases, of course, it may be useful to consider testimonies of persons whom we can trust.

Under this unitary theory, we can thus discern the first signs of factors that were later distinguished with greater precision in order to define the three approaches examined here. For instance, when Jacob (1713) defines probability as a degree of certainty,[1] he is effectively adopting a subjective interpretation, but when he demonstrates the weak law of large numbers, he is choosing an objective (frequentist) interpretation. Similarly, Leibniz (1765 but written around 1703) asserts the need to introduce a new logic for probability:

> And as regards the magnitude of the consequence and the degrees of probability, we still lack the part of logic needed to estimate them, and most of the casuists who have written on probability have not even understood its nature, basing it on authority with Aristotle,[2] instead of basing it on likelihood as they should, authority being only one part of the reasons that contribute to likelihood.

---

[1] Gradus certitudinis.

[2] Leibniz is referring to the logic of probability that Aristotle links to the art of rhetoric in *Rhetoric*.

This is indeed a logicist interpretation of probability, even though Leibniz, who repeatedly insists on its establishment, is barely capable of articulating its foundations. Jacob Bernoulli (1713) had firmly established direct probability with the law of large numbers, which makes it possible to deduce effects from observed causes. By contrast, Bayes (1763) and Laplace (1774), using inverse probability, made it possible to travel the opposite path from effects back to causes (see Chap. 2). Laplace's *Essai philosophique sur les probabilités* (1814) crowns this synthesis:

> One can even say, strictly speaking, that nearly all our knowledge is merely probable; and in the small number of things that we can say we know with certainty, in mathematical science itself, the principal means of arriving at the truth—induction and analogy—are based on probability, so that the entire system of human knowledge is connected to the theory set out in this monograph.

Laplace illustrates this totalizing vision by showing the use of probability in every field, from gambling to astronomy, moral science, testimonies, political decisions, court verdicts, population sciences, economics, and others.

In fact, it is our ignorance of the ties that bind events to the entire system of the universe that explains this importance of probability. As Laplace puts it: 'Probability is partly related to this ignorance, partly to our knowledge'. But all the efforts of humans will bring them ever closer to the knowledge of those ties:

> An intellect that at any given moment knew all the forces that animate nature and the mutual positions of the beings that compose it, if this intellect were vast enough to submit its data to analysis, it could condense into a single formula the movement of the greatest bodies of the universe and those of the lightest atom: for such an intellect nothing could be uncertain; and the future, just like the past, would be present before its eyes.

Laplace recognizes, however, that the human mind will 'forever remain infinitely distant' from such omniscience, and that probability is the only way to approach it.

Throughout the nineteenth century, many criticisms were directed against Laplace's unitary approach, with the ever sharper dualist distinction between subjective probability and objective probability. This eventually led to an objectivist approach to probability, adopted by a large majority of probabilists by the late nineteenth century—although a small number remained staunchly loyal to some of Laplace's principles.

In Chap. 3, we quoted the objections by members of the French Academy of Science to Poisson's use of probability (1836a, b, 1837) to deal with 'moral' issues—specifically, court verdicts (Poinsot and Dupin 1836)—whereas the studies by Condorcet and Laplace on the same subject had not been criticized in their time except by philosophers and clergymen.

Later, Cournot (1843) introduced a number of new notions, while partly preserving Laplace's legacy.

Cournot began by breaking up the unity of probability. He did so by distinguishing between 'the dual meanings of the term *probability*—understood now in an objective sense, now in a subjective sense'. For the objective sense, he sometimes used the term physical possibility:

> The advantage of the term *possibility* (which has already come into use because of the awareness of the truths described here) is that is clearly designates the existence of a relationship that subsists between things themselves, a relationship that is not determined

by our way of appraising or feeling—which can vary from one individual to another, depending on the circumstances in which they find themselves and to the extent of their knowledge.

For the subjective sense, Cournot spoke of 'issues of *probability* that indeed relate in part to our knowledge, in part to our ignorance'. But he did not contrast objective phenomena with subjective phenomena, as later authors would. In fact, he took the human perception of a phenomenon and considered it in both objective and subjective terms. This is very clear from the example he offered:

> When we say that the probability of rolling a double six in backgammon is $\frac{1}{36}$, we can express an opinion based on possibility; and this means that, if the dice are perfectly even, cubic, and homogeneous, so that there is no reason inherent in their physical structure for one side to turn up rather than another, the number of double sixes obtained in a large number of throws, by impulsive forces whose direction is totally independent of the points marked on the sides, will be broadly one-thirty-sixth of the total number. But we can also express an opinion based on mere probability; and without investigating whether that evenness of structure exists or not, we need only be unaware of the direction in which the unevennesses of structure operate—if they exist—for us to have no reason to assume that one side will turn up rather than another.

As we shall see, this reasoning is actually very different from the later reasoning on objective probability. In fact, Cournot devoted several chapters of his book to the probability of court verdicts, in the wake of the studies by Condorcet, Laplace, and Poisson.

Laplace had used his theory of probability to estimate the best average of different observations—by examining, for instance, the case in astronomy in which only three measures are available (1774). Cournot does not seem to follow him on this ground. He writes:

> When the number of trials is insignificant, the formulas commonly given to assess probability *a posteriori* become fallacious: they now indicate merely subjective probability, suitable for setting the conditions of a wager, but incapable of application in the order of natural phenomena.

We can see that in this case he rejects the use of subjective probability, with which 'we shall be unable to determine the ratio of erroneous verdicts to the total number of verdicts pronounced in similar circumstances'.

Boole (1854) goes even further than Carnot in this rejection, describing Laplace's law of succession (see Sect. 4.3) as follows: 'I apprehend, however, that this is an arbitrary method of procedure'. Likewise, in regard to verdicts, Boole writes:

> Laplace makes the assumption, that all values of x from $x = \frac{1}{2}$, to $x = 1$, are equally probable. He thus excludes that a juryman is more likely to be deceived than not, but assumes that within the limits to which the probabilities of individual correctness of judgment are confined, we have no reason to give preference to one value of x over another. This hypothesis is entirely arbitrary, and it would be unavailing to examine into its consequences.

These criticisms were taken up by authors such as John Venn (1866), who noted that the law of succession:

> does not merely mislead us by giving one determinate but incorrect answer; it perplexes us by the offer of several discordant and often contradictory answers, all of them presumably incorrect.

The same attitude was voiced by Peirce (1883) and most researchers who espoused an objectivist approach to probability—despite the previously noted fact that such objections were totally unjustified in regard to epistemic probability. Venn (1866) also stated that 'the distinction between Direct and Inverse probability must be abandoned', thus rejecting a basic distinction in epistemic probability.

These critiques led to the endorsement by most probabilists, in the late nineteenth century, of the objectivist approach described in Chap. 1. Only a few authors still defended some of Laplace's ideas. Francis Edgeworth (1885a), for example, clearly showed—but apparently in vain—the errors of reasoning committed by Boole, Venn, Peirce and many others. This objectivist approach continued with the axiomatization by von Mises (1919, 1928), criticized by many authors, and by Kolmogorov (1933). It culminated in the studies by Fisher (1922a, b, 1956), and others marked the major stages in the development of these methods during the first half of the twentieth century.

By the 1920s, however, some authors were questioning the objectivist view and proposing other approaches to probability, which did not begin to take hold until after World War II.

First, the subjectivist epistemic approach—introduced by Ramsey (1926) but effectively elaborated by de Finetti (1937)—axiomatized this type of probability using the notion of coherence (see Chap. 2). Here, 'each person is free to adopt the opinion (s)he prefers, or, more accurately, the opinion that (s)he *feels*', if it is a coherent opinion. The theory was fleshed out by Savage (1954), who introduced the concept of utility proposed by the economists von Neumann and Morgenstern (1944). This subjectivist approach made it possible to restore a structure of probability of degrees of belief complying with Kolmogorov's axiomatization (Bernardo and Smith 1994) while adding other conditions—in particular certain exchangeability properties—that also need to be met.

However, the psychological basis of this approach led a number of authors to generalize the notion of subjective probability, initially replacing the single probability of a given event by a double probability, upper and lower (Good 1962; Dempster 1967; Suppes 1974; Shafer 1976). Going one step further, Smets (1988, 1990) distinguished between (1) a credal (psychological) level, for which he defines belief functions without resorting to the concept of probability, and (2) a pignistic (decision-making) level, where it becomes possible to define a subjective probability (see Chap. 2). This approach has allowed the use of subjective probability in fields where conventional probabilistic modeling is inappropriate because the statistical data are unavailable, but where quantitative opinions seem very useful.

Meanwhile, the logicist epistemic approach (see Chap. 3), outlined by Keynes (1921) but fleshed out by Jeffreys (1939), largely revived Laplace's approach by axiomatizing it with the aid of the notion of consistency. It was also used by the philosopher Carnap (1950, 1952), but his over-abstract presentation led him to distinguish between frequential probability and logical probability—a distinction that is actually unwarranted in Jeffreys's approach. The logicist epistemic approach was later reworked by Richard Cox (1961)—who showed how the intuitive notion of plausibility can be formalized by the notion of logical probability—then by Jaynes

(2003). Both authors added the notion of entropy as a measure of the information available. They also recognized that Kolmogorov's axiomatization is consistent with their approach, provided that the reasoning concerns propositions rather than sets. Despite many identical algebraic properties, it is necessary to distinguish between the different intrinsic meanings of the two terms (Kardaun et al. 2003).

Classical probability began as a unitary construct, but was replaced by three main approaches often in conflict with one another. However, as we have seen, the same mathematical rules, defined by Kolmogorov's axiomatization, apply equally well to objective, subjective, and logical probability. The only way to tell them apart is to engage in a deeper interpretation of these rules, with the aid of new axioms or by distinguishing between the elements of a set of propositions concerning them (Shafer 1992).

Before examining whether a synthesis of the three approaches is possible, let us see how to move from these forms of probability to the statistics gathered by the various sciences, and what inference we can draw from those statistics.

Inference and Decision-Making Theory

In the previous chapters, we viewed statistical inference as a whole that enables us to move from data on specific units to statements about a new unit. But we did so without distinguishing between the two terms of the expression: 'statistical' and 'inference.' Let us now consider the two terms separately in order to identify their more precise meaning. We shall see whether this examination allows an extension of what we referred to earlier as 'statistical inference'.

First, let us see what is meant by statistics. The term is linked to the notion of the State (Latin: *status*), which collects all the information needed to manage its affairs. As indicated earlier, such information has been used since remote antiquity. In the seventeenth century, William Petty (1690) adopted the notion of the State as the foundation of political arithmetic, an enterprise that involved collecting numerical data from mortality and baptism bills, but also from all other sources. Petty did advocate the collection of observations as the first step in any scientific approach. His goal, however, was to go back—in the Baconian tradition—to the social and economic principles of the observed facts in order to define a political action, such as the massive resettlement of the Irish on English territory (Reungoat 2004). This approach viewed people not as autonomous entities, but as members of a State without an individuality of their own (Porter 1986). By contrast, this vision did not inform the discipline of statistics that emerged in the early nineteenth century. The new science set out to provide information on all aspects of the life of individuals, now viewed as autonomous. The difference is manifest in this introduction by the Council of the Statistical Society of London (1838):

> The Science of Statistics differs from political economy, because, although it has the same in view, it does not discuss causes nor reason upon probable effects; it seeks only to collect, arrange, and compare, that class of facts which alone can form the basis of correct conclusions with respect to social and political government.

Information was to be collected not only on all aspects of people's lives but on all characteristics:

> In fact, as all things on earth were given to man for his use, and all things in creation were so ordained as to contribute to his advantage and comfort, it follows that Statistics enter more or less into every branch of Science, and form part of each which immediately connects with human interests.

Statistics thus came to be seen as a set of methods of mathematical interpretation applied to these phenomena. The result was a new science, with close links to probability from the outset, although it remained distinct from the latter and could even be practiced without relying on probabilistic concepts. March (1908), for instance, wrote:

> It would thus be appropriate, in statistics, to renounce the use of the word probability to express the expectation born of the observation of a frequency; for, while in probability theory the convention on which this expectation rests is wholly trustworthy, in statistics the degree of confidence that this expectation deserves is often altered by the study of connections between facts, and by the lessons of social science.

But such an attitude was rare. Most often, statistics coexisted with probability theory, although it pursued another goal.

First, while probability was axiomatized in the twentieth century along with all the other branches of mathematics, statistics was not. Bernardo (discussant of the paper by Kardaun et al. 2003) had this to say about the issue:

> No wonder that contradictions arose in conventional statistics, and no surprise at the often derogatory attitude of mathematicians to mathematical statistics, too often presented as an 'art' where contradictions could be acknowledged and were to be decided by the wit of the 'artist' statistician.

In talking about statistics, we are therefore already in a field close to that of the social sciences, where axiomatization is also practically non-existent, as we shall see in Part II.

Let us now examine the second term: 'inference.' Again, this notion is general in scope and has been applied to a domain ranging well beyond statistics. The term has been used since the development of logic by the Greek philosophers. In its broadest logical sense, inference is an operation in which a proposition is accepted by virtue of its connection with other propositions deemed to be true. By this definition, Aristotle's syllogism is an inference enabling us to deduce a third premise from two that are held to be true (see Sect. 3.1).

While such an inference may be viewed as deductive in logic and mathematics,[3] it cannot in the other sciences. This is because the researcher will now observe a data set or engage in reasoning, and then deduce via scientific inference the relations that will exist between those data. Another name for this method is amplificative induction, which generalizes an observation or a reasoning from the observation of individual cases.

---

[3] In fact, Gödel (1931) showed that the theories intended to provide a foundation for mathematics, such as Peano's arithmetic and set theory, contain at least one proposition that can be neither proven nor rejected as false by the theory.

Scientific inference lies outside the realm of deductive logic. As shown in Chap. 3, we need to introduce a more general logic to handle it. Also, as its definition indicates, it does not necessarily use statistical data. In some cases, it will use reasonings.

In physics, for example, Einstein (1905) developed his theory of relativity without having access to data that would confirm it. He suggested his two postulates with no experimental basis: '(1) the same laws of electrodynamics and optics will be valid for all frames of reference for which the equations of mechanics hold; (2) light is always propagated in empty space with a definite velocity V which is independent of the state of motion of the emitting body'.[4] Indeed, the German physicist Kaufman (1906), who tried to verify the predictions of relativity theory by experiment, declared: 'these measures are incompatible with the fundamental Lorenz-Einstein hypotheses'.[5] Not until later did more rigorous experiments disprove Kaufmann and confirm the accuracy of Einstein's postulates: these were based not on experiment but only on a thorough analysis of the concepts used, i.e., on reasoning.

This analysis, however, required a deep knowledge of the phenomena studied, a knowledge based in turn on many experiments. We shall see that Pascal and Fermat's findings were not based on statistics either, but on their knowledge of the rules of games of chance. Their findings can, similarly, be verified by experiment or through the use of statistics.

We therefore believe it is useful to extend the results on statistical inference obtained in the previous chapters to the more general case of other scientific inferences (Jeffreys 1931). The goal of such inferences is to make the most of the information available on one or more phenomena in order to infer which hypothesis best explains them (the information may be supplied by the data observed or the properties of the phenomena). The explanation may be very simple: for example, when different balls are picked from a jar of unknown composition, which distribution best explains the draw? More generally, when several hypotheses can account for the connections between various phenomena, which hypothesis best explains them? We may have at our disposal statistics on the phenomena or a detailed observation of their properties.

It is also clear that the resolution of this problem enables us to extend the notion of inference to the broader yet related domain of decision-making theory. Admittedly, using statistical inference, we can, for example, deduce the probability distribution that best represents the final state of our knowledge. But we lack the rule that allows us to transform this representation into action. That is the role of decision-making theory.

To better appreciate the significance and implications of both inference and decision-making theory, and their links with probability theory, we must begin by viewing these issues in a historical perspective.

---

[4] (1) die gleichen elektrodynamischen und optischen Gesetze gelten, wie diese für die Größen erster Ordnung bereits erwiesen ist, (2) das Licht im leeren Raume stets mit einer bestimmten, von Bewegungszustande des emittierenden Körpers unabhängigen Geschwindigkeit V fortpflanze.

[5] Die Messungsergebnisse sind mit der LORENTZ-EINSTEINschen Grundannahme nicht vereinbar.

From the outset, Pascal and Fermat used their observations of games of chance to develop the notion of mathematical mean (expectation), which enabled them to solve the problem they had posed. This consisted in determining the fairest distribution (*parti*) to be inferred from the rules of the game, in the event that the players want to stop the game at a certain point (see General Introduction). The solution suggested by Pascal and Fermat is the mathematical mean of each player's expected gains—a mean that, according to a fairness argument, must be identical for each. This is indeed a scientific inference and not a statistical one, for we do not need to observe a sample in order to make the decision: the information on the rules of the game suffices to infer the gains expected by each player. As noted earlier, there is no collection of statistical data on which to base the inference. All we have is the experience of games of chance. The decision-making criterion adopted here is expectation, i.e., the same one used for inference.

Sometimes, however, it is hard to equate the mathematical mean with the decision to be taken. One example is the Saint Petersburg paradox, submitted by Nicolas Bernoulli to de Montmort in 1713 (see Chap. 2), which consists in tossing a coin until it comes up heads for the first time. If this happens on the nth toss, the gain will be 2^n *écus*. The question is: what initial sum is needed for a player to be able to join the game? The previous solution, which is to take the mathematical mean of the gains, would here require a wager tending toward infinity in order to ensure a fair game. While such an inference is mathematically acceptable, the decision is not. In the words of Nicolas Bernoulli, quoted by Daniel Bernoulli (1738):

> Although the standard calculation shows that the value of Paul's expectation is infinitely great, it has, he said, to be admitted that any fairly reasonable man would sell his chance, with great pleasure for twenty ducats.[6]

To solve the problem, Daniel Bernoulli (1738)[7] assumed that people do not necessarily think in terms of mathematical mean; instead, they introduce the notion of 'emolumentum medium'—today called mean utility[8]—to take a decision. In other words, one needs to take each player's wealth into account:

> the utility resulting from any small increase in wealth will be inversely proportionate to the quantity of goods previously possessed.[9]

This condition yields a linear utility function of the log of the person's wealth. Bernoulli shows that, if a is Paul's wealth, his expected gain is:

$$G = (a+1)^{1/2} (a+2)^{1/4} (a+4)^{1/8} \cdots (a+2^{n+1})^{1/2^n} \cdots - a.$$

---

[6] Quando-quidem calculus dicet, sortem Pauli infintum esse, nec tamen ullus sanae mentis, ut dicit, futurus sit, qui non libentissime spem suam vendiderit pro summa viginti ducatorum.

[7] At the end of his presentation, given in 1731 but published in 1738, he notes that Cramer, the famous Swiss mathematician (1704–1752), had already described a similar theory in a letter to his cousin Nicolas in 1728, which Daniel Bernoulli quotes.

[8] Cramer and Laplace still refer to this notion as moral expectation.

[9] aestimari posse emolumentum lucri valde parvi summae bonorum reciproce proportionale.

Hence:

> If he owned ten ducats his opportunity would be worth approximately three ducats; it would be worth four if his wealth were one hundred, and six if he possessed one thousand. From this we can easily see what a tremendous fortune a man must own for it to make sense for him to purchase Paul's opportunity for twenty ducats.[10]

As in Pascal's wager, we must take the players' wealth into account in order to estimate the utility of the game. If a player with a wealth of 100 ducats bets 50 in a game of heads or tails, his or her wealth after placing the bet will be reduced to 87 ducats. This example also enables Bernoulli to demonstrate the inanity of games of chance.

These ideas introduced by Daniel Bernoulli and Cramer were used by Laplace to examine the application of probability to many decision-making problems.

However, the rejection of classical probability theory in the nineteenth century and the adoption of objective probability led to a very different approach both to inference and decision-making criteria. The objective approach confined itself to events capable of recurring in identical circumstances. In consequence, the notion of the probability of an intrinsically unique event—or, more generally, of the probability that a proposition is true—no longer made sense. But inference should apply precisely to testing the truth of a statistical hypothesis, such as the null hypothesis. The objectivist approach could not draw such an inference directly. It fell back on the probability of obtaining the observed sample if the hypothesis is true and not on the probability that the hypothesis is true (see Chap. 1)—a roundabout answer to an initially clear problem. Moreover, for the objectivist approach to be of value here, we would need to be able to repeat the sample selection an infinite number of times (Jaynes 1976).

Despite these difficulties, Neyman and Egon Pearson succeeded in developing an objectivist theory of statistical tests (Neyman and Pearson 1933a, b; Neyman 1937). For this purpose, they distinguished between two types of errors. The first is the one we commit by wrongly rejecting the hypothesis that the phenomenon studied is due to chance; the second consists in rejecting the opposite hypothesis. If we guard against one of the two, then we necessarily increase the probability of the other, on a constant-information basis. As a result, the inference that should have informed us whether a hypothesis was true or not has been replaced by rules of action whose meaning is not the one we could have expected. For instance, when we calculate a 95% confidence interval for an estimated parameter, all we know is that if we picked a large number of samples, the estimated parameters would lie within the interval in 95% of cases.

The objectivists viewed decision-making theory as totally distinct from inference. This led them to use game-theory concepts to address decision-making. By the 1940s, Wald had developed a complex decision-making theory from game-theory concepts.

---

[10] Si decem habuerit ducatos, proxime tres valebit expectatio, and quatuor cum triente praeter propter si centum habuerit, ac denique sex cum mille habuerit. Facile hinc indicatu is quam immensas quis divitas possidere debeat, ut cum ratione viginti ducatis sortem Pauli emere possit.

However, after a series of articles (Wald 1947a, b), he eventually admitted in 1949 that the admissible decision-making rules were actually Bayes strategies (Jaynes 2003). We therefore do not feel it is useful to describe his theory in greater detail here. Let us see instead what the epistemic approach contributes to these strategies.

The subjective approach, elaborated by Bruno de Finetti and Leonard Savage, not only provided a means to analyze many events about which the approach objectivist had nothing to say, but also brought a totally fresh approach to these issues.

By introducing the notions of coherence and exchangeability, de Finetti (1937) opened up the possibility of calculating an inference with which one could move from data on a set of units to assertions about a new unit (see Chap. 2). The tests that could be conducted under this subjective approach would indicate whether a hypothesis is true or not. Thus, when calculating a 95% confidence interval, we know that the estimated parameter effectively has a 95% probability of lying in that range.

Meanwhile, the introduction of utility by Savage (1954) allowed the development of a decision-making theory whose meaning would merge with that of inference. To the question: 'Is it fruitful to treat inference and decision analysis somewhat separately?' (Kardaun et al. 2003), Bernardo responded in discussion on the paper:

> At a foundational level certainly it is *not*: decision analysis provides the coherent framework which guarantees that no inconsistencies and/or obvious wrong answers (a *negative* unbiased estimate of a probability, or a 95% confidence region for a real-valued quantity which happens to be the *entire* real line, say), will be derived.

By using prior probability to take into account the utilities of the potential consequences of different actions, it thus became possible to provide a solid foundation to Wald's decision-making theory, initially developed in an objectivist framework.

The logicist approach, developed by Jeffreys, Cox, and Jaynes, had many features in common with the subjective approach in its treatment of statistical inference (see Chap. 3) and decision-making theory. However, the personal viewpoint on which coherence was based gave way to independent analysis of the user's personality and to a vision rooted in consistency.

Unlike the objectivists' confidence intervals, the Bayesian intervals proposed by Jaynes (1976) for estimating an inference thus shared the objectives of the intervals advocated by the subjectivists:

> Our job is not to follow blindly a rule which would prove correct 90% of the time in the long run; there are an infinite number of radically different rules, all with this property. Our job is to draw the conclusions that are more likely to be right in the specific case at hand; indeed, the problems in which it is more important that we get this theory right are just the ones (such as arise in geophysics, econometrics, or antimissile defense) where we know from the start that the experiment can *never* be repeated.

At the same time, he rejected the individualist view of subjective probability. Commenting on the use of Bayes's theorem, he noted:

> To recognize these things in no way forces us to accept the 'personalistic' view of probability (Savage 1954, 1962). 'Objectivity' clearly does demand at least this much: the results of a statistical analysis ought to be independent of the personality of the user. In particular, our prior probabilities should describe the prior information; and not anybody's vague personal findings.

This concern for objectivity is shared by all advocates of logicist probability (Berger et al. 2009; Bernardo 2011).

Similarly, this approach recognized a theory of statistical inference resembling that of the subjectivists. Jaynes (2003) showed the rules for solving a decision-making problem that avoided the drawbacks of both the objective and subjective approaches. We shall not describe the rules in detail here, but merely note that they are determined by 'elementary desiderata of rationality and consistency'. Jaynes also pointed out their closeness to those offered in the eighteenth century by Cramer, Daniel Bernoulli, and Laplace.

Does Cumulativity Exist in Probability?

The desire or, on the contrary, the refusal to restore the unity of classical probability has driven many researchers and philosophers of science to offer arguments of variable persuasiveness. We set aside those who have striven to establish the relevance of their theory, such as von Mises (1928, 1932) and Kolmogorov (1933) for objective probability; de Finetti (1937) and Savage (1954) for subjective probability; Jeffreys (1939), Cox (1961), and Jaynes (2003). They naturally defended their points of view and showed the reasons why the approaches followed by the other theories are mistaken or flawed. Thomas Kuhn's term 'scientific revolution' (1962) would seem to suit them well, but as all these schools are still very much alive, can we speak of revolution?

We begin by reviewing the arguments of those who, without really choosing one of the three approaches, have militated against the unity of probability; we then look at the arguments of those who championed unity. However, we shall not aim to be exhaustive.

To begin with, most statisticians, physical scientists or social scientists display little interest in the foundations of probability. They apply statistical concepts to their data pragmatically, without any concern for their basis. Mosteller and Wallace (1964) clearly show the pragmatic attitude of statisticians:

> Even though individual statisticians may claim generally to follow the Bayesian school or the classical school, no one has an adequate rule for deciding what school is being represented at a given moment. When we have thought we were at our most Bayesian, classicists have told us that we were utterly classical; and when we have thought ourselves to be giving a classical treatment, Bayesians have told us that the ideas are not in the classical lexicon. So we cannot pretend to speak for anyone but ourselves.

This approach has prevailed in many applications, although in some cases the paucity of data points has forced the researcher to adopt the Bayesian method.

The philosopher Ian Hacking takes a similar but clearly different attitude. In 1975, he admitted that probability had two facets:

> On the one side it is statistical, concerning itself with stochastic laws of chance processes. On the other side it is epistemological, dedicated to assessing reasonable degrees of belief in propositions quite devoid of statistical background.

Does Cumulativity Exist in Probability?

As we can see, each of these facets does not apply equally to all problems involving probability: the first applies to statistics, the second to personal beliefs. Hacking goes even further when, in discussing an article by Glenn Shafer (1990a), he totally rejects all reunification of the concept of probability, arguing that: 'It is like most of our other concepts, a radial one, not characterized by necessary and sufficient conditions'.

Similarly, the philosopher Patrick Suppes (2002a), after describing in detail the various theories of probability, recommends probability users to adopt a pragmatic attitude:

> Probability is too rich and diversified in application to be restricted to one single overweening representation. The pragmatic context of use will sometimes fix the representation chosen, but more often, a deeper pragmatic attitude will dominate and no explicit choice of representation will be made [...]

This attitude is supported by examples from quantum mechanics as well as 'data mining, adaptive statistics, boosting, neural networks and a variety of other approaches to large-scale data analysis'.

However, in the opposite direction, some authors have sought a more cumulative approach, which would allow a synthesis of the different schools.

Shafer (1985, 1990a, b, 1992), for instance, regrets that the unity of classical probability has been lost because of the split into divergent schools, and he suggests that we should think about ways to restore that unity. He starts with the games of chance studied by Pascal and Fermat, which he calls the 'ideal picture of probability', where the unity between belief and frequency is unquestionably visible. When we use probability theory for an altogether different type of problem, however, Shafer (1985) argues that:

> the different ways Bayesians, frequentists and others use probability should be thought of as different ways of relating problems to the ideal picture.

He notes that what he calls the Bayesian approach is now regarded as subjectivist, and that it is useful to consider the logicist Bayesian approach as well:

> Today, most scholars who understand probability as degree of belief have dropped the adjective 'rational'. These scholars no longer hold, as Keynes did, that given evidence logically determines a probability for a given proposition.

Shafer accordingly believes (1992) that we need to return to the notion of 'fair odds' rather than to that of 'personal odds' if we want to reunify probability:

> There is no reason for a person to have personal odds at which she would bet on either side. But a person can draw an analogy between her evidence and the special situation where fair odds are known. She can say that her evidence is analogous, in it strength and import, to knowing certain fair odds, which are based on long-run frequencies. This recasting of belief interpretation pulls it towards both the frequency and support interpretations.

He sets out to axiomatize these odds differently from Kolmogorov's axiomatization, which concerned the probability of isolated events. Shafer sees the need to reintroduce the repetition of events in this context. For this purpose, he develops a wager-based ('game-theoretic') interpretation of probability (Shafer 1996, 2010) that we have already presented (in part) and criticized in Chap. 2. We shall not examine this approach any further here, for we believe it is too focused on games and finance (Shafer 2001) to supply a fully valid generalization of probability theory. However, the introduction of a time dimension is a valuable feature of Shafer's theory.

More recently, Knuth (2008, 2009, 2010a, b) uses the studies by Shannon (1948), Cox (1961), and Jaynes (1956, 2003) as his starting point to propose a synthesis not only of probability theory, but also of information theory and entropy. He begins by noting that probability and entropy describe our state of knowledge about both physical and social systems, but do not describe those systems themselves. From this observation, he shows that the theory of partly ordered sets (posets) and lattice theory make it possible to unify the frequentist and epistemic logicist approaches (2010a):

> Here the two perspectives of logic and sets, on which the Cox and Kolmogorov foundations are based, are united within the lattice-theoretic framework.

It will be recalled that a partial order, producing a 'poset,' is a binary relation on a set that is reflexive, antisymmetrical, and transitive (see Chap. 3), and that a lattice is a partially ordered set in which each couple of elements admits an upper bound and a lower bound. Having briefly described the approach in Sect. 3.2, let us now examine in greater detail how this unification is possible.

The theory considers three closely interlinked spaces, each of which constitutes either a partially ordered set or a lattice. Knuth (2008) characterizes them as follows:

> the poset that describes the state space gives rise to a lattice of statements, called the hypothesis space, and a lattice of questions, which is called the inquiry space, via order-theoretic exponentiation.[11]

He sets out to work simultaneously in all three spaces to unify probability theory, information theory, and entropy.

Knuth defines a set of logic propositions composed of identifiable elements ('atoms'), which can be combined with the aid of a 'join' operator, $\cup$, which is the equivalent of the logical operation OR and includes the null element $\emptyset$. Also, when we consider two independent systems, their direct-product, $\times$, consists in taking two elements of each system as a new unit. Knuth has shown that both Cox's logicist approach and Kolmogorov's set-theory approach can be based on this more general lattice theory. The latter possesses a set of seven symmetries, leading to seven axioms governing the theory's quantification (Knuth 2010a):

1. $\emptyset$ does nothing[12]
2. $\cup$ obeys strict order

---

[11] Exponentiation relies on generating new lattice elements from old by grouping the old elements into sets called downsets. Downsets are constructed so that they contain their lower bound. That is, given any element in the downset, all elements included by this element are also members of the set (Knuth 2008).

[12] Here we use the notations that we have adopted for the present book and not Knuth's (2010a), which are slightly different.

3. ∪ is associative
5. × is distributive
6. × is associative
7. order is associative

To begin with, axioms 1, 2, 3, and 4 make it possible to define a measure $m(x)$ on the space of states that satisfies:

$$m(x \cup y) = m(x) + m(y).$$

Axioms 5 and 6 enable us to deduce the direct-product rule:

$$m(x \times t) = m(x)\, m(t).$$

The increase in the number of independent measures results in:

a unique form of variational potential for assigning measures under constraints, yielding a unique divergence of one measure from another.

In addition to this measure on the space of states, we can define a corresponding measure for the space of hypotheses, $p(x \mid t)$, which represents the plausibility of question x conditioned by context t. Using axiom 7, combined with axioms 1–4, we can show that the measure satisfies the condition:

$$p(x \mid z) = p(x \mid y)\, p(y \mid z),$$

and that this takes us back to Kolmogorov's probability axioms and Bayes' theorem. Working now on probability distributions, Knuth concludes as follows:

The variational potential defines the information (Kullbach and Leiber 1951) carried by a destination probability relative to its source, and also yields the Shannon entropy of a partitioned probability distribution.

As we can see, this approach not only unifies Kolmogorov's objective probability and Cox's logicist probability, but, at the same time, offers a single quantification of Shannon's information theory and entropy. We should also note that these results are valid for all lattices whereas probability theory was established in the context of Boolean algebra. The new theory therefore paves the way for a true cumulativity of probability.

However, it does not allow the inclusion of subjective probability theory, in which the degrees of belief may no longer be 'rational'—in other words, everyone is free to have his or her own opinion. This is consistent with Shafer's position and confirms it. However, unlike Shafer's approach, Knuth's does not make it possible to introduce temporality into the definition of probability. It is perhaps in this direction that probability theory could still evolve.

Part II
From Population Sciences to Probability

Introduction to Part II

Taking the opposite path to the one followed in Part I, we shall now explore the relationship that developed over time between social science and probability. However, as this book is not an encyclopedia, our analysis will necessarily focus on a selected group of social sciences.

Let us first examine what are the links between social science and probability.

We previously noted the near-simultaneous introduction of a geometry of chance by Pascal and Fermat and of a political arithmetic by Graunt and Petty. While the two approaches were linked, the historical development of social science also left room for the possibility that those connections may have been very weak, and even irrational.

In his course on positive philosophy applied to astronomy (Lecture 27), Auguste Comte rejected the notion of probability (see our Sect. 3.4). Later, in Lecture 49 (vol. 4 [1839] on social physics), he spoke of a 'fanciful subordination to the illusory mathematical theory of chance', for which he castigated Jacob Bernoulli and Condorcet. But he saved his harshest criticism for Laplace, finding it truly impossible to excuse his 'sterile reproduction of such a philosophical aberration, even as the general condition of human reason was already making it possible to glimpse the true fundamental spirit of sound political philosophy […].' Comte pursued his attack as follows:

> Indeed, would it be possible to imagine a more radically irrational conception than that which consists in assigning as a philosophical base for the totality of social science, or as the principal means of its final elaboration, a so-called mathematical theory in which, by routinely taking signs for ideas—in keeping with the customary nature of purely metaphysical speculations—one seeks to submit to calculation the necessarily sophistic notion of numerical probability, whose direct result is to offer our own true ignorance as the natural measure of the degree of likelihood of our various opinions?

This violent condemnation of the use of probability in social science—a practice that, in Comte's view, has registered no tangible improvement in a century—was also endorsed by members of the French Academy of Science (Poinsot and Dupin 1836) hostile to the use of probability for moral issues. Many contemporary philosophers, such as Bordas-Desmoulins (1843), similarly rejected the application of probability to social or cultural phenomena. This refusal, tied to a particular historical period, should, in fact, be viewed in a fuller perspective on the social sciences in order to assess it in its true context.

More recently, Shafer (1990a) has noted the need to consider the limits of probability as a method for dealing with many problems:

> In truth, most problems of inference in science and in the professions do not lend themselves to effective probabilistic or statistical treatment. An understanding of the intellectual content of applied probability and applied statistics must therefore include an understanding of their limits. What are the characteristics of problems in which statistical logic is not useful?

He gives the example of inference methods specific to artificial intelligence, but his argument is broader and notably extends to all the social sciences. What problems in these sciences does probability enable us to address? Is it truly indispensable for the progress of those sciences?

Part II examines first how probability can be incorporated into some of the social sciences, whether it plays a major role there, and, if not, the role it does play.

Let us now see why to restrict the field of social science, too large for a non encyclopedic study, to a more restricted field of research. In Part I, we examined the applications of probability to a large number of social sciences. However, we shall need to confine ourselves to a smaller group of social sciences so that we can examine their development in greater detail and see how they used probability throughout their history. We think that population sciences will constitute an interesting case to consider here, as their development was simultaneous with the development of probability.

We have already pointed out the near-simultaneity of the emergence of probability and political arithmetic. It would therefore be interesting to examine the latter science here. From the seventeenth century to the early nineteenth century, political arithmetic was regarded as an indivisible whole. However, as it eventually split up into different social sciences such as demography, epidemiology, political science, actuarial science, and sociology, we must restrict our scope even further.

In consequence, we shall take a more detailed look at population sciences, particularly: actuarial science, demography, paleodemography, historical demography, etc., without excluding the other sciences when useful.

For example, epidemiology, which was a part of political arithmetic on providing 'a method to quantify the costs of mortality' (Susser 1996), shares many features with population sciences. Our investigation must even include the form of sociology championed by Durkheim (1895, 1897), which adopted approaches similar to those followed by population sciences at the time. Population genetics is also partly in the same field but will not be examined in detail here: the interested reader may refer to Vetta and Courgeau (2003).

Similarly, numerous social sciences today practice common statistical approaches. Regression methods, event-history approaches, and multilevel approaches are used not only in population sciences, but in epidemiology (Greenland 2000), economics (Florens 2002; Heckman and Singer 1984a), sociology (Yule 1895; Tuma and

Hannan 1984), medical statistics (Andersen et al. 1993), education sciences (Goldstein 2003), population geography (Jones 1993), and many other disciplines. When we discuss these methods in population sciences, we shall thus need to show their more general scope of application.

As a result, even restricting our choice of social sciences to population sciences, our examination will be much broader and will enable us, in certain cases, to draw more general conclusions.

Chapter 4
The Dispersion of Measures in Population Sciences

Having presented and discussed the various meanings of the term probability, in Part I of this volume, we shall focus in this chapter on the more detailed links that exist between population sciences and probability. More specifically, we shall examine the concept of *dispersion* in the two disciplines.

The term 'dispersion' derives from the Latin verb *dispergere*, which means 'to spread.' In French, the first edition of the *Dictionnaire de l'Académie Française* (1694) defined the term only as 'action of dispersing or by which one is dispersed', illustrating it by the example of the dispersion of the Jews. Over time, the word has come to designate far more diverse realities, notably in the scientific field.

Today, the *Académiciens* distinguish between two major meanings of the term, and we shall see that these meanings are also used—albeit in more specific senses—in probability and social science.

The first meaning of dispersion is 'action of spreading, of scattering abroad', which has led, among other things, to the probabilistic meaning of the term: a spread of observations around their central value. We can thus define and measure this dispersion using various numerical indicators: variance,[1] standard deviation, confidence interval, variation coefficient, etc.—at the cost, however, of a loss of information on the broader concept of spread (Barbut 2002). Likewise, in population sciences, one can speak of the dispersion of a rate, a probability or an index, in the same sense as above.

The second meaning is 'action of separating elements, of breaking the unity of a set'. For instance, in probability, we can use a single random variable, defined for the total population considered. The variable may be characterized by its mean and variance. After breaking down the population into more specific sub-populations, we find that the variable's mean and variance are strongly dispersed for each

---

[1] In modern Greek, the term for variance is diaspora (διασπορά), which brings us back to the dispersion of the Jewish people.

sub-population. We must then split up the population and examine each sub-group separately. As a result, a set of facts regarded as equivalent in the explanation of a phenomenon may cease to be equivalent when we analyze the phenomenon in greater depth. In population sciences, for example, we shall see that the multilevel approach will examine separately the effect of a characteristic on various groups, whereas, in the earlier approaches, the characteristic was assumed to have a uniform effect on a total population.

We shall also examine the opposite of dispersion, namely, 'homogeneity', which denotes the concentration of observations on a single value, or the characteristic of a set whose breakdown into constituent elements does not seem useful.

We shall review the variations in the meaning of 'dispersion' throughout the history of population sciences, and we shall link it to the paradigms prevailing in each period.

4.1 From Pascal and Fermat's Wagers to Graunt's Bets

This section looks at the works of some authors already discussed in earlier chapters but also new ones; here, however, the focus is on dispersion.

As noted earlier, Pascal and his correspondence with Fermat (1654, 1922), the founders of probability theory, made the hypothesis that the game rests on pure chance, in other words, is fair, i.e., not rigged in any way. Thus, rather than predict the outcome of each future round of play—an impossible task—they can calculate each player's fair share of winnings if the game did not continue. More generally, even when the game is fixed (for example, by using a loaded coin or die), a player's odds of winning or losing always has a determined value. As Bernoulli's theorem states, the value can be calculated from a large number of tosses of the loaded coin or die, with an accuracy that increases with the number of tosses. Objective probability theory will apply perfectly here.

By contrast, when Graunt (1662), as mentioned in the General introduction, writes that:

> it is esteemed an even Lay, whether any man lives ten years longer, I supposed it was the same, that one of any 10 might die within one year,

he is no longer speaking of a known or at least knowable probability—as in gaming—but of a probability that is not only unknown, but whose very existence cannot be established a priori. Graunt's assumption is a heroic one, whereas the death of an individual seemed inherently irreducible to all rational consideration. As Vilquin stated (1977):

> Measuring and quantifying phenomena that were God's secret (birth, illness, death, the games of life and chance)—the only ones who dared engage in these pursuits were experts whose science was too mysterious for them to fear public disapproval: doctors, theologians, or polemicists with scant regard for truth and falsehood.

Vilquin puts the 'games of life' and 'games of chance' on an equal footing here. In our view, however, the games of chance, where the probability of an outcome can

4.1 From Pascal and Fermat's Wagers to Graunt's Bets

be estimated even if the game is not fixed, differ from the games of life, where the probability can be estimated only by making far more heroic assumptions.

Graunt effectively posits such hypotheses to estimate human probabilities of dying. We shall try to identify them in greater detail.

First, he is working on the life of an individual, whose probabilities of dying can vary, in principle, from age to age. In contrast, the probability of an outcome in a game of chance is independent of the instant in which the toss occurs, provided that we use the same coin, die, etc. However, as he does not know the age distribution of deaths, he assumes that the annual probability of dying during the 'ten years longer' period is the same. Then, he extends the probability from ages 10–60—in other words, he regards the probability as age-invariant, i.e., without dispersion or homogeneous, in the second meaning of the word. He therefore treats the study of games of chance and the study of human life as equivalent, which is what Vilquin ultimately argues.

Second, Graunt posits another underlying hypothesis by treating deaths observed in a given year as equivalent to deaths observed over a generation. Admittedly, the distinction between period analysis and longitudinal analysis was still far from established, but only this hypothesis allows him to reason thus. In Sect. 1.3, however, we showed that his calculation was incorrect and that the annual probability, which he estimates at $p = \frac{1}{20} = 0.05$, should be estimated at $1 - \sqrt[10]{0.5} = 0.067$, a value more than one-third higher.

Having made these two hypotheses, he can then deduce the London population exposed to this age-invariant risk from the total 10,000 deaths that he observed for persons aged 10–60. He may be doing so as follows. Writing this age-invariant probability as:

$$p = p_{10} = \frac{D(10,11)}{N_{10}} = p_{11} = \frac{D(11,12)}{N_{11}} = \cdots = p_{60} = \frac{D(60,61)}{N_{60}} = \frac{\sum_{x=10}^{60} D(x, x+1)}{\sum_{x=10}^{60} N_x},$$

where p_x is the probability of dying at age x, $D(x, x+1)$ the number of deaths between ages x and $x+1$, and N_x the number of survivors at age x. We can thus deduce the population aged 10–60 from the deaths observed and the estimated probability p:

$$\sum_{x=10}^{60} N_x = \frac{\sum_{x=10}^{60} D(x, x+1)}{p}.$$

If, like Graunt, we assume a probability of $\frac{1}{20}$, we obtain, from the 10,000 deaths observed, a population of 200,000 people aged 10–60 years and not 100,000, as he states incorrectly, for he takes a multiplier of 10 instead of 20 ('which number being multiplied by 10,…,). But we have also seen that his estimate of p is inaccurate; we must take $p = 0.067$ instead, i.e., a multiplier of 14.925, which yields an estimate of 149,250 inhabitants, or approximately 150,000.

We shall see that the use of the multiplier not only persisted but expanded throughout the following century, when deaths and births were the only figures available to estimate a country's population.

Although Graunt's probabilistic reasoning was still very uncertain, his demographic hypotheses—given his knowledge of human mortality—were clearly defined, even if they are debatable. He took Pascal's reasoning one step further: Pascal spoke only of expected gains, whereas Graunt introduced the probability of dying.

It is Halley (1693) who made it possible to establish a more satisfactory life table, albeit still subject to the second hypothesis noted above. He verified the dispersion—in the second sense of the term—of age-specific probabilities of dying. After him, no-one would regard them as equal.

However, like most savants of the seventeenth century and first half of the eighteenth century, Halley had only birth and death statistics at his disposal, and they were inadequate for constructing an accurate life table. The population categories at risk were missing. The gap was not filled until 1766, when the Swedish astronomer Wargentin produced a true life table. He was able to do so because his country maintained population registers, which give the figures for population at risk, and death registers, which provide the numerators of the rates or probabilities to be computed. The censuses introduced in the nineteenth century made it possible to generalize the calculation of these tables.

In conclusion, we can say that, without the estimation of age-specific deaths, Graunt assumes their homogeneity, at least between ages 10 and 60, in order to be able to estimate the corresponding population. Once these deaths have been measured, the hypothesis becomes useless, for one can now verify its validity and show the dispersion—in the second meaning of the word—of their values by age.

What about the statistical dispersion of population science measurements—in the first meaning of the term—throughout the same period? As we have already seen in Sect. 2.1, Jacob Bernoulli (1713) showed that 'the estimate of a probability can be bounded by two limits, as precise as we wish them to be'.[2] We should therefore be able to estimate the dispersion of population sciences indices, when we have the measured population size to estimate them.

To our knowledge, only one author applied these results to population science data: Nicolas Bernoulli (in de Montmort, 1713). He sought to refute Arbuthnott's argument of Divine Providence (1710) using the observation of children born in London between 1629 and 1710, which he presents as follows:

> if chance ruled the world, it would be impossible for the number of males and females to converge as closely for several consecutive years as they have been doing for the past 80 years […]

---

[2]binis limitibus conclusam, sed qui tam arcti constitui possunt, quàm quis voulerit.

To this end, he sets out to prove:

> that there is a very high probability that the number of males and females lies each year between even narrower limits [sic] than those observed in the past 80 years in a row.

His demonstration actually resembles the one offered by Jacob Bernoulli, as he himself recognizes:

> I recall that my late uncle demonstrated a similar thing in his treatise *De Arte conjectandi*, now printed in Basel [...]

By contrast, none of the other authors working on population until 1774 thought of setting limits within which to perform calculations. Neither Kersseboom (1742), nor Deparcieux (1746), nor Süßmilch (1741, 1761–1762), nor any other contemporary attempted to assess the dispersion of his estimates. Doubtless some of these authors, such as Süßmilch, believed that the immutable causes underlying these phenomena lay in the Divine Order, which a perfect society would illustrate. In that case, the observed dispersion of the indices would vanish. But, as Nicolas Bernoulli showed, using a probabilistic reasoning (see Sect. 1.4), this argument had to be refuted: mythological thought—which substituted for the lack of explanation of the observed phenomena—cannot provide valid research guidelines (Courgeau 2010).

4.2 Introduction of Epistemic Probability in Population Sciences

In Sect. 2.1 we examined the emergence of epistemic probability. We began with Jacob Bernoulli's direct approach (1713), and we ended with the so-called indirect approach initially proposed by Bayes (1763), then expanded and applied by Laplace to many scientific fields from 1774 onward. It is as well applicable to objective events, such as a person's probability of dying, as to more subjective ones, such as trial verdicts.

Laplace (1778) showed very clearly how the probabilities of elementary events can be obtained in three different ways:

> 1. *a priori*, when by the very nature of events, we see that they are possible in a given ratio; [...] 2. *a posteriori*, by repeating many times the trial that may produce the event in question, and by examining how many times it has occurred; 3. lastly, by considering the reasons that may lead us to assert the existence of this event [...]

In the first situation, we know what Laplace calls the absolute possibility of events, as when Pascal (Pascal 1654a) assumes a fair game in which the probability of winning is $\frac{1}{2}$. In the second situation, we obtain the absolute possibility after an infinite number of trials—for example, when Jacob Bernoulli (1713) determines with increasing precision the probability of drawing, with replacement, a token of a given color from an urn of unknown composition. In the third situation, we find only the relative possibility of the event given the state of our knowledge. Laplace decided to focus all his efforts on the analysis of this third situation.

We have already described his approach to the masculinity proportion at birth, absent any prior data on the possibilities of male and female births (see Sect. 3.4). Here, Laplace assumes that all possibilities from zero to unity are equally probable. He can thus provide not only the best estimate of the masculinity proportion but also the probability that the proportion will lie within given limits around the estimated value.

Let us examine his treatment of one of the major population-related topics in the absence of census data, a topic first broached by Graunt (1662) as we have just seen: the estimation of the population from the ratio of the population to annual deaths and births (Laplace 1783b)—also known as the multiplier method.

In the introduction to his article, Laplace underscores the importance of knowing a country's population:

> The population is one of the surest ways to judge the prosperity of an empire; and the variations that it experiences, compared with the events preceding them, are the fairest measure of the influence of physical and moral causes on the happiness or misfortune of the human species. It is therefore of interest, in every respect, to know the population of France, to track its progress, and to determine the law whereby men are distributed across the surface of this great kingdom.

Like most countries at the time, however, France had neither a reliable census nor a good-quality population register that would have allowed an accurate estimate of its population. The only sources available were birth and death registers. Laplace does note that in a country where the number of deaths is roughly equal to that of births—i.e., what we would now call a stationary population—there is a constant ratio of the population to annual births or annual deaths: life expectancy at birth. More generally, this factor, i, by which we need to multiply births[3] in order to obtain the population, is the most subtle and interesting topic in population research.

Laplace did state that the result obtained with this multiplier method can never be strictly accurate and is subject to error. He therefore concentrated his work on the estimator's accuracy, i.e., its dispersion in the first meaning of the term.

For this purpose, he imagines each annual birth as represented by a white ball, and each individual in the population studied as represented by a black ball. To simplify the calculation, their total number is assumed to be infinite, whereas in fact it is very high. Laplace proposes what would now be called a sampling operation, which consists here in selecting a large number of parishes from all of the country's provinces to properly represent its total population.[4] The sampling will involve an enumeration of inhabitants and a tabulation of births recorded in the 10 years prior to the enumeration. This will constitute an initial sampling of p inhabitants and q annual births, so that $i = \frac{p}{q}$. The second sampling will cover the entire country, but we shall only obtain the number q' of annual births; the population p' will be unknown.

---

[3] As Moheau (1778) noted, 'humane mortality is not regulated in the same manner as fertility: there are years that produce a multitude of deaths, there are others that spare our days, whereas the rate of annual newcomers is almost equal and invariable'. That is one of the reasons why eighteenth-century authors preferred to calculate a birth multiplier rather than a death multiplier.

[4] It is interesting to note that such a sampling replaces the exhaustive census in France after 1999.

4.2 Introduction of Epistemic Probability in Population Sciences

To determine this unknown population, we rely on the sampling, which yields a value of $\hat{p}' = \dfrac{pq'}{q}$. But it is important to find the probability that the error in this result does not exceed $a = \dfrac{pq'\omega}{q}$, where a can, for example, be set at 500,000. To this end, Laplace uses the results obtained in two earlier articles (Laplace 1782, 1783a), which present his theory of the probability of future events derived from observed events.

Let x be the unknown ratio of the total population to this total population plus annual births. The probability of the second sampling will be:

$$\frac{(p'+q')!}{p'!\,q'!} x^{p'} (1-x)^{q'}.$$

But as p' is unknown, it can take all the prior values in the interval $[0,\infty]$. These values, however, will be more or less probable, according to whether they make the second sampling more or less probable. We shall therefore obtain the probability for each value of p' by dividing the previous quantity by the sum of all whole values of this quantity taken from zero to infinity, i.e., by the following series:

$$(1-x)^{q'} \sum_{p'=0}^{\infty} \frac{(p'+q')!}{p'!\,q'!} x^{p'} = (1-x)^{q'} \left[1 + (1+q')x + \frac{(1+q')(2+q')}{2!} x^2 + \ldots \right].$$

If we call the quantity in brackets $S_{q'+1}$, we can easily ascertain that the following relation is verified:

$$S_{q'+1}(1-x) = \left[1 + q'x + \frac{q'(1+q')}{2!} x^2 + \ldots \right] = S_{q'},$$

and that when $q' = 1$:

$$S_1 = 1 + x + x^2 + \ldots = \frac{1}{1-x} \quad \text{therefore} \quad S_{q'+1} = \frac{1}{(1-x)^{q'+1}}.$$

Hence:

$$(1-x)^{q'} S_{q'+1} = \frac{1}{1-x},$$

and, as a result, the probability for each value of p', with x assumed known, will be:

$$\frac{(p'+q')!}{p'!\,q'!} x^{p'} (1-x)^{q'+1}.$$

But, as x is unknown, we again need to make its prior vary from zero to unity. These different values are also more or less probable depending on whether they make the first sampling more or less probable. As the probability of this first sampling is:

$$\frac{(p+q)!}{p!q!} x^p (1-x)^q,$$

the probability of x will be:

$$\frac{x^p (1-x)^q dx}{\int\limits_{x=0}^{1} x^p (1-x)^q dx},$$

and multiplying this probability by the one determined earlier for p', with x known, we obtain the final probability for each value of p':

$$\frac{(p'+q')! \int\limits_{x=0}^{1} x^{p+p'} (1-x)^{q+q'+1} dx}{p'! q'! \int\limits_{x=0}^{1} x^p (1-x)^q dx}.$$

The last step is to calculate the probability, P, that the total French population lies between the two values $\frac{pq'}{q}(1+\omega)$ and $\frac{pq'}{q}(1-\omega)$. We shall not describe these approximate calculations based on the previous equation but simply give the final result:

$$P = 1 - \frac{2 \int\limits_{t=V}^{\infty} e^{-t^2} dt}{\sqrt{\pi}},$$

where $V = \omega \sqrt{\frac{pqq'}{2(p+q)(q+q')}}$. This value of P is accurate to within the value of ω.

Let us examine France in early 1782 within its borders of the time. The average number of births between 1781 and 1782 was 973,054.5. Let $a = 500,000$: consequently, the probability $P = \frac{1000}{1001}$, hence $\int\limits_{t=V}^{\infty} e^{-t^2} = \frac{\sqrt{\pi}}{2002}$. This equation therefore determines $V = 2.327$. But as Laplace had not yet performed the enumeration that should have yielded the multiplier to be used, i, he resorted to earlier enumerations that gave him a multiplier of roughly 26. As the exact multiplier did not diverge significantly from this number, he took the values

4.2 Introduction of Epistemic Probability in Population Sciences

$i_1 = 25.5$, $i_2 = 26$, $i_3 = 26.5$, which yielded, respectively, the following populations to be enumerated: $p_1 = 727,510$, $p_2 = 771,469$, $p_3 = 817,219$. We should note that Louis Henry's survey on the population of France from 1740 to 1860 (Henry and Blayo 1975) indicated a population of 27,550,000 in 1780 for 1,053,800 births recorded within the borders of 1861,[5] in other words, a multiplier of 26.14—well within the limits given by Laplace.[6]

Thus Laplace introduced into demography the concept of dispersion of measures, in the first meaning of the term. As we shall see, the concept would later disappear entirely from the discipline, when most European countries introduced exhaustive census-based population counts.

In his *Mémoire sur les probabilités* (1778) Laplace recommended also an analysis of dispersion in the second sense of the term:

> it is here, above all, that we need to have a rigorous method for distinguishing, among the phenomena observed, those that may depend on chance from those that depend on specific causes, and to determine the probability with which the latter [*phenomena*] indicate the existence of the causes.

In our view, his recommended distinction is essential for defining present-day demographic approaches, whether of the event-history or multilevel type.

But it was Gauss (1809) who actually proposed using the least-squares method for a true regression analysis. The main goal of his book was to present his detailed mathematical research on planetary orbits. This research, begun in 1795, enabled him to predict the position of the asteroid Ceres in 1801 from a small number of observations made early that year.[7] Gauss concluded his volume with a more general description of his least-squares method, whose principle consisted in resolving a system of linear equations containing fewer variables than equations. When the number is equal, we can simply resolve the system, which usually has a single solution. When the variables outnumber the equations, there is generally no solution. But when the number of variables is smaller, the problem is overdetermined. For the unknown variables, we therefore need to find values that most closely approximate the true values.

---

[5] The Comté de Nice and Savoie (Savoy) were annexed to France in 1860.

[6] For the record, these figures are currently challenged (Brian 2001). However, our purpose here is not an accurate reconstruction of France's population, but a discussion of Laplace's estimates. Laplace himself revisited the subject in *Théorie analytique des probabilités* (1812), using a different approach from the one described here. In this later study, he worked on an enumeration of 1802 and on the average number of births recorded between September 22, 1799, and September 22, 1802. Laplace also introduced the longitudinal analysis of mortality ('Let us suppose that we have tracked the distribution of mortality among a very large number n of children, from their birth to their total extinction'), and the mean duration of marriages between boys aged a and girls aged a'. He naturally calculated confidence intervals for all these quantities.

[7] Legendre published an application of the least-squares method to a simpler case (1805) before Gauss and claimed precedence for the discovery. However, besides the evidence that Gauss used the method before 1805, the key element of this approach is missing from Legendre's publication: he fails to present it in a clear probabilistic framework.

To pose this problem in terms of probabilities, Gauss assumed μ linear equations:

$$V = ap + bq + cr + ...$$
$$V' = a'p + b'q + c'r + ...$$
$$\vdots$$

where the observations are $a, b, c, ...$, with v unknown coefficients $p, q, r, ...$, which we must estimate using observations of the values $M, M', ...$ of the functions $V, V', ...$ Gauss then assumes that the errors: $\Delta = V - M$, $\Delta' = V' - M'$, ... have probabilities given by the function $\varphi(\Delta)$, $\varphi(\Delta')$, ... Applying Laplace's inverse probability principle, he assumes that all the values of the unknowns are equally plausible priors. This leads him to the values of the unknown coefficients that will maximize the quantity:

$$\Omega = \varphi(\Delta)\varphi(\Delta')...$$

All that is needed now is to derive Ω relative to the unknowns, and to resolve the corresponding equation system. But these equations comprise the unknown function $\varphi(\Delta)$, whose formulation must be determined.

Gauss then starts from the axiom that the arithmetical mean of a variable's observed values is the variable's most probable value. Here, he shows that Δ is normally distributed, in other words:

$$\varphi(\Delta) = \frac{h}{\sqrt{\pi}} e^{-h^2 \Delta^2},$$

where h is a positive constant, which may be viewed as a measure of estimation accuracy. Here, we can write:

$$\Omega = \left(\frac{h}{\sqrt{\pi}}\right)^n e^{-h^2\left(\Delta^2 + \Delta'^2 + ...\right)},$$

a quantity that takes its maximum value when the quantity:

$$L = \Delta^2 + \Delta'^2 + ...$$

is minimal. This leads to what has since been named the least-squares method to supply the values of $p, q, r, ...$

Laplace, however, while recognizing the merits of Gauss's linkage of the least-squares method to probability theory, quite rightly criticized this axiom:

> Mr. Gauss, in his *Theory of elliptical motion*, sought to tie this method to Probability theory, by showing that the same law of observation errors, which generally yields the rule of the arithmetical mean between several observations, accepted by observers, likewise yields the rule of least squares of observation errors, [...] But, as nothing proves that the first of these rules yields the most advantageous result, the same uncertainty exists for the second. (Laplace 1812)

4.2 Introduction of Epistemic Probability in Population Sciences

In the first of two papers (1809a), Laplace showed that if we assume that positive and negative errors are equally possible in each observation, then the probability that the mean error of n observations will lie within the limits $\pm\dfrac{rh}{n}$ is:

$$\frac{2}{\sqrt{\pi}}\sqrt{\frac{k}{2k'}}\int_{r=-\infty}^{+\infty} e^{-\frac{k}{2k'}r^2} dr,$$

where h is the interval in which the errors of each observation can lie, k is the integral $\displaystyle\int_{x=-\frac{1}{2}h}^{\frac{1}{2}h} \varphi\left(\frac{x}{h}\right) dx$, with $\varphi\left(\dfrac{x}{h}\right)$ the probability of error $\pm x$, and k' is the integral

$\displaystyle\int_{x=-\frac{1}{2}h}^{\frac{1}{2}h} \frac{x^2}{h^2} \varphi\left(\frac{x}{h}\right) dx$. In his second paper (1809b), written after seeing Gauss's book (1809), Laplace shows how this theorem can provide a solid basis for Gauss's choice of a normal distribution for observation errors, by using the 'mean error to be feared':

$$\int_{\Delta=-\infty}^{+\infty} |\Delta|\, \varphi(\Delta)\, d\Delta.$$

Laplace combines the μ previous equations so as to obtain a system of ν equations that are linear combinations of the earlier ones, such that for each unknown $p, q, r, \ldots$, the mean error to be feared is minimal.

This supplies a clear demonstration of the least-squares method, without resorting to the axiom that the arithmetical mean of a variable's observed values is the variable's most probable value. However, as Laplace notes:

> When the observations are few in number, the choice of these systems depends on the law of errors of each observation. But, if we consider a large number of observations, which is the most common occurrence in astronomical research, that choice becomes independent of the law, and we have seen, in the preceding discussion, that the analysis will then directly yield the results of the least-squares method for observation errors. (Laplace 1812)

Independence from the error distribution is obtained from what we now call the 'central limit theorem'. However, we can discern in the use of the least-squares method the first step in the abandonment of inverse probability and the move toward objective probability, which is solely concerned with events whose frequency is stable. Furthermore, as early as 1816, Gauss used probability in its frequentist sense to estimate the standard deviation of the parameters of his regressions (Hald 2007). This is equivalent to using the maximum-likelihood method, now standard in frequentist statistics.

Over the years, the least-squares linear regression method was refined: Laplace (1827) applied it in conjunction with correlated errors and a known variance-covariance matrix. Its use was long confined to astronomical, geodesic, and climatological applications, although there was every reason to assume that its application to the social sciences, particularly demography would be highly profitable (Stigler 1986). For example, with regard to mortality, Laplace quite rightly stated:

> So many causes influence mortality, that the Tables representing it must change according to place and time. The different states of life offer, in this respect, significant differences relative to the dangers inseparable from each state, differences that must be taken into account in calculations based on length of life. But these differences have not yet been sufficiently observed. They will be, some day; then we shall know what sacrifice of life each occupation demands, and we shall take advantage of this knowledge to reduce its hazards. (Laplace 1812)

It is interesting to note, however, that Laplace, like many other investigators before him and in his day, worked on certain data drawn from parish registers—which, in France, became 'civil registers' (*registres d'état civil*) in 1793—such as the sex of newborn children. But he did not realize the richness of these sources for the study of many demographic issues, which only recent surveys have made it possible to grasp (for example, the Henry Survey on the population of France from 1670 to 1792, and Dupâquier's TRA survey on persons whose last names begin with the letters TRA from 1803 to 1986).

After Laplace, few researchers pursued this path (Poisson 1837; Bienaymé 1838) and his approach soon came under fire, as we shall see in the next section.

4.3 Toward an Objectivist Approach in Population Sciences

By the mid-nineteenth century, most statisticians and population scientists had begun to reject the epistemic approach to probability—sometimes violently. For the population sciences, there are several reasons for this, which we shall now examine in some detail.

First, the distinction grew sharper between objective probability and subjective probability, a distinction that not all earlier authors since Pascal had made as we have already said. For instance, in his 1812 work, Laplace discussed objective problems of dice tosses and lottery draws alongside the most subjective problems such as the probability of testimonies and convictions. Cournot (1843) draws a very clear distinction between these two meanings:

> Nothing is more important than to carefully distinguish between the dual meanings of the term *probability*—understood now in an objective sense, now in a subjective sense—if we want to avoid confusion and error, both in the exposition of the theory and in its applications.

For Cournot, only objective probability is measurable, when the trials of the same natural phenomenon, physical or moral, can be repeated *ad infinitum*, leading to an ever more precise measure of their probability. Thus we can say that a demographic phenomenon such as the probability of dying at a given age or moment can

4.3 Toward an Objectivist Approach in Population Sciences

be determined with the aid of a population large enough to be regarded as infinite, or rather as being drawn from an infinite population. By contrast:

> When the number of trials is not very significant, the formulas commonly given for measuring posterior probability become fallacious: they now indicate merely subjective probabilities, suitable for setting the terms of a wager, but inapplicable to the production of natural phenomena.

His criticisms focused on what is called the law of succession, identified by Laplace back in 1774. Let us take the case of an urn containing an infinity of white or black tickets. We have previously drawn $p+q$ tickets, of which p are white. The probability that a new ticket drawn from the urn is white is $\frac{p+1}{p+q+2}$. Cournot applies this rule to the demographic example of the masculinity proportion at birth. He remarks that, as we do not know each woman's chances of bearing a child of either sex, Laplace's rule should lead us to conclude that if a woman already had a boy, her chances of having another boy would be $2/3$. But in this case, no-one would be willing to make such a wager, thus demonstrating that the law would have 'only a futile and derisory consequence'. He did not take into account that this rule is verified when we have no other information on the studied phenomenon. More generally, Cournot observes:

> However, people have not feared to make applications with just as little basis, in matters of grave import for society and morals, such as those concerning judicial decisions and testimonies; and we have thus fallen into aberrations unworthy of great geometers.

We can see the depth of the disagreement with Laplace, for Cournot refuses to apply probability theory to cases where the number of observations is small and to subjective events.

Ellis (1849) wrote:

> The principle, on which the whole depends, is the necessity of recognizing the tendency of a series towards regularity, as the basis of the theory of probabilities.

He also sought to show that the estimators supplied by the inverse epistemic probability method are fallacious. As mentioned in Sect. 1.1.2, Venn (1866) discussed in detail the characteristics that a series should possess in order for it to be studied using probability theory. Boole (1854), after examining different applications of epistemic probability, expressed doubts about their use:

> These results only illustrate the fact, that when the defect of the data is supplied by hypothesis, the solutions will, in general, vary with the nature of the hypotheses assumed; so that the question still remains, only more definite in form, whether the principles of the theory of probabilities serve to guide us in the election of such hypotheses.

Boole specifies that he is voicing these criticisms for the benefit of English authors who would want to use Laplace's methods.

Thus, as we have seen, a theory of objectivist or frequentist probability was taking hold—a theory that would prevail for over a century. For example, von Mises (1957), taking a demographic example, was prompted to write: 'The phrase 'probability of death', when it refers to a single person, has no meaning for us'. We can see how strongly his approach to probability differed from that of Bernoulli, who sought to measure the chances of dying of an individual whom he calls Titus (see Sect. 2.1), but especially from that of Bayes or Laplace, who sought to refine the estimate of those chances from a certain number of individuals resembling the person concerned. In this case, we can speak only of the probability of dying in a population whose size may be regarded as being as large as we want, tending toward infinity. Moreover, to speak of the probability of an intrinsically unique event or more generally of the probability that a proposition is true makes no sense for an objectivist.

At the same time, population sciences practically gave up their pursuit of an analysis as refined as Laplace's of the phenomena that they were studying. It is important to realize that the introduction of population censuses took some earlier concerns off the agenda—most notably by supplying exhaustive counts of populations at risk. This avoided the use of civil-registration data to estimate population by means of the multiplier method, already used by Graunt.

In other words, the use of probability was changed by the population censuses—which first appeared in the eighteenth century and were gradually established in Europe all during the nineteenth century—coupled with exhaustive civil-registration data. By collecting data on the total population at a given moment, population scientists can work with an objectivist approach, given the large number of individuals involved. The variance of their probabilities becomes so small that they will no longer even calculate it. In fact, neither the cross-sectional approach, with the concomitant-variation method (Durkheim 1895; Landry 1945), nor the longitudinal approach (Pressat 1966; Henry 1972) ever envisaged the calculation of variance. As shown in Sect. 1.4, the determination of the variance of an annual probability of mortality, assuming a binomial distribution of deaths (i.e., among a homogeneous population), yielded such low values that they were no longer worth computing.

Only very seldom, when calculating probabilities over shorter periods (for example, monthly), would population scientists need to take these variances into account (Hoem 1983). The reason is that the number of individuals experiencing the event was far smaller, even when fully counted. This explains perfectly why classical demography, while preserving the probabilistic meaning of dispersion, refrained from measuring dispersion in the first sense of the term. However, the hypothesis of a population in which the probability of dying at a given age is identical for all its members is utterly unrealistic—as the following section will show.

For dispersion, in its second meaning, the situation differs depending on whether cross-sectional or longitudinal analysis is involved.

As censuses were introduced, population sciences developed what is known as cross-sectional (or: period) analysis, taking civil-registration data contemporaneous with the censuses to calculate 'period' indices. This type of analysis was practiced until the end of World War II.

After a purely descriptive phase, which consisted in compiling population pyramids, crude rates, age-specific rates, and so on, these data underwent a more statistical analysis. In our view, the clearest exposition of the goals of this approach was given by the sociologist Durkheim (1895, 1897). We shall describe it briefly here, referring the interested reader to our more detailed examination elsewhere (Courgeau 2004a, b, 2007a).

Durkheim sought to show that one social phenomenon could be the cause of another. To this end, he used a suitable comparative method, which he called the concomitant-variation method. It was identical to the one proposed later by Landry (1945), who noted that if we want to 'understand a variation across time, a difference across space, we shall need to prove a relationship of obviousness, concomitance, covariation or all other', between the phenomena studied. In reality, this was tantamount to a regression analysis of the kind that Gauss and Laplace had advocated back in the early nineteenth century, simplified here by the fact that it consisted of a linear regression.

The concomitant-variation method actually concerns aggregated data. For instance, to demonstrate the influence of religion on suicide in Bavaria and Prussia, Durkheim used suicide rates showing the share of Protestantism and other religions in the different provinces of the two States (Durkheim 1897). Moreover, Durkheim did not even calculate the parameters of a simple regression, but merely observed that 'suicides are directly proportional to the number of Protestants and inversely proportional to that of Catholics'. As noted earlier, these regression methods were not used in demography until much later—to our knowledge, around the 1950s (Robert Schmitt and Crosetti 1954). But their principle is indeed identical to that of a linear regression. It effectively allows a distinction between sub-groups with different behaviors in a population, so that one can measure the dispersion of a phenomenon across a heterogeneous population.

By contrast, longitudinal analysis, which emerged after World War II, completely disregarded this dispersion in the second meaning of the term. This was because the main goal of the new approach was to remediate the vision of cross-sectional analysis, which assumed that phenomena were determined by the characteristics of the population studied immediately preceding their occurrence. By giving precedence to a time span linked to people's persistence in a given state, the longitudinal approach stressed duration and allowed a clearer separation between long-term and momentary factors. It therefore allowed a more relevant analysis of the timing and intensity of demographic phenomena in a given cohort.

To do this, however, it was obliged to frame various hypotheses on the phenomena that would make the study of dispersion, in the second sense of the term, nearly impossible. Under this approach, the only truly feasible analysis is that of phenomena regarded as *independent* of one another and occurring in a *homogeneous* population (Blayo 1995). The events that determine the entry or exit of the population studied must be viewed as independent of the phenomenon studied. In particular, this makes it impossible to study exits via competing events, such as exit from never-married status via marriage or cohabitation. Most important, the study of dispersion, in the second meaning of the term, which assumes a heterogeneous population, eventually clashed with the hypothesis of a homogeneous population. The solution for breaking out of the impasse by dividing the population into homogeneous sub-groups made the analysis so complex as to become impossible.

We refer the reader to Courgeau (2003, 2004a, 2007a) for a more detailed critical analysis of this approach. The outcome was the impossibility of taking account of dispersion, in the second sense of the term—including both the dispersion created by the other demographic phenomena for the one under study and the dispersion created by the diversity of the population's members.

To sum up, in the period when longitudinal analysis prevailed—from the end of World War II to the early 1980s—population science was able to almost totally ignore the two aspects of the dispersion of the phenomena that it studied, preserving only the age-specific difference in probability.

4.4 Return of Dispersion in the Event-History and Multilevel Approach

In response to criticisms of the longitudinal approach, the event history approach involved setting up ways to analyze individual life histories by examining (1) the characteristics of people at the time when they are about to experience various events and (2) the links between the phenomenon studied and the other phenomena that have marked the person's earlier life. This required access to far more detailed surveys that, unlike census and register data, provide detailed event histories of respondents. For instance, the 'Triple Biographie' (Triple Event History) survey (also known as 3B for short), which we conducted at INED in 1981, was designed to permit an event-history analysis of the interactions between the various aspects of respondents' family life, working career, and migration history (Courgeau 1982).

The event-history approach will therefore consider a set of individual trajectories in all their complexity, generally captured by detailed surveys. The unit of analysis will no longer be the single event, as in longitudinal analysis, but the personal event history, regarded as a complex stochastic process. It will no longer consider the events studied as unrelated; on the contrary, it will analyze the dependences between them. Similarly, it will no longer treat the population as homogeneous, but instead examine its heterogeneity. This approach addresses most of the criticisms directed against longitudinal analysis (Courgeau and Lelièvre 1996).

Let us take a simple example to illustrate how to formulate such a semi-parametric analysis, from which we can show the dispersion (in both senses of the term) of the population. Despite its simplicity, the example is complete enough to include both the interaction between phenomena and the heterogeneity of the population. Suppose we wish to study an initial phenomenon such as exits from agriculture in France, knowing that another phenomenon—nuptiality—can interfere with the exits. At the same time, however, other individual characteristics will influence these phenomena, such as the number of siblings, being the eldest child, and so on (Courgeau and Lelièvre 1986).

Next, we define two random variables, T_1 and T_2, corresponding to the durations at which the two types of events occur: exit from agriculture and marriage. We then define what are known as instantaneous rates of occurrence of each event according

4.4 Return of Dispersion in the Event-History and Multilevel Approach

to whether the other event has occurred previously or not (for more details, see Aalen et al. 1980, and Courgeau and Lelièvre 1989). We can thus write:

$$h_{01}(t) = \lim_{\Delta t \to 0} \frac{1}{\Delta t} P\left(T_1 < t + \Delta t \mid T_1 \geq t, T_2 \geq t\right),$$

which, when Δt tends toward zero, yields the instantaneous rate of the first event when the second has not occurred beforehand. We similarly define a rate $h_{21}(t)$, when the second event precedes the first:

$$h_{21}(t) = \lim_{\Delta t \to 0} \frac{1}{\Delta t} P\left(T_1 < t + \Delta t \mid T_1 \geq t, T_2 = u\right) \quad \text{where} \quad u < t,$$

the second event having occurred at a date u prior to t.

The estimation of these rates and their variance makes it possible to show their dispersion in the first sense. When the confidence intervals are disjoined, the estimation enables us to measure a dispersion in the second sense of the term—namely, that the occurrence of the second phenomenon, at an instant u prior to t, will identify a new sub-population whose behavior differs from the population initially examined.

Now, let us incorporate the fixed characteristics of the individuals studied. We shall represent them by a column vector z before the occurrence of the second event and z' after,[8] and we shall estimate the following model, known as semi-parametric, which is fuller:

$$h_1(t; z, z', u) = h_1' \exp\left[z\beta_1 + H(t - u)(\beta_0 + z'\beta_2)\right],$$

where:

$$H(x) = \begin{cases} 0 & si \quad x < 0 \\ 1 & si \quad x \geq 0 \end{cases},$$

where u is the date of occurrence of the other event, β_0 a parameter to estimate, and β_1 and β_2 two parameter vectors to estimate as well. The first analysis shows that the number of never-married women exiting agriculture greatly exceeds that of married women. But when we incorporate various characteristics of these women initially analyzed separately, we obtain highly significant parameters, whose effect—this time, their combined effect—is given in Table 4.1.

We can see that the more siblings a woman has, the more likely she is to exit agriculture. Conversely, women who are the eldest siblings and those with a farmer father will be less likely to leave agriculture. But once married, while the sibling-number

---

[8] When the second event has occurred, and consists of the marriage of the surveyed individual, it is useful to incorporate the spouse's fixed characteristics in addition to those of the individual.

Table 4.1 Effect of characteristics taken simultaneously in semi-parametric model for women's exit from agriculture

Total characteristics	Main effect β_1	Disturbance β_0	Interaction β_2
Number of siblings	0.012[a]		0.000
Eldest	−0.320[a]		0.296
Farmer father	−0.928[a]		0.806[b]
Married		−0.228	
Farmer husband			−0.359[a]
Farmer father-in-law			−0.126
Farmer at marriage			−1.040

Source: Courgeau and Lelièvre (1986)
[a]Result significant at 5% limit
[b]Result significant at 10% limit

effect is unchanged, the fact of having a farmer father will no longer affect the probability of exiting agriculture. The results also show that having a farmer husband will keep the woman in agriculture. The analysis indicates that the parameter β_0, which measures the effect of marriage on exit from agriculture, is no longer significantly different from zero, although still negative. The characteristics introduced actually make it possible to explain this effect. We shall not elaborate on the analysis of these results any further, referring the reader to our article for fuller explanations (Courgeau and Lelièvre 1986). This brief presentation has shown the essential role of the dispersion—in both senses of the term—of the results of the analysis.

David Cox (1972) initially presented this analysis under a purely objectivist statistical approach,[9] but the potential application of these methods to many other fields—such as epidemiology, industrial reliability and medical statistics—was already raised as distinct possibility. Many other authors who have explored this approach (Kalbfleisch and Prentice 1980; David Cox and Oakes 1984; Courgeau and Lelièvre 1989; Andersen et al. 1993) have also set out a basically objectivist version of this analysis.

Over time, however, some authors have preferred to adopt a Bayesian approach that has, more recently, made it possible to overcome many estimation difficulties and is more consistent with the spirit in which this analysis is performed. One example is the semi-parametric analysis proposed by David Cox, which we applied earlier to a demographic example. Several authors published articles on this analysis in the 1990s using Bayesian processes (Clayton 1991; Sinha 1993). Finally, Ibrahim et al. (2001) wrote a detailed book on Bayesian event-history analysis. This approach was made possible by the critique of the objectivist approach, as discussed in Chaps.

---

[9]It is interesting to note how strongly David Cox criticized the Bayesian approach. For instance, in discussing a later work by Cox and Hinkley (1974), Jaynes (2003) stated that the presentation of Bayesian methods led Cox 'to repeat all the old, erroneous objections to them, showing no comprehension that these were ancient misunderstandings long since corrected by Jeffreys (1939), Savage (1954) and Lindley (1956).'

2 and 3, and by the computational possibilities offered by Monte Carlo Markov Chain (MCMC) methods, which provide acceptable approximations of integrals and other functions that depend on a distribution of interest.

In particular, the Bayesian approach makes it possible to include all information potentially useful for the issue studied, which the objectivist approach did not allow. Similarly, these methods, thanks to Gibbs sampling and MCMC methods (Robert 2006), have made it far easier to solve complex problems without resorting to asymptotic objectivist calculations. It offers many other advantages over the objective approach, thanks to the availability and flexibility of tools for developing models and analyzing data.

As regards dispersion, in the second sense of the term, the Bayesian approach takes it into account as well, by introducing the estimation of the heterogeneity of a population and of the dependence between the phenomena studied. The reasons both internal to population sciences (dependence between phenomena) and external to them (heterogeneity of a population) can thus be identified and their effects on individual behaviors can be analyzed in great detail. However, we run the risk, in this case, of committing what is known as the atomistic fallacy, for by taking only the individual's characteristics into account, we ignore the context in which human behaviors occur. This risk stands in contrast to that of the ecological fallacy under the cross-sectional approach, noted by sociologists (Robinson 1950), which involved assigning to the individual reasons of a more collective nature, pertaining to the groups used in the analysis.

To avoid these fallacy risks, the contextual and multilevel approaches enable us to explain an individual's behavior by bringing several groupings of individuals into the process simultaneously. Using the contextual approach, we can associate an individual's behavior both with his or her characteristics (individual measure) and with the characteristics of the groups to which he or she belongs (aggregated measure). The multilevel approach enables us to go one step further by introducing an internal dependence in each group to individual and contextual characteristics simultaneously. These approaches therefore provide a means of avoiding both the ecological fallacy (for the aggregate characteristics are no longer viewed as a substitute for individual characteristics), and the atomistic fallacy (assuming the individuals' living environment is properly taken into account) (Courgeau 2003, 2004a, b, 2007a).

Of course the Bayesian approach allows an even more satisfactory multilevel analysis (Goldstein 2003; Courgeau 2007b; Draper 2008), as in the case of event-history analysis. In particular, when the number of units in one of the aggregation levels is small, the maximum-likelihood estimation of asymptotic standard deviations may be heavily biased as a result, and the maximum may actually be negative (Draper 2008). The situation is even more delicate with binary variables and, more generally, discrete variables: in certain cases, the classic likelihood methods do not even allow an evaluation of the model's parameters. The use of Bayesian methods then becomes necessary.

In other words, after completely neglecting the dispersion of their measures, population sciences have reverted for nearly 30 years now to the notion that the population is heterogeneous and that phenomena are interdependent. This allows the introduction of a fuller analysis of dispersion. More recently, these sciences have also begun to use a Bayesian analysis, better suited to capturing the refinement of event-history and multilevel approaches.

To show more clearly how only a Bayesian analysis can solve certain problems, we shall now give a more detailed example of an application to paleodemography (Caussinus and Courgeau 2010), a population science where the number of observed individuals is very often small. The purpose of this analysis is to estimate the age structure of a population without measuring it.

4.5 Estimating the Age Structure in Paleodemography

At their very origin, population sciences set out to determine the age structure of human populations (Graunt 1662). As noted earlier, they soon managed to do so with increasing accuracy (Halley 1693; Wargentin 1766) by exploiting data from parish registers, then from population registers. Unfortunately, paleodemography does not have such data at its disposal and is forced to rely on indirect measures of the ages of past populations, mostly thanks to the structure by evolution stage of selected biological indicators (Séguy and Buchet 2011).

We thus need another source of information to link these indicators to the chronological ages of individuals. The source may consist in the observation of a *reference population* for which both the direct and indirect measures exist. These two sources combined should enable us to extract an age structure for the *observed population*. However, various solutions to this problem are feasible and have been discussed extensively among paleodemographers.

First, let us present, in the most general terms, a situation in which we observe l stages tracked by a given biological indicator (stages in which the femurs of individuals are classified [Bocquet-Appel 2005]; stages in which the cranial sutures of individuals are classified [Séguy and Buchet 2011]; etc.) of a reference population and in which we distinguish c age groups. This reference population is shown in Table 4.2, with the number of individuals n_{ij}, by age, j, and by stage, i.

Table 4.2 Matrix of reference population by stage and age group

	Age groups								Totals by stage
Stages	n_{11}	.	.	.	n_{1j}	.	.	.	n_{1c}	$n_{1.}$
	.								
	.								
	n_{i1}	.	.	.	n_{ij}	.	.	.	n_{ic}	$n_{i.}$
	.								
	.								
	n_{l1}	.	.	.	n_{lj}	.	.	.	n_{lc}	$n_{l.}$
Totals by age	$n_{.1}$	.	.	.	$n_{.j}$	.	.	.	$n_{.c}$	$n_{..}$

4.5 Estimating the Age Structure in Paleodemography

Table 4.3 Number of individuals by stage observed in the new population

Stages	1	.	.	i	.	.	l	Total
Nb of individuals	m_1			m_i	.	.	m_l	$m_.$

From this table we can calculate different frequencies: first, the frequency of the age, j, given the stage i: $f_{j|i} = \dfrac{n_{ij}}{n_{i.}}$, and the frequency of the stage, i, given the age, j: $f_{i|j} = \dfrac{n_{ij}}{n_{.j}}$. We can also calculate the marginal frequencies: age frequency of reference population: $f_j = \dfrac{n_{.j}}{n_{..}}$ and stage frequency of reference population: $f_i = \dfrac{n_{i.}}{n_{..}}$.

We now want to estimate the age structure of a new population of which we only know the number of individuals by stage, m_i, given in Table 4.3. From this table, we deduce the stage frequency of the observed population: $\pi_i = \dfrac{m_i}{m_.}$. These consist, therefore, of measures taken either on a reference population (Table 4.2), or on the population whose age structure we want to estimate even though we only know the number of individuals by stage (Table 4.3).

To these measured numbers of individuals and frequencies correspond various unknown probabilities. Let p_{ij} be the probability that an individual taken at random in the population studied will be in stage i ($i = 1,...,l$) and in the age class j ($j = 1,...,c$) of a given indicator; the sum of p_{ij} values on i will be written simply p_j (probability that an individual is of age j), the sum of p_{ij} values on j will be written simply p_i (probability that an individual is in stage i); the conditional probability of stage i given age j will be noted $p_{i|j}$. These various probabilities are positive and satisfy the equations $\sum_i p_i = \sum_j p_j = 1$ and $\sum_i p_{i|j} = 1$ for all j. They are also linked by the following equation:

$$\sum_j p_j \, p_{i|j} = p_i \quad \text{for all} \quad i = 1, \ldots, l \tag{4.1}$$

Depending on the hypotheses posited to estimate these probabilities, we can use different solutions to obtain the age structure of the observed paleodemographic population.

4.5.1 Methods Proposed Earlier

The first method seeks to estimate the matrix for the observed population by using the criterion of greatest closeness between this matrix and that of the reference population. This approach was introduced by Kruithof (1937). In a study of telephone networks, he used a complete table of telephone flows derived from a reference population in order to estimate a new matrix of telephone flows, of which the only elements known to him were the margins for an observed population.

This research was extended in 1940 (Demming and Stephan 1940; Stephan 1942[10]) to estimate the cells of a contingency table subject to various constraints on one or both of its margins, given that we know all the cells of an initial table toward which it must converge as closely as possible. This method is usually called IPFP (Iterative Proportional Fitting Procedure) but, as shown later, paleodemographers have given it other names.

It consists in assuming that the probabilities are correctly estimated by the frequencies derived from Tables 4.2 and 4.3. A simple reasoning is applied: we minimize the distance of χ^2 between each cell of the reference-population table and the cells of the table to be estimated, under the constraint that its stage-specific frequencies are equal to π_i. This yields the following estimate of the age structure of the observed population:

$$\hat{p}_j = \sum_{i=1}^{l} \pi_i f_{j|i} \qquad (4.2)$$

This estimate is identical to the one given in paleodemography by the *probability-vectors method* (Masset 1971) or the ALK (Age Length Key) method. The latter was initially developed by Fridriksson (1934) to determine the age of fish belonging to a given species from a sample taken from the same catch. The method was later proposed in paleodemography by Konigsberg and Frankenberg (1992).

The distribution thus computed is necessarily dependent on the age distribution in the reference sample and is 'flattened by the influence of the reference sample,' as Masset notes (1995). This is self-evident, given the hypothesis that each cell of the estimated matrix must be as close as possible to each cell of the reference matrix. Of course, the stronger the correlation between age and stage, the more satisfactory the estimate. But unfortunately these correlations are rather weak in paleodemography—typically around 0.5 (Bocquet-Appel and Masset 1982)—generating a significant effect of the reference population on the age structure of the observed population.

Moreover, the hypothesis underlying the ALK method—i.e., the reference population must be extracted from the observed population—no longer applies when we have two populations of the same species taken from different catches. This problem has been raised by a number of researchers (Kimura 1977) and is of crucial importance in paleodemography. Indeed, in this case, the two populations are necessarily different, as noted earlier.

Both the probability-vectors method and the ALK method seek to estimate an observed theoretical matrix that is closest, term for term, to the initial matrix. This explains the dependence between the two matrixes. It also disregards the *invariance hypothesis* (Müller et al. 2002)—also called *uniformity hypothesis*—which holds

---

[10] In this second article, Stephan recognized that the results published in the previous article did not coincide with those given by the least-squares method—as the authors had mistakenly claimed—but he argued that they supplied a proxy solution.

4.5 Estimating the Age Structure in Paleodemography   177

that, for any bone of given age at death, the probability that a bone will be classified in a given stage depends only on that age, regardless of the population from which the bone was taken.

This assumption introduces a dissymmetry in the tables considered. Hence the search for another, more satisfactory estimation method, which takes the hypothesis fully into account.

The alternative method will not assume that each cell of the reference matrix must be as close as possible to the cell of the matrix corresponding to the observed stages; rather, each column of the reference table, relative to its margin, must be as close as possible to the identically defined column of the corresponding table for the observed population. In paleodemography, we begin with the stage-specific distribution for each age group of the reference population. We shall then find the weights that enable us, by multiplying them by the various distributions previously estimated, to find the number of individuals per stage of the observed population. These weights will accordingly reflect the number of individuals by age in the observed population. In this case, the invariance hypothesis is perfectly verified. The problem, which differs from the previous one, requires different methods for its resolution (Masset 1982; Konigsberg and Frankenberg 1992; Bocquet-Appel and Masset 1996).

Like the previous one, the problem was first posed to determine the age of an observed population of fish, of which only the size distribution is known. Here, however, the reference population is derived not from the same observed population, but only from a population of the same species, for which we know both the length and age, again measured from otoliths. Hasselblad (1966) supplied an iterative method for this type of estimation, followed by Orchard and Woodbury (1972), then Chikuni (1975). It was developed statistically by Kimura and Chikuni (1987), who proposed the name IALK for the method, to indicate that it involved iterations. Unlike ALK, the only assumption in IALK is that size distributions for each age of the reference population are applicable to the observed population, which is no longer derived from the same total population (Kimura and Chikuni 1987) and may therefore have a very different age structure.

In paleodemography, Masset (1982), in his unpublished dissertation, sets forth a method of successive approximations to avoid the excessively flat result obtained with the probability-vectors method. Masset's method proved very similar to IALK. For this purpose, he wrote an iterative program called *Approx* with Bocquet-Appel, supplied in an appendix to his dissertation. The sample application of the method (pp. 275–276 of the dissertation), on a population comprising seven age groups, yields results that are hard to accept. Although Masset starts with a reference population composed of seven age classes and seven stages, and the stage vector for the observed population of 60 individuals has no null element, he obtains an age structure of the observed population that is implausible:

$$(34.10 \quad 1.72 \quad 0 \quad 24.18 \quad 0 \quad 0 \quad 0),$$

for it includes four null proportions. Faced with these disappointing results, he prefers the probability-vectors method, which is more rustic but, as he notes, truer.

Meanwhile, Konigsberg and Frankenberg (1992), who were searching for a more satisfactory method than ALK, realized that IALK provided a means to avoid the biases above. The authors applied IALK to paleodemography using the maximum-likelihood method.

Bocquet-Appel and Masset (1996) revived their approximations method, which they now wrongly called IPFP, for we have shown that the latter seeks to minimize a distance of χ^2 between each cell of the reference-population matrix and the unknown cell of the observed population, of which we know only one margin. Here, instead, the aim is to minimize the distances between each of the two columns corresponding to the same age. To distinguish the method from the 'original' IPFP, we shall therefore refer to it here as *approximations method*—the initial name assigned by the authors. In their article, the authors continue to point out the difficulties in achieving convergence toward acceptable results with this method. Consequently, they now suggest confining its use to determining the mean age at death of the individuals in the population. But this restriction greatly diminishes the method's value.

The two approaches described above—which, for simplicity's sake, we shall call American and French—stirred many controversies between 1992 and 2002. In the end, however, they proved nearly identical (Konigsberg and Frankenberg 2002; Konigsberg and Herrmann 2002). To show this, let us examine their principles.

We begin with the basic principle of the IALK method. This time, we shall take the frequencies of the distribution of the biological indicator, conditioned by the age group in the reference population, i.e., $f_{i|j}$. We still assume that this frequency gives a valid estimate of the probability $p_{i|j}$. Applying the maximum-likelihood method, we see that we can obtain the age structure, $\hat{p}_j$, through successive iterations from any initial structure, but often set at $\frac{1}{c}$, i.e., uniform:

$$\hat{p}_j^{n+1} = \sum_{i=1}^{l} \pi_i \frac{\hat{p}_j^n f_{ij}}{\sum_{j=1}^{c} \hat{p}_j^n f_{ij}}. \tag{4.3}$$

We then perform as many iterations as needed for $\hat{p}_j^n$ to differ from $\hat{p}_j^{n+1}$ by as small a quantity as we want. This solution is not valid unless all the estimators are positive. In the case where some of them are zero, the solution is no longer that of the maximum-likelihood method. We can also estimate the variances of the estimators (see, for example, Cribari-Neto and Zarkos 1999).

The approximation method differs only in the first iteration. It too starts from an initial age structure that is purely uniform, and not generic as in the previous method. The two basic equations of this algorithm are consequently as follows:

$$\hat{p}_j^n = \sum_{i=1}^{l} \pi_i \frac{\hat{f}_{ij}^{n-1}}{\sum_{j=1}^{c} \hat{f}_{ij}^{n-1}} \quad \text{and} \quad \hat{f}_{ij}^n = \hat{f}_{ij}^{n-1} \frac{\hat{p}_j^n}{\hat{p}_j^{n-1}}.$$

4.5 Estimating the Age Structure in Paleodemography

From the initial values $\hat{p}_j^0 = \dfrac{m}{c}$ and $f_{ij}^0 = f_{ij}$, we deduce, from the first equation:

$$\hat{p}_j^1 = \sum_{i=1}^{l} \pi_i \dfrac{f_{ij}}{\sum_{j=1}^{c} f_{ij}}. \tag{4.4}$$

Note that for this first iteration, the expression (4.4) differs from the general form (4.3). The second equation enables us to calculate:

$$\hat{f}_{ij}^1 = \dfrac{c}{m} f_{ij} \hat{p}_j^1,$$

and again, to conclude, with the first equation:

$$\hat{p}_j^2 = \sum_{i=1}^{l} \pi_i \dfrac{\hat{p}_j^1 f_{ij}}{\sum_{j=1}^{c} \hat{p}_j^1 f_{ij}}.$$

We now return to Eq. 4.3. We also observe that it is not useful to consider $\hat{f}_{ij}^1$ or the population count $\dfrac{m}{c}$, which, being included in both the numerator and denominator, disappears from the equation. At this point, all we need is to see that, if the relation:

$$\hat{f}_{ij}^{n-1} = \dfrac{c}{m} f_{ij} \hat{p}_j^{n-1}$$

is satisfied, then:

$$\hat{p}_j^n = \sum_{i=1}^{l} \pi_i \dfrac{\hat{p}_j^{n-1} f_{ij}}{\sum_{j=1}^{c} \hat{p}_j^{n-1} f_{ij}} \quad \text{and} \quad \hat{f}_{ij}^n = \dfrac{c}{m} f_{ij} \hat{p}_j^n,$$

which is simple to show using the previous algorithm. As we have demonstrated that these equations were true for $n = 2$, they are true for all n. This does indeed take us back to the same formulation (4.3) as with IALK, but only from the second iteration onward. What Konigsberg and Frankenberg (2002) had shown empirically has thus now been demonstrated mathematically in its most general form.

However, the main difference between the two methods is that the first makes it possible to start from any initial structure, provided that its probabilities sum to unity, whereas the second requires us to begin with a uniform structure. This is simply due to the different formulations and therefore the different values in the first stage, as the formulations are identical from the second stage on. If we take a non-uniform initial distribution for the second method, then the solutions found will no longer be maximum-likelihood estimators.[11]

We should also note that another method can be used to estimate the same age structure, but has never been proposed by paleodemographers: the least-squares method. It consists in searching for the p_j values that minimize the following sum of squares:

$$S = \left(\sum_j p_j f_{1j} - \pi_1\right)^2 + \cdots + \left(\sum_j p_j f_{ij} - \pi_i\right)^2 + \cdots + \left(\sum_j p_j f_{lj} - \pi_l\right)^2$$

under the constraint $\sum_j p_j = 1$. To begin with, we see that, when $l=c$, we obtain a Cramer system whose solution verifies the linear relations:

$$\sum_j p_j f_{ij} = \pi_i.$$

When $l>c$, let us assume that we had the p_j values. Applying a variation ∂p_j, the differentials of the previous two equations give us:

$$\tfrac{1}{2}\partial S = \left[f_{11}\left(\sum_j p_j f_{1j} - \pi_1\right) + \cdots f_{l1}\left(\sum_j p_j f_{lj} - \pi_l\right)\right]\partial p_1 + \cdots$$

$$+ \left[f_{1c}\left(\sum_j p_j f_{1j} - \pi_1\right) + \cdots f_{lc}\left(\sum_j p_j f_{lj} - \pi_l\right)\right]\partial p_c = 0$$

and $\sum_j \partial p_j = 0$.

Multiplying the last equation by the arbitrary Lagrangian multiplier λ and adding the two equations together, we obtain:

$$\left(\frac{\partial S}{\partial p_1} + \lambda\right)\partial p_1 + \cdots + \left(\frac{\partial S}{\partial p_c} + \lambda\right)\partial p_c = 0$$

which leads to the Cramer system of $(c+1)$ equations with $(c+1)$ unknowns $p_1, p_2, \cdots p_c, \lambda$:

$$\begin{cases} p_1 \sum_i f_{i1}^2 + \cdots + p_j \sum_i f_{i1} f_{ij} + \cdots + p_c \sum_i f_{i1} f_{ic} + \lambda = \sum_i f_{i1} \pi_i \\ \quad \cdot \quad\quad \cdot\cdot \quad\quad \cdot \quad\quad \cdot\cdot \quad\quad \cdot \quad\quad\quad \cdot \\ \quad \cdot \quad\quad \cdot\cdot \quad\quad \cdot \quad\quad \cdot\cdot \quad\quad \cdot \quad\quad\quad \cdot \\ p_1 \sum_i f_{ic} f_{i1} + \cdots + p_j \sum_i f_{ic} f_{ij} + \cdots + p_c \sum_i f_{ic}^2 + \lambda = \sum_i f_{ic} \pi_i \\ p_1 + \quad\cdot\cdot\quad + p_j + \quad\cdot\cdot\quad + p_c \quad\quad\quad = 1 \end{cases}$$

---

[11] For more details on this comparison, see Courgeau (2011).

4.5 Estimating the Age Structure in Paleodemography

If the invariance condition and the hypothesis that the observed population sizes are error-free are fulfilled, we should find that this equation system is solved by a weighting system whose values all lie in the interval [0, 1]—which will correspond to the age distribution of deaths. But as the data are necessarily affected by uncertainty, given their modest number, and as we obtain the estimate from least squares, some of the estimates may lie outside the interval [0, 1] even if the invariance hypothesis is satisfied.

After this detour toward the various possible estimates in the chosen example, let us return to the estimates proposed by paleodemographers and the further research conducted.

The Americans continued to use the IALK method, introducing a continuous age rather than a discretized age, but without changing the principle. This is impeccably presented in the volume edited by Hoppa and Vaupel (2002a), after a seminar on the topic in Rostock, attended by many English-speaking anthropologists—but with no French specialists invited. Konigsberg and Herrmann (2002) clearly noted the similarity of results obtained with IALK and these more sophisticated methods: 'Our current methods fit fairly comfortably within the approaches taken during the Rostock workshop'.

First, the age distribution of a given stage in the reference population—with age now treated as a continuous variable—is provided by various types of non-parametric or parametric regression models. However, the volume's main originality is the use of a parametric event-history model (Courgeau and Lelièvre 1989) to model the probability density of the observed population's mortality. Provided the model does not include too many parameters (Gompertz two-parameter model, Gompertz-Makeham three-parameter model, Siler five-parameter model, etc.), we can estimate it using the maximum-likelihood method with the previously estimated age distribution of stages. Applying a notation similar to the previous one, we can summarize this formulation in the following form, where the age variable, j, is now continuous:

$$\pi_i = \int_j w_i(j) p(j,\theta) dj,$$

and where $w_i(j)$ is the distribution of the stage, i, by age, j, estimated in the reference population and $p(j,\theta)$ the age-specific probability density of the observed population, whose parameters, θ, we need to estimate using the l similar relations for each stage.

The problem is that these methods introduce a number of additional hypotheses, notably: a stationary or stable population, so that the event-history model can apply to a current population; and continuity in the age distribution of a given stage, yielding different estimates according to the methods used. In principle, therefore, there is no reason why these hypotheses—which we have no way of verifying—should be fully satisfactory. For instance, a past population that has experienced an epidemic cannot be considered stationary or stable. Similarly, to impose on that population a parametric event-history model—ultimately rather simple and verified on current populations—may fail to capture past situations where these models were not verified. Lastly, these methods always regard the reference population as perfectly observed, whereas large sampling errors may occur in paleodemography. Failure to take this into account, as in the IALK method, introduces a major risk of error in estimating the age structure of the observed population.

The French, for their part, not only doubted the validity of the approximation method for calculating a fully reliable age structure—as shown in Masset's example (1982) reproduced earlier—but held that taking a uniform distribution as a starting point to determine the age distribution could produce an unsatisfactory solution. After various tests (Bocquet-Appel 2005), which we shall not describe here,[12] let us conclude with the latest method proposed (Bocquet-Appel and Bacro 2008).

This time, we perform 1,000 equiprobable draws with replacement, using the bootstrap procedure, in each age group of the reference population. The population size is not fixed as before, but can vary according to the draw. The essential reason for introducing the procedure here is to be able to estimate the age distribution's confidence intervals. For each of these reference populations, we then use each of the age-specific prior probabilities—calculated from a mortality model that encompasses standard mortality (through attrition) and crisis mortality (catastrophic)—to determine a distance between (1) the stage composition of the observed population and (2) the composition obtained by performing a calculation that uses each prior probability and stage structure for each age of the reference population.

The authors point out that this procedure does not allow any valid estimation of the terms of the now random matrix f_{ij}, but it does enable us to choose for each draw the age-specific prior probability that supplies the shortest distance to the stage structure of the observed population. They then determine the mean of each of these probabilities and, from the result, compute a 95% confidence interval with the aid of the various bootstrap estimates. Now we know that, while the bootstrap method can be used when the model is properly specified, no theoretical result allows validation of its results when—as here—one uses an empirical model without sufficient specifications.

While the method does indeed introduce a random factor into the reference population, it is still not fully Bayesian, for the observed population is treated here as non-random. In fact, as we shall see later, it is more important to consider the observed frequencies as random than those of the reference population. By choosing the age structure in a parametric mortality model, it introduces—as before—a structure that is not necessarily verified by past populations. If the solution lies outside the proposed list, we have no means to verify it.

4.5.2 A New, Truly Epistemic Approach

All these reasons drove us to find a fully epistemic solution to the problem (Caussinus and Courgeau 2010, 2011). We view the parameters themselves as random, with a

---

[12] Again, we refer the reader to Courgeau (2011) for fuller details.

4.5 Estimating the Age Structure in Paleodemography

probability distribution—called a 'prior distribution'—chosen by the user to reflect his or her knowledge (and ignorance) prior to the observation. We then adjust the distribution on the basis of observations to arrive at the 'posterior distribution', which is the conditional probability distribution of the parameters taking the observations into account. The method involves both a dispersion in the table based on the reference population and a dispersion in the data on the observed population.

It is logical to regard the frequencies m_i ($i = 1, ..., l$), observed on site for the different stages as the observed values of a multinomial distribution whose parameters p_i are linked to the p_j and $p_{i|j}$ values under system [1]. We shall use the latter parameters to continue the modeling.

Let G be the prior density of the parameters $p_{i|j}$, $i = 1,..., l$ and $j = 1,..., c$ (we shall see, shortly, how to express it) and let us suppose that the parameters p_j ($j = 1,..., c$) have a prior density g and are independent of the $p_{i|j}$ values.

With M as the m_i vector, P as the $p_{i|j}$ vector, and p as the p_j vector, the joint density of (M, P, p) will be f, given by:

$$f(M,P,p) = g(p)G(P)\frac{m!}{\prod_i m_i!}\prod_i\left(\sum_j p_j p_{i|j}\right)^{m_i},$$

where the index i always ranges from 1 to l and the index j from 1 to c.

The marginal density of the pair (M, p) is:

$$\int f(M,P,p)dP$$

and the marginal density of M is:

$$\iint f(M,P,p)dp\,dP.$$

The integrals are taken from the variation domains of P and/or p, which are a simplex (for p) or a product of simplexes (for P).

The conditional density of p, given M, is therefore:

$$\frac{\int f(M,P,p)dP}{\iint f(M,P,p)dp\,dP}.$$

This is the posterior density of the p_j ($j = 1,..., c$) values, on which we shall base the Bayesian estimate.

For example, the posterior mean of p_j will be:

$$\frac{\iint p_j f(M,P,p)dp\,dP}{\iint f(M,P,p)dp\,dP}.$$

More generally, the conditional expectation relative to M of a function φ of p will be given by:

$$\frac{\iint \varphi(p) f(M,P,p) dp dP}{\iint f(M,P,p) dp dP}. \qquad (4.5)$$

This gives us, for example, the kth-order moment of p_j with $\varphi(p) = p_j^k$. Taking for $\varphi(p)$ the function equal to 0 for $p_j \geq x$ and 1 for $p_j < x$ (dummy of the event $p_j < x$), we express the posterior distribution function of p_j at point x.

We can evaluate the integrals of Eq. 4.5 using a Monte Carlo method as follows.

Let $X = (X_1, \ldots, X_c)$ be a random vector with a density distribution g and Y a family of c vectors $Y_j = (Y_{1j}, \ldots, Y_{ij})$ ($j = 1, \ldots, c$), whose joint distribution is independent of X and admits density G. We verify that [5] is equal to:

$$\frac{E\left(\varphi(X) \prod_i \left[\sum_j X_j Y_{ij}\right]^{m_i}\right)}{E\left(\prod_i \left[\sum_j X_j Y_{ij}\right]^{m_i}\right)}.$$

Let us generate S independent sets of such random vectors (X, Y), with s $(s = 1, \ldots, S)$ denoting the different repetitions. By virtue of the law of large numbers, if S is large enough, the expression above is proxied by:

$$\frac{\sum_{s=1}^{S} \varphi(X_s) \prod_i \left(\sum_j X_{js} Y_{ijs}\right)^{m_i}}{\sum_{s=1}^{S} \prod_i \left(\sum_j X_{js} Y_{ijs}\right)^{m_i}}.$$

This supplies, in particular, the posterior expectation of each p_j $(j = 1, \ldots, c)$, which can be taken as a one-time estimate, or the posterior variance that is useful to characterize the accuracy of the estimate. If we wish, we can use the same method to evaluate cross-moments, such as the covariance matrix of the posterior distribution of the p_j values. Lastly, the posterior distribution function of a p_j enables us, for example, to calculate intervals containing p_j with a given probability. Called 'credibility intervals' in the Bayesian context, they are equivalent to 'confidence intervals' in the classical approach.

We must now consider the choice of prior distributions.

Our only source of information on the conditional probabilities $p_{i|j}$ is the reference data. If they consist of raw data obtained simply by recording the frequencies of stages in a sample of skeletons of known ages, it is logical to admit that, for each

4.5 Estimating the Age Structure in Paleodemography

age j $(j = 1,..., c)$, the frequencies n_{ij} are the observed values of a multinomial distribution with a total n_j and probabilities $p_{i|j}$ $(i = 1,..., l)$. Adopting a prior distribution for the $p_{i|j}$ probabilities, we deduce a posterior distribution, conditional upon the reference data. This distribution, in turn, is taken as a prior distribution of the $p_{i|j}$ probabilities in the final model. Given the scarcity of additional information on the $p_{i|j}$ probabilities beyond what the reference data provide, it makes sense to adopt a uniform distribution as the prior distribution of the $p_{i|j}$ probabilities for each j. For a given j, the posterior distribution of the $p_{i|j}$ probabilities will accordingly consist of a Dirichlet distribution with parameters $\alpha_{ij} = n_{ij} + 1$ $(i = 1,..., l)$. The density G is the product of these c Dirichlet densities:

$$G(p) = \frac{\prod_j \Gamma(\alpha_j)}{\prod_i \prod_j \Gamma(\alpha_{ij})} \prod_i \prod_j p_{i|j}^{\alpha_{ij} - 1}.$$

In practice, the raw reference data can be 'worked' in different ways (for example, to apply a suitable weighting to a sample of men and a sample of women), so that their distribution is no longer strictly multinomial. But the prior distribution G defined above appears to remain suitable, as the multinomial property mentioned was merely a notional means of obtaining this distribution.

We can, however, consider refining the choice of G. One option seems of interest for addressing the practical problem in paleodemography. The reference data provide an instrument for modeling conditional probabilities by relying on the invariance hypothesis, but there are grounds for not placing excessive confidence in this hypothesis. If we want to guard against excessive confidence in the reference data, we can multiply the α_{ij} values by a 'reducing' coefficient r $(0 < r < 1)$ and choose $\alpha_{ij} = r(n_{ij} + 1)$. This does not affect the prior means of the $p_{i|j}$ probabilities, but it increases prior variances, thereby expressing our lack of confidence. These variances are approximately multiplied by $1/r$. We can observe that it is virtually equivalent to assume the n_{ij} values multiplied by r. This is yet another way to reduce the information contained in the reference data, since we proceed as if the relative frequencies observed for the reference data were preserved but obtained on a smaller sample.

The choice of the prior distribution of the parameters p_j is more delicate. We shall state our preference, using it throughout our presentation, but we shall also briefly mention other possibilities.

As the 'class' of distributions in which we should search for the prior distribution does not appear to stand out in any particular way, the most logical choice is a Dirichlet distribution, well suited to probability vectors. We are left with the problem of choosing the distribution parameters, say $(\beta_1,..., \beta_c)$. Absent specific information, we can, as above, opt for a uniform distribution and take $\beta_j = 1$ for all j. Such a choice, coherent with a logicist interpretation, allows us to stay 'neutral' and may be justified in certain cases. We shall see that it yields reasonable results on simple examples. However, in paleodemography, other choices are presumably more appropriate, as certain indications are naturally available. We can, for example, start from a 'standard' mortality distribution for which we calculate probabilities by age class. We then take these as the means of the prior distribution, giving us the β_j parameters to within one proportionality coefficient, i.e., the β_j/β values, where β is the sum of the β_j values for $j = 1,\ldots, c$. We are left with the choice of β values, i.e., in practice, the variances of the prior distribution. We should bear in mind that the variances must be relatively small in order to express the fact that the prior means are not very reliable and that the prior distribution should not play a dominant role—in other words, that the family of options considered covers a broad field. It seems, therefore, that the β values should be fairly small—for example, below or barely above unity. We shall see that this is indeed the case in the simulated examples studied below.

The prior means may be viewed as 'test' values: if the data are few in number and the estimates consequently imprecise, it will be interesting to use the posterior distribution in a qualitative manner, and observe which way the means move—i.e., how the data 'adjust' the prior values.

We can extend the above principle for choosing the prior distribution in different ways. For example, instead of picking a standard mortality distribution as a base for constructing the prior distribution, we can choose a mixture of two 'standard' distributions, which will yield a mixture of two Dirichlet distributions. The mixture could consist (in judiciously chosen proportions) of a standard mortality distribution (attrition) and a catastrophic mortality distribution.

As noted earlier, system (4.1) is undetermined when we use frequentist methods, if the number of rows (stages) l is smaller than the number c of columns (ages). In other words, the parameters of interest are not identifiable, in the sense that several values yield the same distribution of observable samples. The Bayesian method enables us to circumvent the difficulty since we start from a prior distribution and need only to make it change by means of the data. The posterior distribution will, in that case, direct us toward a distribution of the unknown parameters, which is entirely compatible with the fact that they are not fully determined. We can therefore use this method with $l < c$. Obviously, the posterior distribution will be relatively dispersed to take account of the indeterminacy inherent in the situation.

We shall now use an observed example to illustrate more clearly the advantages of a Bayesian method.

4.5.3 *Example of Archeological Application*

Using an actually observed data set, we shall now report the results obtained with the various methods. This will clearly demonstrate the need for a fully Bayesian method in order to achieve satisfactory results.

The data set concerns a population of nuns at the Maubuisson abbey (France) observed during the seventeenth and eighteenth centuries. To check the quality of the various estimates, we have the actual age structure for the entire population of nuns.

4.5 Estimating the Age Structure in Paleodemography

Table 4.4 Deceased Maubuisson nuns in the table (per 1,000)

Age group	20–29	30–39	40–49	50–59	60–69	70–79	80+	Total
Deceased	12	25	87	170	289	210	207	1,000

Table 4.5 Stage-specific distribution of sample of Maubuisson nuns

Stage	0–4	5–7	8–12	13–18	19–23	24–34	31–40	Total
Nb of individuals	6	2	4	5	3	9	8	37

Table 4.6 Distribution (combined synostosis coefficients) by stage and age group observed in female reference population (compiled from three Portuguese data collections)

	20–29	30–39	40–49	50–59	60–69	70–79	80+	Total
0–4	85	40	45	26	11	7	5	219
5–7	6	13	12	10	2	4	5	52
8–12	6	11	15	14	6	11	5	68
13–18	5	2	11	13	13	11	11	66
19–23	3	6	6	11	7	12	8	53
24–30	3	5	6	12	19	13	12	70
31–40	1	6	2	14	12	8	23	66
Total	109	83	97	100	70	66	69	594

In most cases, of course, the age structure will never be available, and we shall have only the estimate obtained with one of the methods available. Table 4.4 supplies the age breakdown.

We also have a sample of nuns whose cranial sutures have been measured as an age indicator. Table 4.5 gives the numbers by observed stage.

As a female reference population, we shall use a compilation of three Portuguese data collections (Séguy and Buchet 2011). The reference population is shown in Table 4.6.

We are now in a position to use all the methods proposed earlier to estimate the age structure of the Maubuisson nuns from these two populations and then compare it with the actual structure.

We should begin by noting that a χ^2 distance is an approximate distance between distributions, especially when the population observed is small (as here), but it already supplies information on the adjustment quality for a calculated distribution compared with a theoretical distribution. We proxied this theoretical distribution by that of the total population of Maubuisson nuns applied to the 37 skeletons examined. The results are given on Table 4.7.

The probability-vectors method (Prob. Vect.) gives results far removed from the expected distribution: the value of χ^2 with six degrees of freedom is 31.64, signaling the method's poor quality. In fact, its results are far closer to those of the reference population with a χ^2 of 2.299.

Table 4.7 Proportions estimated using alternative methods, theoretical proportions, and distances between them

Method	Age groups							χ^2 distance
	20–29	30–39	40–49	50–59	60–69	70–79	80+	
Prob. Vect.	0.11	0.11	0.13	0.18	0.16	0.14	0.17	31.64
IALK ML	0.00	0.11	0.00	0.00	0.61	0.000	0.28	42.50
IALK LS	0.09	−0.50	−5.02	11.56	1.71	−3.39	−3.45	–
Bocquet	0.02	0.04	0.07	0.13	0.21	0.27	0.26	2.15
Bayesian	0.03	0.04	0.08	0.15	0.31	0.23	0.16	1.53
Theoretical	0.01	0.02	0.09	0.17	0.29	0.21	0.21	

The IALK method using maximum likelihood (IALK ML) yields even worse results, with a χ^2 of 42.496. The population of four age groups is zero, an unbelievable result. If we try IALK with least squares (IALK LS), the result is worse yet, for many age groups now post negative population figures, although they still sum to unity. In this case it is not even useful to estimate a χ^2 distance between the estimated and theoretical distributions, given that the estimated distribution is flawed.

The method proposed by Bocquet-Appel and Bacro (2008) yields a now acceptable χ^2 of 2.153. The estimates for young age groups are very accurate, but the number of older nuns is overstated.

The Bayesian method (Bayesian) gives an even better χ^2 of 1.528. For this estimate, we have used information on the population concerned. In particular, as the population consisted of a nuns' convent, it was immune to the then very high risk of maternal mortality in childbirth. We modified the prior-probability vector to take account of this important information, which cannot be factored into the Bocquet-Appel and Bacro method.[13]

For the Bocquet and Bayesian estimates, we can calculate a more satisfactory distance than the χ^2 distance. Recall that the mean quadratic deviation of an estimator X of the real parameter θ is equal to the mathematical expectation of the square of the deviation $X-\theta$, i.e. $E\left\{(X-\theta)^2\right\} = Var(X) + E\left\{(X-E(X))^2\right\}$.

It therefore takes into account (1) the variance of the estimator, the first term of the sum above, and (2) its bias via the second term of the sum. We shall study the quality of the results obtained for these two larger examples by means of a total criterion for the distance between the vector of the actual probabilities and the vector of the estimated probabilities. In fact, we shall test two criteria: (1) the sum of the mean quadratic deviations obtained for the various age groups ('total MQD') and (2) a comparable sum, weighted by the actual probabilities ('relative MQD'). Here, these quantities are respectively as follows:

for the Bayesian method : 0.003 and 0.040

for the Bocquet – Appel and Bacro method : 0.014 and 0.078.

---

[13] For more details on this estimation, see Caussinus and Courgeau (2010).

The results give a significant advantage to the Bayesian method. However, if we had simply performed the Bayesian analysis without taking into account the information supplied by the nuns' specific status, we would have obtained the distances 0.007 and 0.199. The method would thus have preserved an advantage in terms of total deviation but would have lost it in terms of relative deviation because of an excessively large 'error' on small probabilities.

From our examination of this archeological case, and its analysis provided by the Bayesian method compared with the other methods, we can conclude that the first method gives a better estimation of the age structure of past populations—an instance where records of age at death are lacking, but are replaced by measures of biological indicators.

4.6 Conclusion

Throughout this chapter we have observed an alternation between homogeneity and dispersion that, however, occurred between different units, depending on the period considered. The reason is that, like any scientific discipline, population science concerns itself not with population reflecting the full complexity of the individuals who compose it, but with selected aspects of it. These may become complex over time, but they are always characterized by a small number of measures, deemed essential for understanding the phenomena that affect the population.

At the outset, our investigation was confined to probabilities concerning a total population. We tested the hypothesis of their dispersion—in the second sense of the term—by age. The conclusion is that we need to consider different probabilities for each age. We also tested the dispersion—in the first sense of the term—of certain indices. However, we find few examples of its use, particularly in eighteenth-century works of political arithmetic.

Laplace continued to observe a population in its entirety but, this time, from a Bayesian standpoint. He started from the hypothesis of prior probabilities uniformly distributed in the interval [0,1] in order to obtain posterior probabilities, whose dispersion—in the first sense of the term—he could estimate accurately. He also noted the usefulness of analyzing dispersion—in the second sense of the term—by bringing in causes that would affect certain sub-populations and not others: manners, climate, food. However, the regression methods proposed by Gauss were not used in population sciences at that time.

The spread of censuses in the nineteenth century and a critique of the foundations of Bayesian calculus led to a rejection of Laplace's approach. The cross-sectional and longitudinal analyses used until the early 1980s left aside all evaluation of the dispersion of demographic measures, in both senses of the term.

Dispersion then made a comeback in population sciences through the event-history and multilevel approach, whose inclusion of individual and aggregate characteristics broke up the analytical framework. Dispersion came into play in the first sense of the term, for it became essential in estimating the variance of the estimated effects in order to assess their validity. Dispersion also came into play in the second sense of the term, because of (1) the simultaneous introduction of units of different levels into the analysis and (2) individual characteristics affecting a different sub-population each time.

In the final section of this chapter, we examined the estimation of the age structure of a population for which the only data available to paleodemographers were biological indicators measured on skeletons. We were able to show the power of Bayesian methods as a tool for preparing such estimates.

To conclude this chapter, we need to address one final issue. Throughout our discussion, we have contrasted the points of view of those who apply probability to an individual case and those who reject this possibility altogether. To put it simply: in one camp, Jacob Bernoulli, when he examines Titus's chances of dying; in the other, von Mises, when he states that the probability of dying makes no sense for him if it refers to a single individual. We are dealing here, in fact, with the distinction between subjective probability and objective probability, applied to population sciences data.

For instance, de Finetti (1937), one of the leading representatives of the subjective approach, clearly stated:

> The degree of probability assigned by an individual to a given event is revealed by the conditions in which he would be willing to bet on the event.

Later, he specifically expresses his belief that an 'event is always a singular fact'. In contrast, von Mises—one of the leading representatives of the objective approach—refuses to speak of the probability of a singular fact, which does not exist for him. It is important to see where population science stands in regard to these two extreme positions.

For the classical approach, objective probability seems entirely suitable, under the hypothesis that the exhaustively observed population can be regarded as drawn from an infinite theoretical population with the same probabilities as itself of experiencing different events. The variances of the probabilities estimated under this hypothesis are weak enough, as we have shown, to sustain it perfectly.

But once we move to event-history or multilevel approaches, which often rely on non-exhaustive data from surveys, the choice of objective probability can be called into question—although it remains a possible alternative and is used by demographers. Similarly, when we want to address a paleodemographic problem, with a limited amount of data, the continued recourse to objective probability yields results that are often totally erroneous and even impossible to accept. In such conditions, subjective probability seems better suited to incorporating all the theoretically useful information on the phenomena studied, which objective probability did not allow. Indeed, the use of subjective probability is often indispensable because of the extreme dispersion of probabilities (in the second sense of the term 'dispersion') according to individual characteristics and due to interactions between the phenomena studied. But subjective probability allows only very approximate individual forecasts, for people have many other characteristics besides the ones analyzed, and those other characteristics can strongly modify the forecast (Courgeau 2007b).

Chapter 5
Closer Links Between Population Sciences and Probability

We could extend the analysis in the previous chapter to the other social sciences. In Sects. 1.4, 2.4 and 3.4, we have already given some examples of the use of probability in economics, sociology, education science, and political science. Equally well, we could have described its use in human geography, anthropology, and other disciplines. The topic deserves a full volume for each of the social sciences, given the number of such connections and the extent of their development throughout the history of these disciplines. The exercise has already been conducted, for example, in econometrics (Louçã 2007). Here, however, we prefer to explore in greater depth the close ties between probability and the population sciences. This calls for a detailed analysis of the basic concepts and symbols used, and the various paradigms that have supported the probability-based approach to social phenomena. We shall try to show how the axioms that have marked the history of probability are intimately linked to those paradigms.

The concepts of population and individual range well beyond the boundaries of population sciences, and their history has already been recounted by many authors (Landry 1909; Vidal 1994; Charbit 2010). Our discussion will largely focus on a new angle: what is their relationship with probability? What is the position of an individual in a population, and can we analyze these literally individual cases one by one, or must we resort to a more abstract concept of fictitious individual?

Population sciences also use other primitive symbols and concepts, and we shall examine them in detail later, again from the standpoint of their ties with probability. What events should we consider, and what individual or more general characteristics of groups—such as family, household, firm, and religion—should we introduce into the analysis? What role does time play in these sciences, and how should we apply stochastic temporal processes to deal with these issues?

Once we have clearly framed all these prior questions, we shall be able to examine in greater detail the paradigms that have successively prevailed in population sciences (Courgeau 2004a, b, 2007a, b; Courgeau and Franck 2007). Our notion of paradigm differs slightly from those proposed by Thomas Kuhn (1962). Recall that Kuhn, who had used many different meanings of 'paradigm' in the first edition of his book (1962),[1] ended up distinguishing two main meanings in the expanded 1970 edition:

> On the one hand, it stands for the entire constellation of beliefs, values, techniques and so on shared by the members of a given community. On the other, it denotes one sort of element in that constellation, the concrete puzzle-solutions which, employed as models or as examples, can replace explicit rules as a basis for the solution of the remaining puzzles of normal science. (Kuhn 1970)

The slightly different version offered here addresses the following question: how do we go from experienced phenomena to the scientific object as defined by the philosopher Granger (1994)?

> The complex real-life captured in the experience of sensible things has become the *object* of a mechanics and a physics, for example, when the decision was taken to reduce it to an abstract model, which initially incorporated only spatiality, time, and "resistance" to movement.

Granger also admits that the content of this object has not been assigned a broad, explicit definition from the outset. Sciences such as physics and biology have produced successive elaborations of their object, as evidenced by the transition from Newtonian physics to Einstein's general relativity. Similarly, population sciences have spelled out their object in successive paradigms, each of which provides a different way of relating observed phenomena to the scientific object (Courgeau and Franck 2007; Courgeau 2009).[2]

Here, we do not distinguish this notion from what are more generally referred to as programs. These are defined in purely literary terms and offer schemas of the observed system that highlight the social system's relevant entities with their properties and basic relationships (Walliser 2009). We shall not extend our discussion to axioms, which would allow us to characterize with greater precision the principles of the social properties capable of guiding empirical investigations and underpinning our proposed explanations (Franck 2009). The reason is that, unlike probability, axioms are still far from totally defined in these sciences, particularly demography. The history of their establishment deserves a full volume for each of these sciences[3] but would be a digression from the subject of this chapter: the links between social science and probability. However, we shall try to approach axioms as best as possible.[4]

---

[1] See Masterman (1970), who has collected 21 different meanings of the term 'paradigm.'

[2] This notion is similar, in our view, to what Gonseth (1975) proposed under the term 'referential' (*référentiel*). In particular, he noted that 'the referential may be described as holographic, in the sense that the aspect under which it manifests itself to the subject may change according to the relationship that the subject establishes with it.' We shall see that this is what occurs for the various population sciences paradigms.

[3] The best example we can give is Pratt's book (Pratt 2010), *Modeling written communication*, whose aim is to discover the principles or axioms of written communication, considered as a social property (Franck 2002).

[4] I wish to thank Robert Franck and Xavier Bry for our many exchanges on this topic, which allowed me to better define the goal of this chapter.

5.1 The Framework: The Population and the Individual

As just noted, the first concept to consider is population, which is closely connected to the concept of individual. We shall try to see how these have evolved through history, and if they can be treated as scientific objects.

We cannot describe the points of view of all the authors who have discussed these concepts, but we shall give a few examples,[5] chosen for their particular relevance to the present volume.

Plato gives a detailed and even quantitative description of the establishment of a political society (Republic, II, 369 ff.), that is, of a City, πόλις, (Laws, V, 737 ff.). One could view this as the origin of demography. Using a political criterion to define it, he sets up a unit that appears to be suitable for this discipline:

> Then, as we have many wants, and many persons are needed to supply them, one takes a helper for one purpose and another for another; and when these partners and helpers are gathered together in one habitation the body of inhabitants is termed a State. (Republic, II, 369)

But Plato goes on to use criteria that cause him to move away from demography. To begin with, his unit of account for the components of a State is not the individual but the head of the family:

> The number of our citizens shall be 5,040—this will be a convenient number; and these shall be owners of the land and protectors of the allotment. (Laws, V, 737)

Let us set aside the figure of 5,040, which has generated many hypotheses (Charbit 2010), to focus on the definition of the unit of account. By taking the number of families as his starting point, Plato effectively rules out any measure of individual mortality, which population sciences must take into consideration, whereas the family can far outlive its members. Similarly, for the generation of children, Plato lays down rules to ensure that there will always be 5,040 family houses (Laws, V, 740) when there is more than one heir per family, but the arrangement allows no measurement of fertility. Plato also envisages emigration and immigration (Laws, V, 740), again without quantifying them. Lastly, he totally ignores slaves, whose number per family can be very large, but who have been 'separated from royal and political science' (Stateman, 289).[6]

In sum, by placing the discussion directly at the aggregated level of the family, Plato lacked the material needed to institute a truly analysis in population sciences.

Aristotle (Politics, I, 2) goes further than Plato in defining the groups that make up a population:

> As in other departments of science, so in politics, the compound should always be resolved into the simple elements or least parts of the whole. We must therefore look at the elements of which the State is composed, in order that we may see in what the different kinds of rule differ from one another.

---

[5] The work by Charbit (2010) offers a fuller view, although he is forced to choose from among the many historical examples.

[6] The website http://classics.mit.edu/Plato/stateman.html display incomplete: see 'text only version' for quotation in context.

He begins by showing that 'the family is the association established by nature for the supply of men's everyday wants.' He goes on to state that 'when several families are united, and the association aims at something more than the supply of daily needs, the first society to be formed is the village.' Lastly, 'when several villages are united in a single complete community, large enough to be nearly or quite self-sufficing, the State comes into existence.' Aristotle also ranks these units by order of precedence:

> The State is by nature clearly prior to the family and to the individual, since the whole is of necessity prior to the part; for example, if the whole body be destroyed, there will be no foot or hand, except in an equivocal sense, as we might speak of a stone hand; for when destroyed the hand will be no better than that.

Aristotle thus clearly defines all these aggregation levels, and may thus be characterized as a precursor of today's multilevel analysis. There are, however, major differences between the two approaches. First, the order of precedence is reversed: whereas multilevel analysis centers on the individual to determine the effect of more aggregated levels on individual behavior, Aristotle focuses on the State to deduce behaviors at less aggregated levels. Second, he proposes no scientific approach to the individual, as population sciences do. As he clearly indicates in *Rhetoric*, the individual cannot be the object of any science:

> But none of the arts theorize about individual cases. Medicine, for instance, does not theorize about what will help to cure Socrates or Callias, but only about what will help to cure any or all of a given class of patients: this alone is business: individual cases are so infinitely various that no systematic knowledge of them is possible.

We should begin by noting that Aristotle often uses the term 'art' (τέχνη) as a substitute for 'science' (επιστήμη). Most importantly, he does not identify the concept of statistical individual, which, as we shall see, is the foundation of population sciences.

In other words, unlike multilevel analysis, which starts with this statistical individual in order to show the effect of higher aggregation levels on his or her behavior, here this individual behavior is deemed unknowable. We can therefore conclude that the modern idea of a science of man has not yet taken shape in Aristotle's mind (Granger 1976).

In fact, the hegemony of Aristotelian thought in the Western world did not begin to falter until the sixteenth century. The *Revolutions* of Copernicus (1543) attacked Aristotle's astronomy head-on by replacing its Earth-centered system by a Sun-centered one. Francis Bacon (1605) contested his authority in every field:

> so knowledge derived from Aristotle, and exempted from liberty of examination, will not rise higher than the knowledge of Aristotle.

Bacon wrote *Novum organon* (1620) to elaborate an inductive method and oppose it to the *Organon*, the name given by Aristotle's disciples to the set of his six works on logic. Contrary to the then standard approach, the inductive method could be reached as follows:

> There are and can be only two ways of searching into and discovering truth. The one flies from the senses and particulars to the most general axioms, and from these principles, the

5.1 The Framework: The Population and the Individual

truth of which it takes for settled and immovable, proceeds to judgement and to the discovery of middle axioms. And this way is now in fashion. The other derives from the senses and particulars, rising by a gradual and unbroken ascent, so that it arrives at the most general axioms at last. This is the true way, but as yet untried.[7] (Bacon 1620)

Thus, for Bacon, observation comes first and underpins an inductive approach that enables us to follow the trail back to axioms.[8]

This is precisely the same induction that allowed Pascal to develop a geometry of chance (1654a, b), using the player's uncertain luck as a starting point. The purpose was to demonstrate, if not an axiomatics of probability at this early stage, at least a hidden logic that enabled Pascal to grasp the facts of the matter. In his own words (1654b): 'but now, having held out against experience, it [i.e., the logic of probability] was unable to escape the empire of reason.'

Similarly, Graunt (1662) notes the following in his preface addressed to Robert Moray, the main founder of the Royal Society:

> because Sr. Francis Bacon reckons his Discourses of Life and Death[9] to be Natural History;
> … I am humbly bold to think Natural History also, and, consequently, that I am obliged to cast in this small Mite into your great Treasury of that kinde Sir Francis Bacon.

Faithfully obeying Bacon's precepts, he begins with an observation that, while detailed, is still far from complete. In particular, it does not mention of the age of the deceased, supplied by the mortality bills of the time. Graunt's objective is to produce a full picture of mortality in London during the period considered, with the few characteristics that he was able to glean from the mortality bills.

In truth, to establish a science—whether mathematical, physical or human—one must select only a few aspects of the phenomena studied, aspects that can be characterized with great precision (Franck 2002). Graunt's merit was therefore to gloss over most differences between individuals, concentrate on what was recorded in the mortality bills of his time, and offer some clear tables summarizing the information contained in these registers. The information that he takes into account regards the deaths recorded with the mention of their cause (plague, accident, illness, etc.), the person's sex, and the place and date of death. These are some of the characteristics examined in population studies. Moreover, from the outset, Graunt addresses the important question of the generality of this information. If he had been dealing with a small number of bills collected here and there, the results would be of little value. But Graunt took care to

---

[7] Duae viae sunt, atque esse possunt, ad inquirendam and inveniendam veritatem. Altera a sensu and particularibus advolat ad axiomara maxime generalia, atque ex iis principiis eorumque immota veritate judicat and invenit axiomate media; atque haec via in usu is. Altera a sensu and particularibus excitat axiomata, ascendendo continenter and gradatim, ut ultimo loco perveniatur ad maxime generalia; quae via vera is, sed intentata.

[8] See Franck (2009) for a detailed description of Bacon's proposed method for seeking the form of a property.

[9] (Bacon 1623). In this book, however, he resorted mainly to compilation, without truly applying his induction principle, and without always relying on experience, as Graunt did.

peruse and record every possible bill gathered in London over a certain number of years.[10]

The very same year as Graunt published his pioneering work, the logicians of Port Royal[11] (Arnauld and Nicole 1662) elaborated two aspects of the same concept (which they termed 'universal idea') that were essential for defining a population:

> I call *comprehension* [*compréhension*] of the idea the attributes that it contains and that cannot be removed without destroying it, for instance, the comprehension of the idea of the triangle encompasses extension, figure, three lines, three angles, the fact that these three angles sum to two right angles, etc.

> I call *extent* [*étendue*] of the idea the subjects to which this idea is suited; this is also called the inferiors of a general term, which is called superior with respect to them; for instance, the idea of the triangle in general extends to all the various kinds of triangles.

Today, we would speak of intension or significance to denote the comprehension of a concept, and extension to denote its extent (Nadeau 1999). If we look more closely at how these aspects appear in the definition of the concept of population, we arrive at the following definition: the intension of the term 'population' establishes its properties and characteristics, whereas its extension consists of the set of individuals who satisfy these properties. Thus, if this property or characteristic consists of nationality alone, we can define the population of French nationality and determine all the individuals who belong to it: they are located around the world, although most reside in France; they will also change continuously over time. If the property or characteristic is both the nationality and the country of residence, we can define more precisely the population of French nationality living in France.

More generally, therefore, a population can be characterized as an 'aggregate of individuals which conform to a given definition. This definition is at least spatial and temporal in specificity' (Ryder 1964). The criteria used for it may, of course, be of many different kinds: political (such as nationality, discussed above), economic (people working in the same firm or industry), religious (church parishioners) or social (persons related through specific family ties). We can take a single criterion, or several. In consequence, these various populations are not distinct but partially overlapping.

As Aristotle very rightly points out, however, individuals belonging to a given population will have an unlimited number of different characteristics other than those assigning them to that particular population. But a true social science cannot be implemented except to explain a finite number of individual characteristics.

For this purpose, we need to deprive individuals of their unlimited and unknowable character, reducing them to a limited number of aspects that will allow the establishment of a social science. This small set of characters can form a usable scientific object. The researcher will set aside an infinity of other characters regarded as

---

[10] He used the bills prepared continuously since December 29, 1603, for earlier bills had been compiled and preserved haphazardly.

[11] Pascal had close ties with the Port Royal authors in this period, when many studies were conducted on a cooperative basis and published anonymously.

secondary to the study to be conducted. Naturally, the choice of characters is essential and must be made with a very detailed knowledge of the phenomena studied.

In population sciences—as in many other social sciences—we shall indeed start by observing the individuals in a population, but we shall observe only a small number of phenomena and of characteristics of these individuals. The characteristics will be examined in greater detail in the next section. For the time being, we shall be vague about them, apart from noting their small number. It is on this reduced set that the discipline will operate. In this case, we shall create an abstract fictitious individual, whom we can call statistical individual as distinct from the observed individual. The statistical individual will experience events that obey the axioms of probability theory chosen to treat the observations. Under this scenario, two observed individuals, with identical characteristics, will certainly have different chances[12] of experiencing a given event, for they will have an infinity of other characteristics that can influence the outcome. By contrast, two statistical individuals, seen as units of a repeated random draw, subjected to the same sampling conditions and possessing the same characteristics, will have the same probability of experiencing the event. We can now see more clearly how the use of observed phenomena and characteristics—which constitute the statistical reality of human facts—can now be transformed into an abstract description of human reality. This is achieved by means of concepts deliberately stripped of those concrete circumstances that, in the researcher's view, can be left aside. We shall explore this notion of statistical individual more fully when examining the different paradigms of population sciences.

Lastly, it is useful to note that, while each individual in a population has a limited life span, their aggregation (which constitutes a population) has an indefinite existence, which can greatly exceed that of its members. The demographic events discussed in the following section introduce a process of creation of new individuals, through birth and immigration, while others disappear, through death or emigration.

We have thus succeeded in narrowing the field of our work in order to make it usable for social-science research. After describing the framework and content, we must highlight the phenomena that will alter them over time.

5.2 The Object: Not Individual Behaviors but More Abstract Concepts

Many authors see population sciences as the study of 'the behaviors of human populations, from the individual level to the social level' (Tabutin 2007), thereby assigning it a field that is both indefinite and infinite. If it were to follow that path, these sciences would end up covering everything. We believe, instead, that it is useful to

---

[12] We deliberately refrain from using the term 'probability' here, for it cannot be estimated in these circumstances.

refocus research on a more specific object of population sciences (Courgeau and Franck 2007).

On the face of it, one might think that its object is the study of births, deaths, and migration flows, which affect the population. But, like the concept of population itself, such an object has been envisaged since Antiquity without ever producing a truly approach in population sciences, as noted earlier. Vilquin (1977) expresses this idea very clearly in his introduction to Graunt's work:

> Measuring, quantifying phenomena that were God's secret (birth, illness, death, the games of life and chance): the only ones who dared to venture there were scholars whose science was mysterious enough for them not to fear public censure: physicians, theologians, or polemicists caring little for truth and falsehood. The multiplicity of the numbers reported without mystery and with assurance in Graunt's work was therefore bound to provoke astonishment, if not scandal.

He shows that Graunt goes beyond quantification by envisaging—as we have seen—wagers on the numbers ('it is esteemed an even lay [...]'), an exercise that requires introducing the notion of probability. Similarly, Lotka (1939) writes:

> the large number of variables, and the random nature of their connections, give demographic analysis a special character.

Setting aside the fact that the variables far outnumber the phenomena of classical physics, the main point here is the randomness not only of the events themselves but also of the connections between them.

Therein lies the originality and deep foundation of population sciences, which we can thus describe as the study of the probability of fertility, mortality, and migration—a definition that clearly introduces probability as a fundamental concept. For instance, fertility is a measure of the probability of having children, provided we simultaneously define the population at risk.

Interestingly, most researchers in the mid-twentieth century regarded the link between probability and population sciences as so obvious that they no longer bothered to point it out, or viewed it as one of several possibilities. The term 'probability' hardly ever occurs in Henry's major demographic treatise (1972).[13] He speaks far more often of frequency in the sense of the ratio of a part to a whole. Even more strikingly, in the *Multilingual Demographic Dictionary* (1981), the same author notes:

> The *relative frequency* of a non-renewable event is often regarded as an empirical measure of the *probability* of occurrence of that event. This presumes that all the individuals who appear in the denominator have been exposed to risk in some way, i.e. there must have been a chance or risk that the event in question could happen to them.[14]

He therefore views the interpretation in terms of probability as a possible interpretation of frequencies, which he does not deem indispensable. This is totally

---

[13] However, he uses it in the expressions 'probability of family enlargement' and 'probability of survival.'

[14] The English edition uses 'often' whereas the original term in the French version is 'parfois' (sometimes), which attenuates the link with probability.

5.2 The Object: Not Individual Behaviors but More Abstract Concepts

consistent with the attitude described in Chap. 4, which prevailed in population sciences from the mid-nineteenth century through most of the twentieth: to cope with the large numbers of persons observed in censuses, an objective approach sufficed, without even introducing the notion of probability.

However, in Chap. 4, we also noted the importance attached by Laplace in *Théorie analytique des probabilités* (1812) to the analysis of population sciences phenomena, which he viewed as an integral part of his book. We also showed the need to use probability in the event-history and multilevel approaches, now prevalent in population sciences analysis. Our proposed definition is consistent with this approach, which we believe is essential in today's population sciences.

We must also show that, from the outset, events are situated in a time frame that is always present, even though the cross-sectional approach tends to minimize it. The frame may consist of the date at which the events occur, of age or, more generally, of the time elapsed since a founding event taken as the origin of time. On the latter definition, age becomes the time elapsed since the individual's birth.

In sum, although this individual characteristic was missing from his *mortality bills*, Graunt (1662) did see its crucial value very clearly. He tried to estimate the age-specific probability of dying, under hypotheses that were rough but the only conceivable ones (see Sect. 1.4). Later, all political arithmeticians, then all researchers in population science, have used age by connecting birth certificates to the other certificates for named persons recording the occurrence of a demographic event for the same individual.

Time can also play its conventional role as the instant at which an event occurs. We shall see that this notion of time is the one prevailing in cross-sectional analysis, where time is static, since we are looking at a given moment. But time can also play a role as a marker for events affecting personal lives, such as wars and crises.

We can now try to see more precisely what demographers mean by the study of the probability of fertility, mortality, and migration, which are the object of their discipline.

To begin with, we should note that an explicit definition of the object is not necessary. Granger (1994) states this clearly:

> In the case of social sciences [*sciences de l'homme*], the transmutation into a scientific object of the complex and changing life experience that consists of the human fact remains problematic, even in its aspects commonly recognized as public. This is not because a prior and general explicit definition of the object of a science is a prerequisite for its development. Quite to the contrary, physics and biology are, if anything, progressive explanations of their object. But it is no less true that, in these conditions, every objectivated aspect of sensitive experience, every phenomenon, is identified by means of criteria accessible to all those who possess a specific physical and intellectual tool set. These criteria, however refined, mediated, and abstract they may seem, nevertheless 'salvage' phenomena, in the sense that they always allow, within the limits of a recognized tolerance, the establishment of a univocal correspondence of the phenomena perceived, and consequently experienced by a private conscience, with their publicly intelligible schematization.

This text therefore encourages us to spell out the meaning of the object of population sciences as we gradually examine the various proposed interpretations, and it spares us from having to provide a prior and general explicit definition of the object.

In contrast, we note the object's close ties with the phenomena known to all: births, deaths, and migration flows, which are common occurrences in a population. It is perfectly legitimate to establish a univocal correspondence between these phenomena and the schematization involved in speaking of fertility, mortality, and migration, but it is vital to distinguish between the two here. The term 'probability' allows us to show that we are dealing with a property of the populations that we can use to explain and try to predict their changes.

For instance, fertility makes it possible to increase the size of a population by introducing new individuals who have just been born. Immigration will have the same effect, but by introducing new individuals at any age. By contrast, mortality will reduce population size in a similar manner to emigration. Population sciences will study the changes in the population due to these events, taking different points of view of the time frame in which they occur. Moreover, these phenomena cannot be studied without reference to the factors that can influence their occurrence. The examination of such factors can thus be introduced into the field of population sciences, but at a later stage and in a limited number. The purpose will be to determine whether they influence the three phenomena and, if so, how.

We must now explore in greater detail the various approaches followed by population sciences. Only by examining these different points of view can we attempt to define their object more precisely.

5.3 The Cross-Sectional Approach

This approach largely prevailed in the early days of population sciences and consisted in taking a period point of view to study phenomena. By period we mean a short duration, typically 1 year, in which we can measure the numbers of events occurring in a given population.

We have already described (Courgeau 2003, 2004a, b, 2007a) the paradigm of this approach. Let us recall it briefly here. The social facts of the period exist independently of the individuals who experience them. We can explain them by the economic, political, religious, social, and other characteristics of society. Furthermore, these phenomena are independent of one another. Let us look in greater detail at the consequences of the independence posited between period phenomena and the consequences of their dependence on the characteristics of the society in which they occur.

5.3.1 Independence Between Phenomena

Euler (1760), for instance, already set three hypotheses in his work on mortality and the multiplication of the human species:

- The first hypothesis is based on 'the vitality or power of life that is specific to humans.' It leads to equating this vitality with the probability of dying at each

age for members of a given population. Therefore, the first hypothesis will indeed be represented by a probability, assumed identical for all persons of the same age. Moreover, as Euler was studying very large populations, he did not need to take the estimated dispersion of the probability into account.
- The second hypothesis rests on 'the principle of propagation, which depends on marriages and fertility.' For Euler, it consists in identifying the principle with the fact that 'the number of children born every year is always proportional to the number of all living persons.' Once again, therefore, he introduces a probability: that of the birth of a child in a given population. Admittedly, this is a rough approach to the fertility of a population. All present-day population scientists would reject it as utterly inadequate, for the ratio is calculated not with respect to the female population of childbearing age, but to the total population. However, the notion of identifying the principle with the fertility of a population, conceived as a probability, is already well established in Euler's work.
- The third and final hypothesis is that 'the two principles of mortality and propagation are independent of each other.' Therefore, Euler does not need to take account of possible interactions between the two probabilities.

From these principles, he can thus calculate all the other probabilities that population scientists would want to estimate, such as: 'given a certain number of men, all of the same age, find how many will probably be alive after a certain number of years'. Likewise, what is 'the probability that a man of a certain age will die during a given year' [?], and also 'find the point in time that a man of a given age may hope to reach, such that it is equally probable that he will die before the point as after.' [and so on].

Euler also recognizes the accidents of history that introduce discontinuities in population change, as well as migration phenomena, which he does not address. However, he notes the following:

> For places subjected to such irregularities, one should keep accurate registers of all the living as well as of the dead, and then, following the principles that I have just set out, we would be able to apply the same calculation. Everything always comes back to these two principles, that of mortality and that of fertility; once these are well established for a particular location, it will not be difficult to resolve all the questions that may be raised on this topic, of which I have merely described the main ones.

It follows that, if we have detailed data on these phenomena (population registers are an example for migrations), we can treat them in a manner similar to mortality (e.g., famine, emigration) or fertility (e.g., immigration, colonization). The first set of phenomena remove members from a population, the second add new ones, not only at the age of birth but also at all ages for immigration. Indeed the propagation principle implies the possibility of entering a population otherwise than through birth.

In fact, we could speak here not of an axiomatization in the full sense, but of a proto-axiomatization of population sciences. Let us show why.

First, the probabilities on which this proto-axiomatization is based were not axiomatized themselves until more than 160 years later (Kolmogorov 1933). But Euler built his approach on these probabilities. Next, let us examine the successive stages that Franck (2002) proposes for modeling in the other social sciences as well:

> (1) Beginning with the systematic observation of certain properties of a given social system, (2) we infer the formal (conceptual) structure which is implied by those properties. (3) This formal structure, in turn, guides our study of the social mechanism which generates the observed properties. (4) The mechanism, once identified, either confirms the advanced formal structure, or indicates that we need to revise it.

We shall show that this approach resembles Euler's. Regarding the first point, he clearly tells us:

> As this task is most difficult to perform, we must be very grateful to Mr. Süssmilch, Councilor of the Higher Consistory, who, after overcoming nearly insuperable obstacles, supplied us with so great a number of such observations that they seem adequate to settle most of the questions arising in this research.

Euler indeed relies on a very large corpus of observation data to determine their properties: the data from birth and death registers. He then seeks out the various properties of the corpus and identifies the basic ones, which will allow him to reconstruct everything that can be said about the data, whatever they are. However, he does not address these phenomena in all their complexity. He selects a small number of aspects, which he sees as essential for his approach: the age-specific probability of dying in a given year, the probability of giving birth to children in that same year, the independence between these two probabilities, and their constancy over time and for the location studied (city, province, country, and so on), which he adds to the three previous properties in order to extend his period results to the flow of time.

He can state that these four properties of a population allow him to reconstruct the set of characteristics of each population and test their validity on the many examples provided by Süssmilch.

We can consider the forces acting on a population as altering its mortality and fertility rates, as Euler defines them (Bourgeois-Pichat 1994). If the rates stay constant over time, then the population remains stationary or stable, depending on the rates' respective values. Equally well, if these formerly variable rates become identical at a given moment, the stationary or stable population is not reached immediately. As Lotka showed (1939), it will serve only as a limit.

We can draw a parallel between these results and the first axiom (or law) of Newton's theory (1687), known as the inertia principle:

> Every body perseveres in its state of rest, or of uniform motion in a right line, unless it is compelled to change that state by forces impress'd thereon.[15]

Similarly, any population whose fertility and mortality are assumed to become constant from a given instant on will tend toward the stationary or stable population that meets these conditions. But, whereas the physical body immediately acquires its uniform motion when no force acts upon it, the stable population is not affected

---

[15] Projectilia perseverant in motibus suis nisi quatenus a resistentia aeris retardantur & vi gravitatis impelluntur deorsum (English transl. by Andrew Motte, 1729 ed., p. 19).

at once, since it is a limit. However, if certain conditions are met, this population can also promptly reach its stable state.

Bourgeois-Pichat (1994) showed that, in reality, many observed populations—such as most developing-country populations at the time of his writing, in which fertility varied little—meet these conditions and can thus promptly reach a stable state while preserving an invariable age structure over time. Such populations are called semi-stable populations. This is less true today, however, with the downtrend in fertility in these countries.

It is also easy to flesh out this model of population change with age-specific emigration and immigration rates, expressed as a net emigration rate. This yields a basic relationship between age structure, mortality, fertility, and migration at a given point in time (Preston and Coale 1982).

From a cross-sectional population sciences study giving the probabilities of events in the period, we can thus deduce the future change in the population or social group considered, if conditions remain the same. However, the conditions will depend on the social, economic, religious, political, and other characteristics of the society or group in which the events occur. We must therefore identify any links between these characteristics and the phenomena studied.

5.3.2 Dependence of Characteristics on Society

In Chap. 4, we showed that the least-squares method, introduced by the probabilists Legendre, Gauss, and Laplace in the early nineteenth century to study planetary orbits, was long confined to the examination of astronomical or geodesic phenomena in the physical sciences. Its use in the social sciences for dealing with population characteristics required a long process during the nineteenth century, whose main stages are described below.

In astronomy and, more generally, the physical sciences, it was possible to reduce the complexity of events captured in experiments to an abstract model—for example, by incorporating a small number of facts into a regression equation. The biological and social sciences, by contrast, lacked the tool for choosing, among myriad possible causes of the phenomena studied, those whose impact could be distinguished while disregarding the rest.

We should also bear in mind that the social sciences observe phenomena whose variability is far greater and, more importantly, very different from those of physical phenomena, which could be dealt with by means of the least-squares method. The uncertainty of physical phenomena is solely due to the quality of measurement. For human phenomena, it is far more complex; above all, it is related to the diversity of the societies in which we observe them.

In Chap. 3, we examined some of Quetelet's studies on court convictions. Let us now examine in greater detail his work on subjects more related to population sciences. The first volume of his monograph on 'social physics' (1835) is an attempt to explain phenomena by what the author describes as natural causes: age, sex, and occupational differences between individuals, climate and place, affiliations with civil and religious institutions, and so on. He takes the following statement by Laplace as his epigraph (1814):

> Let us apply to the political and moral sciences the method based on observation and calculation, a method that has served us so well in the natural sciences.

In fact, however, Quetelet makes no explicit use of probability calculus or the least-squares method. He simply notes that:

> the larger the number of individuals observed, the more individual characteristics, whether physical or moral, fade away, leaving as the dominant factor the series of general facts by virtue of which society exists and is preserved.

While physical measures are genuine mathematical quantities, non-physical measures—such as 'the age at which the average man' in a given country 'passes away'—are based on time, which is 'amenable to as much precision as that which we use in physics.' To study demographic phenomena, Quetelet proposes the use of a large data set. He clearly states:

> We must, above all, lose sight of individual man, and view him as merely a fraction of the species. By stripping him of his individuality, we eliminate all that is only accidental; and the individual particularities that have little or no effect on the mass will vanish of their own accord, enabling us to grasp the general results.

He thus implicitly assumes the existence of laws to which humans are subject—laws that a frequentist statistical analysis should reveal, once we work on large populations.

Unfortunately, Quetelet does not always follow these principles. In discussing the influence of the sexes on the number of births by parents' age, he uses Sadler's data (1830) on very small populations without performing the significance tests suggested by Arbuthnott as early as 1710 and elaborated by Laplace (1778), which would have allowed Quetelet to demonstrate the inanity of the hypothesis. Likewise, he uses Hofacker's data (1829), which are completely biased as well.[16] While his method is fairly suitable for anthropometric data, it is less appropriate for population sciences measures.

Most important, Quetelet's method also requires a test of the homogeneity of the groups examined. His theory of the *average man* is valid only if the individuals have a normal distribution around a single value for the characteristic considered. But many human characteristics are not distributed normally. In addition, there are as many average men as there are ways to categorize various sub-populations. For instance, when examining mortality and birth rates in Belgium's 19 provinces (1827), he cannot help observing how widely they differ, without offering a means to aggregate them into larger regions that could be regarded as homogeneous.

The truth is that Quetelet lacked a sufficiently elaborate statistical theory to decide whether a particular human group could be viewed as consisting of units

---

[16] For more details, see the full description by Brian and Jaisson (2007).

homogeneous enough for a regression analysis to be applied to it. Quetelet never even uses regression analysis in this work. He merely provides simple tables showing the proportionality of effects usually taken in pairs.

The statistician Lexis tried to generalize Quetelet's methods, which are valid for anthropometric measures, by attempting to apply them to population sciences quantities from 1876 to 1880. In so doing, he showed more precisely the conditions that a series of statistical rates had to meet in order to be regarded as derived from the same probability distribution.

The topic had already been addressed and partially resolved by Dormoy (1874), but Lexis (1877, 1879) gave a fuller and more general presentation. Lastly, trying to use the least-squares method, he explained the reasons why this method of averages is of little use in population sciences. He concluded as follows (1880):

> It teaches us that there are very few statistical ratios that actually behave as mathematical probabilities, i.e., that vary roughly in the same manner as empirical expressions of a constant probability. By dividing the number of deaths between ages 0 and 1 by the corresponding number of births, one believes that one has obtained the probability of dying for this age group. But by calculating this probability for twenty successive annual cohorts, we find variations that greatly exceed the values consistent with the average annual number of births, under the hypothesis of a constant probability of dying.

Among all the ratios and rates considered, he finds that the ratio of male births to births of both sexes combined is the only one that can be viewed as a mathematical probability, although it was later shown to vary as well (Brian and Jaisson 2007). Clearly, the method proposed by Lexis is too strict for analyzing population sciences data.

The late nineteenth century saw the introduction of more efficient methods, largely thanks to British statisticians—in particular, Galton, Edgeworth, Pearson, and Yule—who also addressed topics in social sciences such as genetics and psychology. But let us first examine the approach of the French sociologist: Durkheim. His subjects of study—for example, mortality from suicide (1897)— were close to population sciences, and his sources were very different from those used by the British.

Durkheim (1895) advocated what he called the concomitant-variation method, the only valid one, in his view, for the social sciences among the five methods proposed by Mill (1843) for the physical sciences.[17] Durkheim wrote:

> The reason is that, for [*the method*] to be demonstrative, there is no need to strictly exclude all of the variations that differ from the ones being compared. The simple parallelism of the values taken by the two phenomena, provided that it has been established in a sufficient number of sufficiently varied cases, is the proof of a relationship between the two. The method owes this privilege to the fact that it arrives at the causal relationship not from the outside, as with the previous methods, but from the inside.

---

[17] In fact, Mill had declared that experimentation, even indirect, was inapplicable to the social sciences. Durkheim, however, showed that of the five methods proposed by Mill (*method of agreement, method of difference, joint method of agreement and difference, method of residue, method of concomitant variation*), the method of concomitant variation is, in fact, perfectly applicable to these sciences.

Actually, Durkheim and Mill, like the demographer Landry (1945), had not made the connection between the concomitant-variation method and the least-squares or linear-regression approach—to which it is, in fact, equivalent. Far more important, the fact that it is not necessary to exclude all the variations differing from those being compared has been mathematically demonstrated in the case where these characteristics are independent of one another (for a more general discussion of the effect of omitted variables on generalized linear models and non-linear regressions: Gail et al. 1984; Neuhaus and Jewell 1993), whereas Durkheim asserts it without truly demonstrating it.

By examining separately each of the various social, political, religious, and other characteristics capable of influencing a phenomenon, it thus became possible to dissect their effect on the phenomenon. Durkheim (1895) admits that statistics provides a means to isolate these social facts:

> They are, indeed, represented, not without exactitude, by the birth rate, the marriage rate, the suicide rate, i.e., by the number obtained by dividing the total average annual number of marriages, births, and self-inflicted deaths by the number of persons old enough to marry, procreate, and commit suicide.

He then uses the concomitant-variation method—i.e., ultimately, linear-regression methods—to show the effect of religion (Protestantism, Catholicism or Judaism) or family status (single, married, widow (er) or divorced) on suicide (1897).

As noted earlier, these regression methods were further elaborated to allow their use in biological and social sciences: Galton (1875, 1886a, 1888), Edgeworth (1885, 1893a, b, 1895), Pearson (1896), and Yule (1895, 1897, 1899) crafted the effective integration of probability into these sciences. Let us look at their core contributions.[18]

Galton introduced the notion of correlation between two variables—which he termed 'co-relation' between 'co-related' variables. Having isolated two variables, he published an initial article (1886a) in which he showed in mathematical terms, with the aid of Dickson, the probabilistic relationship between them. In another article (1888), he described how to measure their correlation. In this article, Galton noted:

> It is not necessary to extend the list of examples to show how to measure the degree in which one variable may be co-related with the combined effect of n other variables, whether these be themselves co-related or not.

In fact, however, his discussion was confined to the case of two normally distributed variables with a linear relationship between them. It was Edgeworth and Pearson who generalized these results to the case of n variables, also normally distributed.

As early as 1892, Edgeworth clearly stated the problem in its most general form:

> What is the *most probable* value of one deviation x, corresponding to assigned values x'_1, x'_2, & c. of the other variables? And What is the dispersion of values of x, about its mean (the other variables being assigned)?

---

[18] For a more detailed analysis of their contributions, see Stigler (1986) and Hald (2007).

5.3 The Cross-Sectional Approach

Note that he was still working on normal variables. However, he did not supply their general solution (1893a) in the form of a determinant until after writing four other articles on the subject, then showing how its application to the social sciences (1893b) was a logical consequence. Later, Pearson (Pearson 1896; Pearson and Filon 1898) provided estimates of the standard deviation of the variances and correlation coefficients between the variables, under the same hypotheses.

Yule sought to determine the normality condition of the variables. In 1895, he wrote:

> Though, as we have said, no great stress can be laid on the value of the correlation coefficient (the surfaces not being normal), its magnitude may at least be suggestive.

He spelled out his idea (1897) with examples from biology and population sciences:

> The only theory of correlation at present available for practical use is based on the normal law of frequency, but, unfortunately, this law is not valid in a great many cases which are both common and important. It does not hold good, to take examples from biology, for statistics on fertility in man, for measurements on flowers, or for weight measurements even on adults.

To emancipate himself from this normality condition, he turned to the relationship that exists between the variables, irrespective of their frequencies. For example, when there are two variables, we can construct a regression line for one relative to the other. When the regression line is close to a straight line, we can calculate its parameters by means of the least-squares method. Yule felt confident enough to assert:

> The exponential character of the surface appears to have nothing whatever to do with the result.

This may seem a somewhat extreme conclusion in our time, but in 1897 it freed the regression from all normality conditions. Above all, it allowed the biological and social sciences to use the least-squares method, which had been devised for the astronomical and physical sciences a century earlier.

These advances ended with Fisher (1922a, b), who developed the maximum-likelihood theory and a theory of statistical inference based on the objective approach to probability discussed in Chap. 2.

In sum, objective probability played a crucial role in the cross-sectional approach in population sciences and, more generally, in the social and biological sciences. The approach involves the observation of phenomena in a short period—typically 1 year—and of the characteristics of the population concerned at the start of the period, with the aid of a census conducted in a set of parts of the population's territory. This provides a firm foundation for the objective approach used to analyze the phenomena and the characteristics.[19]

---

[19] One can estimate the linear regression model with purely Bayesian methods (Lindley and Smith 1972), but this solution is of little value in demography, given the exhaustive observation of the population.

We end this section with a simple cross-sectional analysis of the effect of a characteristic on a demographic behavior. The exercise will serve for comparison purposes with the later sections of this chapter.

Our example concerns migrations between regions of Norway of men born in 1948 and observed over a 2-year period after the 1970 census.[20] We calculate the emigration rates for the 19 regions and try to see whether farmers are more or less likely to migrate than other occupations. A regression on the aggregate data shows a significant effect of farmer status on the probability of emigrating from each region: the estimated parameter is 0.597 for farmers and 0.119 for other occupations and men with no occupation, and the share of variance explained (0.24) is quite significantly different from zero at the 5% confidence limit. We can therefore conclude that 22-year-old Norwegian farmers are almost six times more likely to emigrate than other occupations, if the hypothesis that social facts exist independently of the individuals who experience them were met. This result may seem surprising given the financial and personal cost of changing regions for farmers. The only way to settle the issue is through a comparison with results obtained under other paradigms.

In sum, the statistical individual in the cross-sectional approach will actually be a group of individuals defined either by age or by one or more social, economic, family, or other characteristics specified in a census or cross-sectional survey. We observe their aggregate behavior in given units (such as regions or districts) in a short period. This is indeed an aggregate period approach.

5.3.3 Problems Posed by the Cross-Sectional Approach

We shall now discuss various problems that arise when using the cross-sectional approach.

The first problem, pointed out by Robinson (1950), generates what is known as the ecological fallacy. Robinson showed that correlations measured on individual characteristics generally differed from those measured on the same characteristics aggregated by region. This discrepancy may, in fact, apply to the example given at the end of the previous sub-section: the probability of migrating rises with the proportion of farmers in the region, but is not necessarily higher for farmers than for non-farmers. Galton (1886b) encountered the problem when showing that the relationship between the weight of pea seeds and their average diameter is linear: the relationship between weight and average diameter does not imply the same connection between the weight and diameter of an individual seed.

The second problem stems from the grouping of results of a cross-sectional analysis 'as if they were those of a fictitious cohort, experiencing the conditions of the year or period considered throughout its life' (Henry 1959). This use of composite

---

[20] For more details, see Courgeau (2004a, b, 2007a, b).

indices in cross-sectional analysis to summarize a series of age-specific rates can raise serious objections in certain cases. For instance, difficulties arise in the study of phenomena comprising periods of deferral followed by periods of recovery after an economic crisis or a war. As Henry explains (1966):

> in a recovery period, behavior is influenced by the earlier delay; assigning to a fictitious cohort a series of indices observed in a recovery period is thus tantamount to postulating the existence of a cohort that—from one end of its life to the other—would strive to close a gap that had never opened.

That explains why the sum of the age-specific probabilities of marrying (first marriages), which measure marriage intensity and should always be below or equal to unity in an actual cohort, can take on far higher values in a fictitious cohort. In France, just after World War II, the sum exceeded 1.5. Similarly, Whelpton (1946) showed that the sum of first births for a woman in the United States in 1947 was 1.38.

More generally, restricting the analysis to period events does not suffice to explain demographic behaviors, which need to be viewed in the context of individual lives.

A third problem is that the value we are trying to estimate with the least-squares method is, in fact, a probability. The value must therefore lie between 0 and 1. But there is nothing in regression models that forces the estimated coefficients to comply with these constraints. Moreover, when several characteristics are included, a strong correlation between some of them can lead to incorrect results.

Lastly, so long as we are working on a sufficiently large population, the objectivist approach is perfectly valid—but if the population diminishes, the approach can yield aberrant results. In Chap. 4, we observed this outcome in paleodemography, where objectivist methods led to incorrect age-specific probabilities of dying. In such cases, we must adopt an epistemic approach to obtain acceptable results.

Researchers have tried to resolve these difficulties by taking into account the time experienced by individuals in a given population, then combining population sciences with an epistemic approach to probability.

5.4 From a Longitudinal Vision to a Full-Fledged Event-History Approach

The approach using individual life experiences gained ground steadily after World War II. It consisted in taking a longitudinal point of view, then an event-history (i.e., biographical) point of view to study phenomena. The term 'longitudinal' refers to individual life courses aggregated into a generation or cohort, in which we study the evolution of each phenomenon, treated separately. The term 'event history' also refers to individual life courses, but the unfolding of phenomena is now studied in the aggregate.

This section tells the history of the approach—with its pitfalls and successes—through its connections with probability.

5.4.1 An Initially Longitudinal Approach

Because it worked on period data, cross-sectional analysis could not take account of individuals' past histories. It therefore had to aggregate its period results as if they applied to a fictitious cohort. As seen in the previous section, this fiction raised serious objections. To move beyond it, we must work on cohorts or generations actually observed.

Although some earlier voices suggested that period analysis might not be the best approach (Delaporte 1941), it is mainly after World War II, that population scientists set up this new form of longitudinal analysis, which introduces individuals' 'lived time' (Whelpton 1949; Ryder 1951, 1954). The main theoretician of the approach was Henry (1959, 1966).

The observation period is no longer a given point in time, but the life-span of a cohort that can be defined as 'the set of persons who entered into a given population category during a given period, such as the calendar year' (Henry 1959). When the period corresponds to the persons' birth, it is called a birth cohort (generation). The time spent by individuals in the category is reckoned from their entry: it is called their seniority (duration) in the group.

A member of the cohort may experience several events during the period examined. However, from the outset, the analysis will focus on a particular phenomenon that may include one or more studied events of the same type (a woman's first marriage or, instead, her successive births). These events can be affected by others, called disturbing events.

Longitudinal analysis will seek to determine the frequency of the phenomenon studied and its temporal distribution absent disturbing events. The goal is to obtain the phenomenon in a pure state, just as the chemist separates simple bodies whereas most substances do not exist in unalloyed form in nature (Henry 1972).

For this purpose, we must make two hypotheses about the phenomena involved, i.e., both the phenomena studied and the disturbing events.

First, we assume that each cohort member is characterized by a probability[21] of occurrence of the studied event(s) and disturbing events that is identical for all individuals of the population at each seniority. We therefore suppose that the cohorts are homogeneous with respect to each event.

But this condition is not sufficient to estimate the probability of the studied phenomenon (a), after eliminating the effect of disturbing events. We must additionally

---

[21] Let us note that we are speaking here about annual or multiannual probabilities. When later we will introduce instantaneous rates or hazard rates, we will no more use the term probability as these rates or functions are defined on continuous time and did no more depend on the length of the age interval. However they have the dimension of a time frequency, because of the time interval in their denominator (time^{-1}). They have no upper boundary and so may be greater than one, in contrast to probability: this is observed at extreme old ages for human mortality, when survival times are measured in months and these monthly rates are transformed to yearly rates by multiplying them by a factor of 12 (Gavrilova and Gavrilov 2001).

5.4 From a Longitudinal Vision to a Full-Fledged Event-History Approach

assume that the first individuals to experience the disturbing phenomena had the same response to the studied phenomenon as the individuals who did not experience it. In other words, we must assume independence between studied phenomena and disturbing phenomena.

Having posited these two hypotheses, and using vital statistics we can estimate a probability of occurrence for each seniority (Henry 1959; Pressat 1966) and a survivor function for the studied and disturbing phenomena.[22] Under the homogeneity and independence hypotheses, the values would be identical to those calculated for population not exposed to the disturbing phenomena. Their distribution over time gives the calendar of the studied phenomenon; their sum up to the age when the phenomenon can occur gives its intensity.

These estimates are based on objective probability and it is not even useful to estimate the variances of the probabilities or survivor functions, as the populations observed are generally very large. As noted in Sect. 1.4, the calculation of a variance of an annual probability of dying—assuming a binomial distribution of deaths (i.e., in a homogeneous population)—yielded values so low as to become meaningless.

The results of such an analysis, which observes an actual cohort, do not lend themselves to the objections voiced against the composite (synthetic) indices of the cross-sectional approach. For instance, the intensity of marriage (first marriages) or the sum of first births will never exceed unity.

For longitudinal analysis, we can also use data from successive censuses. If so, however, we must set an additional condition: the fact of having experienced the studied event(s) must not influence the probability of disturbing events occurring after them (Henry 1966). This condition, called a continuity hypothesis, must therefore be met for such data to yield satisfactory results.

In sum, longitudinal analysis entails a denial of all specificity of individual lives in order to focus on the occurrence of a type of event, independent of other phenomena, in a population that remains homogenous over time as it is composed of interchangeable units. While the notion of probability underlies the approach, it is secondary in this analysis, which concerns exhaustive populations and focuses on a small number of characteristics that may influence the rate values.

As in the cross-sectional approach, the statistical individual is a homogeneous group of individuals of the same age or having experienced a founding event at the same time—the probability of experiencing the studied event(s) being the same for all individuals. Unlike in the previous analysis, we observe these groups throughout their lives. By contrast, we no longer have the equivalent of regression methods to show the effect of various characteristics on the groups' behavior.

This form of analysis prevailed until the early 1980s, but raised a number of methodological problems, some of which had already been identified by Henry back in 1959.

The first problem is that, in practice, 'demographers very often proceed without concerning themselves with homogeneity' (Henry 1959), or, rather, that this

---

[22] For more details on this estimation, ser Henry (1959, 1972).

hypothesis is very seldom mentioned. Henry clearly addresses the issue by introducing heterogeneous cohorts, each composed of homogeneous cohorts of infinite size. His main goal, however, is not to measure the influence of observed heterogeneity on the studied phenomenon. His first priority is to look for formulas similar to the ones obtained by assuming a homogeneous cohort and, if not, to see how the formulas differ.

For this purpose, he calculates the survivor functions for the observed phenomenon and the disturbing phenomenon of the cohort formed by combining the sub-cohorts. He shows that the number of members who have not experienced the observed event and have not been affected by the disturbing event depends on the correlation coefficient between the survivor functions for the seniority considered and on the variation coefficients of the two survivor functions. These coefficients between groups can now be estimated, unlike the coefficients between individuals of the same group. Consequently, the formulas are only identical to those obtained for a homogeneous cohort when either (a) the correlation coefficient between survivor functions is zero, or (b) the variation coefficient of the observed phenomenon is zero, or (c) when the variation coefficient of the disturbing phenomenon is zero. Henry notes, however, that the error may be negligible, even in the presence of rather significant heterogeneity. For example, using past data on family histories, he shows that in the study of marital fertility by marriage age and duration, one can neglect age heterogeneity without adverse effects.

He then observes that differential demography ought to allow the study of 'differences between different categories (ethnic, religious, social, etc.),' but that, as practiced at the time, differential demography does not seem to him to be of much potential help. He merely states:

> Thus, to solve one of the basic problems of demographic analysis, differential demography ought to be renewed and extended. It should display greater concern for correlations between the differences that it observes; it should cease to confine itself to the study of groups defined by demographic or sociological criteria and should turn its attention to physical or psychological criteria. (Henry 1959)

He goes further in a footnote: 'Given the practical difficulties, the question is bound to arise as to whether the problem posed is soluble. From a certain standpoint, one can conclude that the answer is no.'

In sum, the solution that consists in choosing sub-cohorts treated separately in order to analyze, by means of differential demography, a phenomenon dependent on different characteristics would soon lead to working on groups so small as to preclude longitudinal analysis. The use of objective probability, without even estimating probability variance, is no longer justified either, in this instance where the importance of the variances becomes decisive.

The second problem is that of independence between probabilities. In the same article, Henry writes that 'ordinary observation and reflection lead us, moreover, to believe that in most cases there is no independence between hazards.' But he does not analyze more fully how this dependence between hazards can arise, or how it may influence the probabilities calculated. On the face of it, this hypothesis does not seem very likely when we consider events such as union formation and marriage,

5.4 From a Longitudinal Vision to a Full-Fledged Event-History Approach

entry into the labor market, and the move into one's first independent dwelling—events that must strongly influence one another. More generally, it becomes impossible to study exits by competing events, and

> for the same reason [...] one must give up the idea of conducting the study in a population to which several events allow entry. (Blayo 1995)

In many cases, therefore, the independence condition rules out any possibility of analysis. It is interesting to note that this paper, whose main purpose was to offer a critique of the event-history approach that we will present in the next section, points out the main problems in longitudinal analysis, to which event-history analysis precisely offered solutions with a new paradigm (Courgeau and Lelièvre 1996, 1997).

When longitudinal analysis attempts to treat more complex problems than the separate analysis of individual phenomena, it runs into serious difficulties. Unlike cross-sectional analysis, it offers no method precise enough to deal with the heterogeneity of populations—apart from a rough differential analysis that would require such detailed breakdowns of the population studied as to invalidate any serious calculation. It imposes such constraints on the studied events as to exclude an entire sector of population sciences analysis: analysis of competing or interacting events, analysis of events in a population with entries and exits, and so on.

We therefore need to alter the assumptions on which the analysis is based in order to put the reasoning process on a firmer footing.

5.4.2 An Event-History Approach

Instead of concentrating on the study of homogeneous sub-populations, the event-history approach will examine individual trajectories between any number of statuses (to take the example discussed in the cross-sectional approach: never married vs. married; working in agriculture vs. working in other sectors or being economically inactive). The unit of analysis will not be the isolated event, as in longitudinal analysis, but the individual event history, viewed as a complex stochastic process determined by personal characteristics. This change of perspective removes the requirements of homogeneity for the populations studied, and of independence between longitudinal-analysis phenomena. It introduces also a continuous time dimension.

By tracking an individual's life events over time, we see that the main way for him or her to escape observation will be to leave the sample at the survey date or—if we are using date from population registers—the study date. As there is no reason for these dates to be tied to the life of an individual, the independence condition is fully met here. The observation is called non-informative and there is a way to factor the exits into the instantaneous rates[23] estimations. On the other hand, there is

---

[23] See note 21 of this chapter for the difference between an instantaneous rate and a 1-year probability. We will see in the following pages how to give a more precise mathematical definition of this intuitive concept.

no longer any reason for the events experienced by an individual to be independent of one another. On the contrary, event-history analyses will largely concern themselves with the dependence between events.

We must now distinguish between interacting events and competing events. Longitudinal analysis did not do so, referring to them collectively as disturbing events.

An interacting event will change the rate of occurrence of the studied event. Let us take the example discussed earlier in connection with cross-sectional analysis: the probability of migrating for a Norwegian depending on whether he or she is a farmer or not. Using a simple analysis (Courgeau 2004a, b, 2007a, b), we can estimate the instantaneous rates at 0.095 for farmers, with a standard error of 0.007, and at 0.150 for non-farmers, with a standard error of 0.002. These values are therefore significantly different, and contradict the ones obtained on aggregate data. However, as noted earlier, the newer result is more consistent with expectations. In Sect. 5.5 below, we shall take a closer look at the significance of these divergences.

When we speak of competing events, we refer to different variants of an event that produce the same outcome: mortality by cause, union formation through marriage or cohabitation, and so on. These events, as well, are fully suited to event-history analysis (Courgeau and Lelièvre 1989, 1992).

Next, when attempting to understand individual behavior, we shall need to factor in the person's social origins and entire past history. Behaviors are not inborn, but change during a lifetime thanks to personal experiences and successive acquisitions of information. Thus event-history analysis effectively addresses the heterogeneity of populations, from a dynamic standpoint rather than a static one as in cross-sectional analysis. In Chap. 4, we saw that the event-history approach allows a regression not on period data, but on complex temporal processes.

Its paradigm can be stated as follows: individuals moves on complex trajectories in the course of their lives; these trajectories depend at any given instant on people's earlier trajectories and on information that they have acquired in the past (Courgeau and Lelièvre 1996, 1997).

The event-history approach began with the non-parametric estimation of the survivor function of an event (Kaplan and Meier 1958), and was then generalized into a semi-parametric approach (Cox 1972) involving individual characteristics. Since the early 1980s, it has proved to be particularly suitable for the analysis of social phenomena in demography (Menken and Trussel 1981; Courgeau 1982), as well as for applications in epidemiology, sociology, psychology, medicine, and other fields.

Unlike in Chap. 4, which examined the event-history approach from the standpoint of population sciences, we shall now try to show its closer links to probability theory. This will enable us to address new aspects of probability not yet described in the present work: martingale theory and counting processes.

While the article by Kaplan and Meier (1958) and the articles by Cox (1972, 1975) represented major advances toward event-history analysis, the probabilistic theory underlying the model had not yet been sufficiently elaborated. One could not offer very clear answers to questions such as: What are the asymptotic properties of

5.4 From a Longitudinal Vision to a Full-Fledged Event-History Approach

the Kaplan-Meier estimator? How and why does the Cox model yield robust estimates? Consequently, the partial likelihood proposed by Cox has been the subject of many articles in statistical journals aimed at justifying its interpretation and the conditions for its validity (Prentice 1978; Kalbfleisch and Prentice 1980).

Aalen (1975) was the first to show the close ties between these methods and martingale theory. Andersen and Gill (1982) incorporated them into counting-process theory. Let us see how this occurred.

In France, the term *martingale* appeared in the 1798 edition of the Dictionary of the Académie Française. Its classic definition was given in the 1832 edition: 'manner of gambling that consists in staking, in each round, twice the amount lost in the previous round.' This strategy, which yields a small identical gain throughout the game, has some drawbacks. It can lead players to exceed the amount that they can wager, and hence lose their entire fortune. The term was later incorporated into the vocabulary of probability when Ville (1939) used it to denote any strategy applied by a player, here in a mathematical sense (we referred to Ville in Chap. 1 in connection with the axiomatization of von Mises's notion of collective).

Ville begins by considering the player's capitalization process determined by the martingale strategy in a game of heads or tails. Let us write the capital as $X_n = m(x)$, where x represents the player's strategy, in the form of a sequence of n zeros or ones. The player defines a martingale by the condition:

$$m(x) = \frac{m(x0) + m(x1)}{2}, \tag{5.1}$$

valid for any finite string. Consequently (Bienvenu et al. 2009):

> Any function m satisfying (5.1) for every finite string x is a capital process arising from a strategy and from some initial capital, and uniquely determines that strategy and initial capital. Because of this one-to-one correspondence, and because capital processes play the most direct role in his theory, Ville transferred the name martingale from the strategies to the capital process.

Ville was thus able to show that the collectives proposed by von Mises and later by Wald did not contain all the zero-measure sets, which his martingale concept defined in their totality.

Doob (1940, 1953)—followed by Hunt (1966), Neveu (1972), and Meyer (1972)—essentially extended this definition to ever more complex stochastic processes and demonstrated the theory's basic results, which we shall merely outline here.

Let us consider what probabilists call a filtration on the probability space (Ω, B, P). It consists of a string $\{B_n : n = 0, 1, 2, \cdots\}$ of sub-spaces of additive measures such that for all n: $B_n \subset B_{n+1}$. A filtration, also called a history, represents our knowledge at successive instants of the process studied, which increases with time. In t_n the player's strategy brought his or her capital to X_n. In these conditions, a process $X = \{X_n, B_n, n = 0, 1, 2, \cdots\}$ is a martingale if for all n:

(i) $\{B_n : n = 0, 1, 2, \cdots\}$ is a filtration and X is adapted to (B_n);
(ii) for each value of n, X_n is integrable;
(iii) for each n, $E\{X_{n+1} \mid B_n\} = X_n$.

Through the filtration concept, this theory, based on Kolmogorov's axioms, introduces time, which was initially missing from the latter.

Meanwhile, Doob (1949) showed that the convergence of the Bayesian estimator, in logical probability,[24] is a consequence of martingale theory. However, this consequence is not observed in zero-measure sets. Space precludes a fuller account here of discussions over this problem in the more general case where the models or prior distributions are incorrectly specified (see Ghosal 1996 and Shalizi 2009).

After formalizing martingales, we shall show that the counting processes are based on the same notion. The processes were studied mathematically by Bremaud (1973) and, later, Aalen (1975). Regarding Aalen, we can even say that the theory was initiated to answer questions raised by demographers:

> Aalen was influenced by his master thesis supervisor Jan M. Hoem who emphasized the importance of continuous-time Markov chains as a tool in the analysis when several events may occur to each individual (e.g. first occurrence of an illness, and then maybe death; or the occurrence of several births for a woman). (Aalen et al. 2009)

A stochastic process of this kind, with values in N (the set of whole numbers), makes it possible to model a random number that changes over time. Let us show how to formalize the intensity process that characterizes it in the basic case of the study of a unique event.

The intensity process is the conditional probability that an event will occur in the interval $[t, t+dt)$, given all that has been observed previously, divided by the interval's duration. Let us write $\lambda(t)$:

$$\lambda(t) = \frac{1}{dt} P\big[dN(t) = 1 \big| past\big],$$

where $dN(t)$ represents the number of jumps (basically 0 or 1) in the time interval $[t, t+dt)$. We can rewrite the formula as:

$$E\big[dN(t) - \lambda(t) \big| past\big] = 0.$$

As $\lambda(t)$ is only a function of the past, we can remove the expression from the conditional mean. Let us now introduce the new process:

$$M(t) = N(t) - \int_{s=0}^{t} \lambda(s)\, ds.$$

We can accordingly rewrite the previous formula as:

$$E\big[dM(t) \big| past\big] = 0,$$

---

[24] As noted in Chap. 3, Kolmogorov's axioms are consistent with this logical probability (Jaynes 2003).

5.4 From a Longitudinal Vision to a Full-Fledged Event-History Approach

which is the definition of a martingale. In other words, the intuitive concept of an instantaneous hazard rate is the same as saying that the counting process minus the integrated intensity process is a martingale (Aalen et al. 2009).

We can thus easily generalize this result to the study of censored data, under certain conditions, or multiple events. Most counting processes can be shown to contain components that are martingales.

Now let us look at the models that, like regressions in the cross-sectional approach, will allow us to introduce population heterogeneity. For these models, we need to consider different counting processes for each individual. Let us suppose that for a given individual i we have at instant t a vector of variables $x_i(t)$ whose components can be time-dependent or time-independent. The intensity process $\lambda_i(t)$ of the counting process $N_i(t)$ can be written as:

$$\lambda_i(t) = Y_i(t) h\left[t \mid x_i(t) \right], \tag{5.2}$$

where $Y_i(t)$ is an indicator equal to 1 if individual i is at risk of the studied event just before instant t, and equal to 0 otherwise; $h\left[t \mid x_i(t) \right]$ is the instantaneous rate of individual i, conditional upon the values of the variables in t (Aalen et al. 2008). This process assumes that all individuals with the same characteristics $x_i(t)$ have the same instantaneous rate of experiencing the event. Similarly, the survivor function $S\left[t \mid x_i(t) \right]$ can be written:

$$S\left[t \mid x_i(t) \right] = \exp\left\{ -\int_{s=0}^{t} h\left[s \mid x_i(s) \right] ds \right\}. \tag{5.3}$$

To obtain a regression model, we must specify in greater detail how the instantaneous rate or survivor function depend on the characteristics considered. We can use a Cox semi-parametric model of the multiplicative type, or any other formulation, for example of the additive type. For fuller details on these options beyond our discussion in Chap. 4, see Andersen et al. (1993) and Aalen et al. (2008).

We now have a much clearer picture of the close ties between the event-history approach and probability theory, and even of its pioneering role in the development of certain probabilistic approaches such as the theory of counting processes. Outside of population sciences, the event-history approach has now been adopted by many social sciences such as economics, epidemiology, sociology, and bio-statistics.

How do we introduce the notion of statistical individual into the event-history approach? Here, we cannot view an individual trajectory as the outcome of a process specific to each person. As we observe only a single outcome (the individual's trajectory), the process is not identifiable. We must therefore adopt a collective point of view: all individuals are assumed to follow the same random process, whose parameters we can estimate from the observation of a sample of individuals with their own characteristics. At first glance, this may seem a very heroic assumption. However, it is important to realize that it is not a hypothesis about observed persons, but about the construction of a process underlying a set of trajectories. In this case, two observed individuals have no reason to follow the same process, whereas two statistical individuals do so automatically, for we see them as random-sampling units (being subject to identical selection conditions) that display identical characteristics.

Before discussing various problems posed by this approach, let us note its high compatibility with a Bayesian point of view. Earlier, we saw that martingale theory was used to show the convergence of the Bayesian estimator in logical probability (Doob 1949). Similarly, the analysis of life tables has long used a Bayesian approach (Fergusson 1973; Kalbfleisch 1978; Kalbfleisch and Prentice 1980). We shall not give a detailed description of the approach—as in Florens et al. (1999), Ibrahim et al. (2001) and Florens (2002)—but simply an outline. We shall focus more specifically on the non-parametric estimation model for a survivor function and the Cox semi-parametric model, which are largely used in population sciences.

Let us begin by examining how to estimate a non-parametric survivor function. To do this, we must introduce the notion of Dirichlet process (Fergusson 1973):

> Let $\mathscr{X}$ be a space and $\mathscr{A}$ a σ-field of subsets and α be a finite non-null measure on $(\mathscr{X}, \mathscr{A})$. Then a stochastic process P indexed by elements A of $\mathscr{A}$, is said to be a Dirichlet process on $(\mathscr{X}, \mathscr{A})$ with parameter α if for any measurable partition $(A_1, \cdots, A_k)$ of $\mathscr{X}$, the random vector $(P(A_1), \cdots, P(A_n))$ has a Dirichlet distribution[25] with parameters $(\alpha(A_1), \cdots, \alpha(A_n))$.

We can see that such a process can be defined as a probability about a probability. Fergusson (1973) shows what happens when we take a prior Dirichlet process:

> If P is a Dirichlet process on $(\mathscr{X}, \mathscr{A})$ with parameter α, and if $X_1, \cdots, X_n$ is a sample from P, then the posterior distribution of P given $X_1, \cdots, X_n$ is also a Dirichlet process on $(\mathscr{X}, \mathscr{A})$ with parameter $\alpha + \sum_1^n \delta_{x_i}$, where δ_{x_i} denotes the measure giving mass one to the point x.

These results apply to the Bayesian estimation of a survivor function with right-censored data (Susarla and van Rysin 1976).

When we introduce individual characteristics into a semi-parametric proportional-risk model, the use of a prior Dirichlet process is more delicate, and requires complex numerical processing, particularly when dealing with equal durations (Florens et al. 1999). For cumulative rates, we may therefore prefer a Gamma process (Kalbfleisch 1978), which is fairly similar to a Dirichlet process. A brief description follows.

The Gamma process is a special case of a Levy process (stochastic process with independent increments, right-continuous and left-censored) with a Gamma density.

---

[25] The random vector $X = (X_1, \ldots, X_k)$ follows a Dirichlet distribution of parameter $a = (a_1, \ldots, a_k)$ if it admits the density $d(x) = \dfrac{\Gamma(a)}{\prod_{i=1}^{k} \Gamma(a_i)} \prod_{i=1}^{k} x_i^{a_i - 1}$, on the simplex D defined by $x = (x_1, \ldots, x_k) \in D \Leftrightarrow x_i > 0$ for all $i = 1, \ldots, k$ and $\sum_{i=1}^{k} x_i = 1$, Γ being the Euler's gamma function. For the properties of this distribution, see, for example, Robert (2006).

5.4 From a Longitudinal Vision to a Full-Fledged Event-History Approach

Let $G(x \mid \alpha, \lambda)$ be the Gamma distribution[26] with the parameter of shape $\alpha > 0$ and scale $\lambda > 0$. Let $\alpha(t), t \geq 0$ now be a continuous left-continuous increasing function such that $\alpha(0) = 0$, and let $H(t), t \geq 0$ be a stochastic process with the following properties:

(i) $H(0) = 0$;
(ii) $H(t)$ displays independent steps in disjoint intervals;
(iii) for $t > s$, $H(t) - H(s) \approx G\big(c\big(\alpha(t) - \alpha(s)\big), c\big)$.

The $\{H(t) : t \geq 0\}$ process is called a Gamma process, where c is a weight or a confidence parameter about the mean (Ibrahim et al. 2001). For event-history analysis, Kalbfleisch (1978) considers a semi-parametric model whose survivor function may be written as:

$$P\big(T_i \geq t_i \mid z_i, H\big) = S\big(t_i \mid z_i, H\big) = \exp\big[H(t_i)\exp(z_i\beta)\big], \qquad (5.4)$$

where the random variable T_i is the duration in which the studied event occurred for individual i with the characteristics vector z_i, $H(t_i)$ being the corresponding cumulative probability. The latter is conditional upon the stochastic process H, which we can accordingly define as a Gamma process. If we start with a prior function $H^*(t)$, associated with the weight c, we can write the process examined as:

$$H(t) \approx G\big(H^*(t), c\big).$$

We can then estimate the posterior distribution of $H(t)$ and the values of the β parameters. If the estimated values of these parameters remain stable when c varies from zero to infinity, we can conclude that the model's hypotheses are effectively met.

These results have, of course, been the subject of many generalizations, which we shall not discuss at greater length here: see Ibrahim et al. (2001) for more details on Bayesian event-history models.

5.4.3 Problems Posed by This Approach

The first important point is to determine whether the exclusion of all the characteristics influencing the studied events will affect the results of the analysis. In Sect. 5.3.2 on the cross-sectional approach, we noted that the results of linear regressions were not influenced by variables omitted from an analysis (unobserved

---

[26] The random vector $X = (X_1, \cdots, X_n)$ obeys a Gamma distribution of parameters α and λ, if it admits density
$$G(x \mid \alpha, \lambda) = \frac{\lambda^\alpha x^{\alpha-1}\exp(-\lambda x)}{\Gamma[\alpha]} \propto x^{\alpha-1}\exp(-\lambda x).$$

heterogeneity), when they are uncorrelated with those introduced in the model. Do event-history models display the same property?

Bretagnolle and Huber-Carol (1988) have shown that the property was not met by a Cox semi-parametric model with censoring. In fact, the omission does not affect the signs of the estimated parameters, but it does reduce their absolute values. Thus, if the effect of a characteristic seemed significant when others were left out, their introduction into the model will merely strengthen the effect of the first characteristic. By contrast, some characteristics that appeared to have no significant effect may become quite significant when the initially unobserved characteristics are introduced. This is an important finding that could be usefully compared with results from other types of models.

The following section on hierarchies examines in greater detail the solution recommended by certain authors (Vaupel et al. 1979; Heckman and Singer 1984a, b; Manton et al. 1992): modeling unobserved heterogeneity as a latent distribution, often called frailty.

Another problem is that individual characteristics are now used to explain the behavior of individuals themselves, whereas in cross-sectional analysis the aggregate characteristics explain behaviors that are themselves aggregated according to the same division of the country into provinces, regions, and so on. We showed the risk of ecological fallacy in the cross-sectional approach, but now another risk emerges, usually called the atomistic fallacy. By concentrating on individual characteristics, we disregard the context in which human behaviors occur. This context can be defined in very many ways. It may consist of the individual's family environment, or, more generally, a contact circle around the individual: neighborhood, town, network of family relations or a larger network including friends, and so on. Context will clearly influence individual behavior, and it seems misleading to isolate individuals from the constraints imposed by these networks or their living environment.

Sociologists have long recognized this risk (Lazarsfeld and Menzel 1961). They have shown the need for a precise definition of the various kinds of groups, communities, organizations, and so on. A group is composed of members possessing common features that make it possible to differentiate the group from others. More generally, what may be treated as a group in one study may be viewed as a member of a broader aggregate in another study. This property is very important for it shows the relativity of the individual, whom the event-history approach regards as the prime unit.

But the formalization of this problem—notably from the probabilistic angle of interest to us here—is complex and required the development of a new approach, which we shall now examine.

5.5 From a Latent Hierarchical Vision to a Multilevel Approach

In probability theory, the hierarchical vision dates from the early twentieth century and was linked to psychological research on intelligence. It was later used in decision-making theory (Good 1952), first mainly in economics, then in sociology, psychology, and political science, but only more recently in demography. It also constituted a major topic of discussion among proponents of epistemic probability (both subjective and logical), who were still in a minority.

It is worth describing here, for it was generalized in the multilevel approach in the mid-1980s. As discussed below, hierarchies represented a special case for multilevel analysis. The approach was introduced and used by the social sciences, which had to deal with different aggregation levels: education science, for example, needed to distinguish between student level, class level, school level, and so on. It then spread rapidly to many other social sciences, including population sciences, with multilevel event-history analysis.

We shall attempt to relate this development to probability theory and the various social sciences.

5.5.1 An Initially Hierarchical and Latent Vision

The hierarchical vision emerged when the psychologist Spearman (1904) introduced factor analysis. With his unifactorial model, he sought a latent level above the quantitative characteristics that he was measuring: this level made it possible to summarize observations, although it was not itself observed. More generally, this research rested on the following notion:

> If a latent variable underlies a number of observed variables, then conditionalizing on that latent variable will render the observed variables statistically independent. (Borsboom et al. 2003)

The goal became to seek a more general set of latent variables. This was done by Garnett (1919) and Thurstone (1927, 1938). The latter developed a multifactor model, in which the factors were always situated at the same level and summed up quantitative characteristics.

In the 1940s–1960s, the sociologist Lazarsfeld introduced the latent-class theory (Lazarsfeld and Henry 1968), now using qualitative variables, most often binary. The theory postulates the existence of a latent variable, also qualitative, displaying a certain number of categories. More generally, the factor analysis of correspondences designed by Benzécri et al. (1973) makes it possible to determine all the dependencies between rows and columns in a cross-table. One can also add supplementary data, not used in computing the inertia of the table(s) but projected on the axes. However, these hierarchies were simple constructs comprising two levels: that of the observed variables and that of the factors extracted by the analysis. In principle, the two factors were orthogonal to each other, i.e., independent of each other.

Factor methods, initially developed in psychology, came into widespread use in most social sciences in the 1970s.

In demography, Le Bras (1971) determined total fertility rates from the 1954 census for seven age groups and for each French *département*, a principal plane representing 97.93% of the inertia of the cloud of points. Noting that the main axes have only a physical significance, he sought the classic demographic indices that would give a clearer vision of fertility. He ended up with the gross reproduction rate and the mean age of mothers at childbearing, which cross-sectional analysis had identified long ago. In his own words: 'the identification is hardly surprising for a demographer.' Its analysis, however, does make it possible to show the dominant role of the two dimensions. Next, the change that Le Bras observed in the standard deviation of the mean age of mothers, which decreased steadily from 1921 to 1962, prompted him to state: 'one could predict that by 1993 any difference due to mean ages will have disappeared.' Unfortunately this prediction did not come true: by 1975, the standard deviation of the mean age of mothers was growing (Courgeau and Pumain 1993), particularly owing to the increase in the mean age, which differed from one *département* to another. While factor analysis makes it possible to summarize past changes, it proves too rudimentary to predict future changes in this case.

Meanwhile, Reichenbach—in his 1935 doctoral thesis in German, translated into English in 1949—introduced a series of nested levels in his objectivist theory of probability. He clearly states:

> We find instances in which we do not know for certain which probability exists in a given sequence. We speak, therefore, of *probabilities of the second level*; they are employed in probability statements concerning the existence of a probability. The iterations may be further continued: it is possible to make a probability statement about the existence of a probability of the second level, so that a probability of third level results and so on. The operations that are carried out with probabilities of a higher level constitute the *theory of probability of a higher level* or the *theory of the hierarchy of probabilities*. (Reichenbach 1937)

Reichenbach developed this theory to answer critics who claimed that objective probability made unfalsifiable predictions. By introducing hierarchy, he believed he would make the theory falsifiable. But unfortunately, even if a limit exists, the speed at which it is reached is unknown. Salmon, after attempting to defend his teacher's thesis, conceded defeat:

> Reichenbach's attempt to vindicate his rule of induction cannot be considered successful. […] My attempt to vindicate Reichenbach's rule of induction cannot be considered successful. (Salmon 1991)

However, the introduction of several hierarchical levels, which he recommended, allowed progress in this area.

The factor analysis described earlier sought to discover the underlying structure of a large set of variables, in the absence of prior hypotheses available to the researcher: this is known as exploratory factor analysis (EFA). The factors extracted are independent of one another, as noted previously. But when a pre-established theory is available, the task is to verify whether the structure developed beforehand is compatible with the observed variables—an approach called confirmatory factor analysis (CFA). In this case, the theory may comprise several hierarchic levels. The factors extracted from observed characteristics, for example, will no longer be independent of one another, and it will be possible to conduct a new factor analysis of these factors leading to a second level (Thurstone 1947; Schmid and Leiman 1957).

For instance, again in the field of psychology, Vernon (1950) sought to generalize the models described earlier by introducing Spearman's general factor, g, as a second-order factor, followed by less general primary factors. In this case, the tests that are supposed to measure the intelligence quotient (IQ) seek to approximate the value of g as closely as possible. Horn and Catell (1966) introduced three secondary factors—fluid intelligence, crystallized intelligence, and visualization intelligence[27]—without a more general third-order factor to correlate these three second-order factors. Statistical procedures, such as the analysis of linear structural relations (Jöreskog and Thillo's LISREL program (1972)), allow a selection of the model with the best data fit.

However, a number of later models failed to elicit consensus. Gustafsson (1984), for instance, showed that when a third level corresponding to g is introduced in the Horn-Catell model, fluid intelligence displays a unity correlation with g, demonstrating their identical roles. But Colom et al. (2004) found that a different factor—working memory[28]—can be viewed as identical to g. These contradictory results can be shown to be due to the ambiguity of the models at several hierarchical levels (Gignac 2007, 2008).

Gould (1981) brought heavy arguments to bear against the measurement of IQ and Spearman's g factor, in which he saw 'our tendency to convert abstract concepts into entities.'

The social sciences, especially demography, seldom used this hierarchical approach when they were working on objective probability. The situation changed with the adoption of a subjective or logical epistemic approach, as we shall see later.

A third class of models consists in treating an observed empirical distribution as a combination of unknown theoretical distributions, which one will then attempt to estimate.

In Sect. 5.3.2 on the cross-sectional approach, we noted Pearson's contribution to the development of regression models. We shall mention him here as the initiator of the method for estimating the components of a combination of two normal distributions (Pearson 1894). Pearson observed many cases where the frequency curves for the sizes of various organs of a given species were normally distributed. When encountering an asymmetrical curve in such cases, he speculated that selection may recently have caused the species to split into two or more new species. This is precisely what he observed in the values of the ratio of the forehead to body length among crabs from Naples. Choosing a method that uses the moments of the distribution, and with the aid of a ninth-degree equation, Pearson managed to show that

---

[27] Fluid intelligence is the ability to find new solutions to new problems, independently of earlier knowledge; crystallized intelligence is the ability to apply past experience to a present situation; visualization intelligence is the capacity for mental handling of two- or three- dimensional figures.

[28] This factor is the ability to keep active in the mind the information needed to perform complex tasks such as reasoning, understanding a concept or learning a concept.

the population can indeed be regarded as a combination of two normally distributed sub-populations, for which he estimated the parameters.

Because of the massive calculations required at a time when computers were not yet available, the method was barely used in the following decades. Only in 1965 did Wolfe offer a program for estimating the maximum likelihood of mixed distributions, reviving research on the issue. Today, many monographs have generalized the method and applied it to a wide range of social sciences (Lindsay 1995; McLachlan and Peel 2000). Let us elaborate on its application to the event-history approach, which—as the reader will recall—is used in demography, epidemiology, economics, sociology, and other disciplines.

In Sect. 5.4.3 above, we pointed out the problem of unobserved heterogeneity. This section takes a closer look at a solution considered as a combination of distributions that differ among individuals.

In contrast to the applications that we have just discussed, these models introduce a temporal aspect of the phenomena studied.

As seen earlier, all individuals with the same characteristics $x_i(t)$ in t obeyed the same intensity process $\lambda_i(t)$. It may be useful to ease this condition by introducing an unobserved heterogeneity to express the variations among individuals with the same characteristics. Analysts speak here of frailty, denoting that some individuals are more likely than others to experience the studied event. This entails a distinction between an analysis at individual level and an analysis at population level, the population being regarded here as a mix of individuals obeying different intensity processes. It is important to see how this can be expressed more clearly in terms of event-history analysis.

Let us assume the frailty of an individual measured by a number z ranging, for example, from zero to m, with a distribution $g(z)$ at the start of the observation. For an individual with frailty z, we can write the survivor function as $S(t,z)$ and the instantaneous rate as $h(t,z)$. One should keep in mind that the distribution of population frailty will vary over time, precisely because of the different survivor functions. It can be shown (Vaupel and Yashin 1985; Aalen et al. 2008) that the survivor function for the total population in t is equal to:

$$\overline{S}(t) = \int_{z=0}^{m} S(t,z) g(z) \, dz.$$

In other words, it is measured by the mathematical mean of individual survivor functions—a mean calculated from the variable's initial distribution. Similarly, we show that the instantaneous rate in t, estimated for the total population, is the mathematical mean of individual probabilities, conditional upon the fact that the calculation is confined to individuals still present in t.

The problem is that we know nothing about the distribution $g(z)$, since it represents an unobserved situation. Admittedly, we can use a random distribution to (1) observe whether its introduction will alter the effects of the observed characteristics, in a manner not significantly dependent on the distribution, and (2) compare the likelihood of a model without and with heterogeneity. But this does not suffice,

5.5 From a Latent Hierarchical Vision to a Multilevel Approach

as shown by the instability problems discussed by Heckman and Singer (1982, 1984a) and Manton et al. (1992). These authors observe that the effects of the observed variables were strongly influenced by the presumed distribution of the unobserved variable. These results were more than confirmed by Trussell and Richards (1985), who showed that the choice of distribution could even change the sign of certain parameters. Given the instability of these results, we can question the usefulness of introducing a mixture model in event-history analysis.

In fact, we can show that, to analyze non-repetitive phenomena, there is only one model estimatable without unobserved heterogeneity, but when we try to introduce an unobserved heterogeneity, there is an infinity of models that fit the data identically with different estimated probabilities (Trussell and Rodriguez 1990; Trussell 1992). In this case, we find that the choice of a distribution to represent unobserved heterogeneity, with no biological or other information on its form, is of little use and can actually be harmful.

However, if we analyze repetitive events, we can better distinguish each individual's contribution even if we have no information on his or her biological characteristics. We can apply another type of model comprising an event level nested in an individual level, as discussed in the next section (Lillard 1993).

Let us now turn to the use of hierarchies in models based on either subjective or logical epistemic probability.

This approach is derived from the work of Good (1952, 1980, 1983), who was among the first to introduce hierarchical epistemic models:

> Once we have decided to objectify a rational degree of belief into a credibility it begins to make sense to talk about a degree of belief concerning the numerical value of a credibility. It is possible to use probability type-chains (to coin a phrase) with more than two links, such as a degree of belief equal to ½ that the credibility of H is ⅓ where H is statistical hypothesis such that $P(E|H) = ¼$. It is tempting to talk about reasonable degrees of belief of higher and higher types, but it is convenient to think of all these degrees of belief as being of the same kind [...] by introducing propositions of different kinds. (Good 1952).

Note that Good defines probability theory as the logic of the degrees of subjective or objective belief, which we can objectivate in the form of credibilities. The simplest approach will consider the existence of hyperparameters and use given prior distributions to estimate the parameters of the resulting Bayesian hierarchical model (Bhattacharya et al. 1992; Younes et al. 2007; Kumar 2010). More generally, however, a hierarchical epistemic approach will model the prior information by breaking it down into several conditional prior distribution levels. Later, we shall see its application to multilevel models when we have specific information on the higher levels, such as the class in which a student is enrolled. Here, we take the case in which we have only sketchy information on the form of the prior distribution, which may itself depend on a prior distribution of its parameters, and so on. In other words, we assume that the observed population can be decomposed into a usually unknown number of latent heterogeneous sub-populations, for which no specific prior information is available.

Mixture models—elaborated most fully in hierarchical Bayesian analysis—prove very useful for addressing this type of problem (Titterington et al. 1985;

West et al. 1994). Richardson and Green (1997) propose a prior distribution composed of normal mixtures. However, while such a distribution may often prove entirely suitable, it may turn out to be dangerous and misleading in other instances, as Bernardo points out in the discussion on Richardson and Green's paper.

There are many applications of mixture models to the event-history approach (Hanson and Johnson 2002; Hanson 2006; Karlis and Patilea 2007). Let us examine the use of artificial neural networks (Ripley 1994), which may be viewed as a special form of normal mixture models.

This approach was inspired by the functioning of biological neurons (McCulloch and Pitts 1943) and has since been applied in many other fields, notably probability. It allows the introduction of the ultimately non-additive effects of the various characteristics into models that were initially of the proportional-risk type.

How do such models operate? Let us take the conventional proportional-risk model, in which, on the basis of equation [4] in Sect. 5.4.2, the log of the instantaneous rate for individual i is written:

$$\log h(t_i | z_i) = \log h_0(t_i) + z_i \beta = \log h_0(t_i) + \sum_{j=1}^{p} z_{ij} \beta_j. \quad (5.5)$$

The observed characteristics, z_{ij}, are assumed to be binary for simplicity's sake. We now introduce weights, w_{kj}, where $k \neq j$, which will produce a linear transformation of the z_{ij} values. This function is called a transfer function, $\varphi(x)$, often regarded as logistical. We also make the β_j parameters time-dependent: this will be indicated by the notation $\beta_j(t)$ (Ripley 1998; Ripley and Ripley 1998). The new model is written:

$$\log h(t_i | z_i) = \log h_0(t_i) + \sum_{j=1}^{p} z_{ij} \varphi\left(\sum_{k \neq j} w_{kj} z_{ik}\right) \beta_j(t). \quad (5.6)$$

We choose the transfer function φ such that $\varphi(0) = 1$. Consequently, when it is regarded as logistical, it will take the form:

$$\varphi(x) = \frac{2}{1 + \exp(-x)}.$$

We can thus interpret the parameter $\beta_j(t)$ as the additive effect of the jth explanatory variable, when all the others are null. The reason is that, if $z_k = 0$ for all $k \neq j$, then the model reduces to:

$$\log h(t | z) = \log h_0(t) + \sum_{j=1}^{p} z_j \beta_j(t),$$

which enables us to return to the initial model (5.5), with $\beta_j(t)$ instead of β_j, in the case where all weights are zero.

In this model, therefore, the network input comprises an individual's characteristics, and the output is the log of the instantaneous rate. The input-output link will depend on the weights w_{ij} and a transfer function φ. This neural arrangement is commonly called a feed-forward network. In essence, it consists in turning the initial model (5.5) into a non-proportional-risk model with non-additive effects of the characteristics (5.6), while preserving the possibility of interpreting its parameters.

It would be outside the scope of this book to examine in detail the various stages in the estimation of a hierarchical epistemic model using the neural-network method. In specific cases, this process enables us to estimate the parameters introduced with the aid of prior distributions located at several levels. These distributions will, for example, promote weak deviations from the additivity and proportionality of effects, which the proportional-risk model provided: for more details, see Gustafson (1998) and Ibrahim et al. (2001).

However, neural networks—and, more generally, the mixture models described above—supply a 'black box' that allows us to predict a given distribution accurately but is incapable of offering a genuine explanation of the phenomenon studied (Ripley and Ripley 1998).

We must now move on to truly multilevel models, where levels are not necessarily ranked in a hierarchy, and where their meaning is more explicit.

5.5.2 A Contextual Then Fully Multilevel Approach

Statisticians generally do not regard these approaches as different from hierarchical ones. However, in population sciences, and even more broadly in the social sciences, they reflect a very clear paradigm that incorporates into the analysis other levels besides the individual level that was the focus of the event-history approach. The analytical process no longer introduces latent hierarchical levels whose links to an explanation in practical terms may leave something to be desired. Now, the analysis operates on concrete aggregation levels whose existence and meaning are clear—levels that can arguably influence individual behaviour.

These levels correspond to different types of groupings of individuals found in all human societies: social groupings, such as the nuclear or extended family, the network of contacts, etc.; economic groupings, such as the firm or organization where a person works; education-related groupings, such as the classes in a school, the school itself or the university; healthcare-related groupings, such as a hospital ward or clinic; groupings of events experienced by the same individual (successive births or migrations during a lifetime); and so on. Other geographic or administrative groupings may be used, such as a building, a town, a county, a region, or a country (for an international study). Their effect may be less direct, but they may serve as viable proxies for other groupings that would be more relevant but are inaccessible to the researcher.

The levels are no longer necessarily nested in a hierarchy, although such an arrangement remains important. There will be cross-classifications that will make it possible to introduce non-hierarchic levels. For example, people may be involved in their family and workplace circles without our being able to establish a hierarchy, as some members of the family circle may also form part of the workplace circle. There may also be both hierarchical classifications and cross-classifications. For instance, the family/work cross-classification may be combined with a hierarchical classification of places of residence by town, county, region, and so on.

Lastly, in an event-history analysis, where several events may occur in the life of the same person, we can introduce a cross-classification in which the first level is no longer individual but records the events experienced by the same person; a second level will be individual, while other cross levels or hierarchical levels may be added. The problem here is that events may occur at different levels. Consider, for example, a woman's successive childbirths: when she gives birth to a new child, the event is recorded at level 1, but when she migrates from her region of residence, or changes households after a divorce, these events are recorded at higher levels.

Let us begin with contextual analysis, which uses aggregate characteristics to handle these different levels. It is the simplest solution to the problem posed by the atomistic fallacy, which can arise in event-history analyses of individual data. Contextual analysis also eliminates the risk of ecological fallacy, which we may encounter in an analysis confined to the aggregate level. The aggregate characteristic will measure a construct that differs from its equivalent at individual level. Here, the aggregate characteristic serves not as a substitute, but as a characteristic of the sub-population that will affect the behaviour of a member of that sub-population.

Consider the example of the migration of Norwegian farmers, used throughout this chapter. We now introduce into the analysis, simultaneously, their farmer status, the percentage of farmers in the region where they live, and the interaction between these two characteristics (Courgeau 2004a, b, 2007a, b). We observe that, when the percentage of farmers rises, the probability of migrating remains identical and consistently below non-farmers' probability of migrating. By contrast, non-farmers' probability of migrating will rise sharply when the percentage of farmers in the region increases. Thanks to this finding, we can combine the results of the analyses at the aggregate and individual levels, by clarifying the apparent paradox between the two analyses. A plausible explanation is that the relative lack of non-farm jobs will make non-farmers more likely to emigrate than farmers, all the more so if the percentage of farmers is rising.

However, the use of these contextual models imposes restrictive conditions on the formulation of relative risks as a function of characteristics. The models notably assume that individuals belonging to a given group behave independently of one another. The more likely assumption is that the risk incurred by the member of a group depends on the risks encountered by other members of the same group. Ignoring this intra-group dependence may yield biased estimates of the variances of contextual effects and so produce overly narrow confidence intervals.

Multilevel models offer a satisfactory means of addressing these issues. Proposed by Harvey Goldstein (1986, 1987, 1991), they were first used in the education sciences then generalized to the other social sciences (Goldstein 2003; Courgeau 2004a, b, 2007a, b). They comprise not only linear models but also non-linear ones,

5.5 From a Latent Hierarchical Vision to a Multilevel Approach

in particular the multilevel models for event-history analysis, which complement the approach described earlier.

Let us briefly examine the proportional-risk event-history model, which comprises a single aggregation level—for example, the region—in addition to the individual level. Using Eq. 5.4, we can write the instantaneous rate for individual i located at aggregation level j and exhibiting the characteristic k as:

$$h\left(t_{ij} \mid z_{kij}\right) = h_0\left(t_i\right) \exp\left(\sum_{k=1}^{p} \beta_k \, z_{kij} + \sum_{k=1}^{p} u_{kj} \, z_{kji} + u_{0j}\right), \quad (5.7)$$

where β_k is the region-independent parameter for characteristic k, u_{kj} is the random parameter for the same characteristic but region-dependent, and u_{0j} is the random parameter that makes the underlying probability dependent on region j as well. This model (5.7) is, of course, rather simple, and we can introduce more complex dependences between the underlying probability and the region or between the explanatory characteristics and the region. We can also include a greater number of levels. As we can see, this type of model introduces no new concepts in demography, but it generalizes the models used if we want to examine different aggregation levels.

What is the status of the multilevel model with respect to (1) the contextual mode, which imposes many constraints, and (2) constraint-free models estimated for each group observed? We return to the previous example, in which contextual analysis used the percentages of farmers living in each region (Courgeau 2004a, b, 2007a, b). It will be useful to begin with a separate analysis of each region to determine the local probability of migration by farmers and non-farmers. Because of the small number of people observed in certain regions, the difference between these probabilities is far from diverging consistently and significantly from zero. The results however, show a higher dispersion of regional results than what we found by applying a contextual model. While not yielding a clear-cut conclusion, the analysis shows us that the contextual model cannot explain all the variations in regional results. We should thus seek a compromise between a model that imposes no constraints but allows few if any significant estimates, and a contextual model with excessive constraints whose validity is almost impossible to test.

The multilevel solution consists in introducing random effects at regional level in addition to individual variance, so as to generalize the logistic regression methods considered here. We continue with the example of Norwegian farmer and non-farmer migrations.

The simple multilevel model supplies fixed parameters close to those obtained with the logit model estimated in Sect. 5.4.2: the probabilities of migrating are 0.092 for farmers and 0.153 for non-farmers, versus 0.095 and 0.150 respectively. By contrast, their standard deviation is greater: 0.011 versus 0.007 for farmers; 0.009 versus 0.002 for non-farmers. This is because we longer assume that all members of the population are equally likely to experience the event. Instead, we now suppose that the probability can vary with the region. The random parameters at regional level point to the same conclusion, as their variance for non-farmers differs significantly from zero.

The contextual multilevel model, which now uses both aggregate and individual characteristics, also yields parameter estimates consistent with those of the contextual model described above. Most strikingly, the variance in the random parameter for non-farmers is nearly one-half of the value found with the simple multilevel model. This decrease is due to the introduction of aggregate characteristics. However, it does not make the parameter differ significantly from zero.

In sum, under the paradigm of the multilevel approach, individual behaviour is always determined by the person's past history, seen in all its complexity. But the paradigm also states that behaviour can also depend on external constraints, whether or not the person is aware of them. Our society is composed of many social, economic, political, religious, educational, and other groups, and any given person is involved in a number of these groups that can shape his or her actions throughout his or her life.

The resulting statistical individual is even more complex than the protagonist of the event-history approach, for (s)he will be involved in different levels, both hierarchical and cross-level. However, the two statistical individuals will remain identical in nature.

5.5.3 *Problems Posed by the Multilevel Approach*

While the multilevel approach offers solutions to certain problems, such as those posed by the ecological and atomistic fallacies, a significant number of issues remain for which it provides only partial answers.

The first issue is the relevance of some of the aggregation levels customarily used. While some levels, such as the household, seem thoroughly relevant, others may be far less so. Certain levels often reflect geographic or administrative divisions, such as municipalities and regions, whose effects on given behaviours may once have been more visible but are now more doubtful. Although an effect of this kind may show up in a multilevel study, it may originate in other levels that are not individualized in the analysis because of their greater complexity. The observed effect thus serves as a proxy for the more fundamental levels. Hence the importance of using appropriate surveys to better identify the latter, which would be a more accurate reflection of our current social organization. We are still a long way from implementing such a program.

The second issue is the choice of an ultimately individual approach here. In other words, the effects of aggregation levels are always defined with respect to the individual. How, then, can we explain the changes observed in the rules prevailing at higher levels—changes that may, however, be due to individual actions? For example, a series of isolated actions in a given community may foster awareness of a problem that concerns the entire community. This may lead to political measures, taken at a more aggregated level. These measures will naturally affect individual behaviours, generating new actions to offset their perverse effects, and so on. The multilevel approach as described here does not allow inclusion of this two-way flow.

5.5 From a Latent Hierarchical Vision to a Multilevel Approach

The third issue is the complexity of the social structure of the groups studied. Even small groups such as the family or household are hard to view as homogeneous. The head of household, his or her adult dependents, and the dependent children all play perfectly distinct roles. Given these circumstances, how should we analyze the interactions between group members? A study by Bonvalet et al. (1997) has made progress in addressing the issue in small groups but without providing a full solution. Its generalization to even more complex groups will require the implementation of new observation tools.

We can conclude from this examination that the multilevel approach will surely entail the establishment of a new paradigm that will enable us to answer at least some of the questions posed above.

Conclusion to Part II

After this detailed examination of the links between population sciences, statistics, and probability, we can now provide clearer answers to some of the questions underlying Part II of our book. First, what is the intensity of the ties between population sciences and probability, partly mediated by statistics? Second, what is the nature of the connections between probability, social sciences, and causal inference? Third, does cumulativity exist in these sciences, and, if so, what form does it take?

Intensity of Ties with Probability

The connection between population sciences and probability was vital from the outset. One can say that population sciences could not exist without statistics.

Ever since 1662, Graunt has rightly been regarded as the first statistician in the social sciences. Using all the precautions needed to obtain the most accurate figures possible, he tabulated the bills of baptisms and burials in London and the provinces over nearly 60 years. He rejected certain data, such as baptisms prior to 1642, on grounds of shoddy record-keeping; in other cases, he adjusted the data, such as the number of deaths from the plague in 1625, which, he plausibly argued, exceeded the official number by one quarter. From these statistics, Graunt drew pioneering conclusions about many demographic and epidemiological phenomena: infant mortality, the effects of the plague, neonatal mortality, the link between the number of births and insalubriousness in a given year, migration to London, the proportion of male births, causes of death, the imbalance between burials and baptisms in London and the provinces, and so on.

At the same time, Graunt established ties between the social sciences and probability. His wager on the annual probability of death clearly shows the role of probabilistic reasoning in his approach. In seeking to connect these statistics to quantities of a more theoretical kind, he assumes an identical probability of dying for all individuals in a 10-year age group and even in a 50-year age group. Although he failed to produce an actual life table, his attempt marked the first step in that direction.

Throughout the eighteenth and early nineteenth centuries, probability played a major role in the social sciences. Laplace (1814) contended that all our knowledge is merely probable, and he applied this precept in remarkable fashion to the social sciences, particularly population sciences, by using Bayesian probability.

Another priority was the quest for ever more exhaustive statistics. One example is the life table compiled by Wargentin (1766), whose access to population registers of excellent quality gave him both the numerators and the denominators of the rates to be calculated.

The introduction of censuses—with their exhaustive enumerations linked to civil-registration data—caused probability to lose much of its appeal for population scientists, who began to concentrate on statistics. Cross-sectional analysis in demography focused on the rates themselves, i.e., on frequencies. Their variance, under the hypothesis of a homogeneous population, was so weak as to make its estimation pointless, and contemporary demographers never mentioned it. Many statisticians working in demography and economics in the late nineteenth and early twentieth centuries rejected the very notion of probability as useless (Armatte 2005). They preferred the notion of frequency to that of probability, and the notion of trend to that of distribution. Even population scientists turned away from probability until the early 1980s. The advent of longitudinal analysis after World War II enabled them to continue working on very large populations, and the underlying hypotheses of a homogeneous population and independence between the phenomena studied deprived probability of its usefulness (Henry 1957). Relative frequency was sometimes viewed as an experimental measure of the probability of occurrence of an event, but the issue was of little concern to population sciences.

The shift to the event-history paradigm in the early 1980s brought a total change of attitude. The examination of large sets of characteristics made it essential for population scientists to return to probabilistic logic. Even when the data are exhaustive, the event-history approach—whether objectivist or Bayesian—requires probability theory. The abandonment of the homogeneous-population and independent-phenomena hypotheses made it impossible to ignore the essential role of probability in demographic analysis.

But statistics were to change as well. Obtaining individual life histories with their many characteristics most often requires surveys of small numbers of persons. Sampling methods are now well tested, but we must ensure that non-responses do not bias the results and that the information provided by respondents is reliable. The test conducted with the aid of an event-history survey (3B bis) and the Belgian population register has allowed a consolidation of this array of collection methods, which is specific to population sciences analysis (Poulain et al. 1991, 1992; Courgeau 1991, 1992). The conclusions of the test are clear:

> Even if errors in the *dating* of past events are frequent, apparently these do not affect their *logical sequence*, or only very slightly so. This sequence is correctly memorized, and the errors only form a kind of background noise, which does not prevent coherent information from being drawn from all sources. Thus, memory seems to be reliable where the analysis needs it to be. (Courgeau 1992)

The multilevel paradigm intensified this need to think in terms of probability and conduct surveys to obtain results. Given the small size of the population observed at a given aggregation level, it is usually necessary to use Bayesian probability. The latter is also mandatory in fields such as paleodemography, which studies populations comprising only a few dozen individuals.

Many other social sciences use both the event-history and multilevel approaches. The conclusions we have just drawn therefore apply to those disciplines as well, including epidemiology, economics, sociology, medical statistical studies, and education sciences.

By contrast, the population sciences and even social science as a whole are barely concerned about which type of epistemic probability they apply—subjective or logical. However, the prior distribution they use is seldom adopted without incorporating the information available on the phenomenon studied (Ibrahim et al. 2001). When no such information is at hand, a uniform distribution is often chosen. In other words, social scientists tend to opt for a logical rather than subjective epistemic approach, without being genuinely aware of the fact.

Links Between Probability, Social Science, and Counterfactual Causality

In this section we will consider more generally the links between social sciences and causality, as population sciences do not differ largely from social science on this topic.

This subject has already been explored in many studies, and space precludes a full discussion of it here (see especially: Franck 1994; Illari et al. 2011; Russo 2009; Williamson 2005, 2009). In the following pages, we take a more detailed look at the notion of counterfactual causality, introduced by Lewis (1973a, b) and elaborated by Holland (1986). It is routinely used by many statisticians, population and social scientists, in particular in the U.S.

The approach is based on a hypothesis that is intrinsically non-testable. Let u be an individual to whom, for example, we can apply a course of treatment to be assessed, t, or a conventional treatment, c. Let $Y_t(u)$ be his/her response to the treatment and $Y_c(u)$ his/her response to the control (i.e., conventional) treatment. Treatment's effect relative to c, which we can assume to be causal, is:

$$Y_t(u) - Y_c(u).$$

Unfortunately it is impossible to apply both treatments to the same person, and if we apply them in sequence (when feasible), we generally can no longer speak of causality. We thus cannot observe the effect of the treatment on a given individual. The hypothesis, therefore, is not testable.

However, statisticians will try to circumvent the difficulty by working on a population that they will subject to strict conditions. Suppose we form two groups by random sampling, such that for two persons taken in each of the groups, u_1 and

u_2, $Y_t(u_1) = Y_t(u_2)$ and $Y_c(u_1) = Y_c(u_2)$. In other words, apart from the treatment, the two groups are strictly comparable. We can thus say that the causal effect of the treatment is:

$$Y_t(u_1) - Y_c(u_2).$$

Consequently, if a laboratory can form the two groups by randomization, it may be able to demonstrate a causal effect of the treatment—not with certainty, for that is impossible, but with plausibility.

While such an experiment can sometimes be performed in epidemiology, ethical and material considerations preclude the same experiment in population sciences or any other social science. This has not stopped many researchers from applying the procedure to non-randomized studies. In their opinion, however, the looser conditions in which these studies are performed make it possible to address the causality issue (Rubin 1974, 1977).

Let us take two recent examples of such an analysis, which has in fact been used in a very large number of comparable cases.

First, an article by Randall Kuhn et al. (2011) concerns the effect of internal migration of children on their parents' health. The article uses a pseudo-randomization in which a sample is formed with a set of characteristics (age, sex, and number of children aged 15+) such that each migrant is linked to another, non-migrant individual in the counterfactual control group. Kuhn shows a positive effect of children's migration on the health of non-migrant parents.

Second, an article by Torche (2011) seeks to assess the effect of a massive earthquake that struck northern Chile in 2005 on the weight at birth of children born after the event. As an earthquake may be regarded as independent of the other characteristics influencing the birth of the persons affected, we may consider that the randomization condition is met. Torche shows an average decrease of 51 g in the weight of children of mothers located in the quake-struck areas relative to the weights in areas that were spared.

The studies by Herbert Smith cast doubt on these findings. In many instances, such as the first example above, he notes:

> How does the investigator know when the proper specifying variables have been incorporated in the experimental design? As with the specification of models for the analysis of data obtained from observational studies, theory is the ultimate guide (1990).

The characteristics taken into account in this example are not necessarily the ones that should be considered, for the phenomenon studied has not been sufficiently theorized. For the second example, Smith cites a condition described by Holland (1986) as crucial: 'no causation without manipulation'. In such circumstances, it is hard to see what actions could avert earthquake-related risks. More generally, the number of characteristics found in these types of models makes the manipulability criterion impossible to enforce:

> The manipulability criterion for causal inference has been difficult to assimilate in a discipline that routinely reports measurements of the causal effects of sex, race, and age, inter alia, on various phenomena (Herbert Smith 1997).

It seems important to avoid these characteristics as far as possible in analyses. For instance, in an event-history analysis of internal migrations (Courgeau 1985), I showed that the age effect, so popular among many demographers, vanished entirely when the person's family-related, social, economic, and other characteristics are factored in.

In our view, however, Herbert Smith goes too far when he says: 'We measure at the micro level, but we intervene—manipulate—at some higher level' (2003). He concludes that analyses at individual level are of little practical value in deciding a course of action.

The example of the analysis of farmer migration discussed throughout this chapter shows the need to qualify Smith's assertion. When looking at the macro level of regions, and performing a standard regression analysis, we find that the probability of migration rises with the percentage of farmers in the local population. We could interpret this finding—as Durkheim would have—as evidence that farmers are more likely to migrate. But a micro analysis shows the exact opposite: farmers are less likely to migrate than other occupational groups. Only a multilevel analysis can reconcile these two apparently contradictory results. While farmers are less likely to migrate whatever their percentage, it is non-farmers who are more likely to migrate when the proportion of farmers in their region rises.

Thus, if we want to promote farmer mobility, the macro-level analysis would prompt us to increase migration bonuses, or any other comparable benefit, for farmers living in areas where they constitute a small proportion of the population. But multilevel analysis shows us that such a measure would have no effect, since their probability of migrating is the same regardless of where they live. Examining the different levels, therefore, gives us a clear understanding of why the macro-level analysis has misled us in our policy-making.

To conclude on this topic, let us note that Dawid (2000) has gone further in the critique of the counterfactual approach, when applied under a strict test protocol. His arguments are very convincing, and we shall recall some of the stronger ones here.

First, he shows that these analyses, in seeking the effects of causes, are based on an attitude he calls 'fatalism':

> This considers the various potential responses $Y_i(u)$, when treatment i is applied to unit u, as predetermined attributes of unit u, waiting only to be recovered by suitable experimentation. [...] Note that because each unit label u is regarded as individual and unrepeatable, there is never any possibility of empirically testing this assumption of fatalism, which thus can be categorized as metaphysical.

Second, in seeking the causes of effects, attention will shift to the following question: for a given individual u_0, did the application of the treatment t cause the observed response $Y_t(u_0) = y_0$ or not? Dawid shows that this inference is even more problematic than the previous one:

> It appears that, to address this question, there is no alternative but to somehow compare the observed valued y_0 with the counterfactual quantity $Y_c(u_0)$, the response that would have resulted from application of c to u_0. Equivalently, inference about the individual effect $\tau(u_0) = y_0 - Y_c(u_0)$ is required. However, the fact that such an inference may be desirable does not, in itself, render it possible.

Dawid concludes that this counterfactual approach can, in most cases, lead to inferences that are not justified by empirical data and are therefore unscientific.

In our General Conclusion, we shall examine whether other approaches to inference can yield more satisfactory results.

Does Cumulativity Exist in the Population Sciences?

In this chapter, we described the paradigms that population sciences have encountered throughout their history. We shall now examine in greater detail, as we did with probability, whether cumulativity exists in these paradigms or not.

Let us begin by taking a more detailed look at the shift from the cohort paradigm to the event history paradigm, and let us try to compare it with the move from Newtonian physics to Einstein's general relativity.

As we indicated, the longitudinal approach rested on the notion of a homogeneous population and mutually independent events. Likewise, Newton's physics rested on the notion of a space that was homogeneous and isotropic in its dimensions. The two approaches can be regarded as roughly comparable if we consider that physics is interested in space (we deliberately set aside time, which complicates the comparison, without altering its conclusions) and population sciences in population (again, setting aside time). Both are homogeneous and the entities that we can regard as the dimensions of population sciences—the phenomena studied—are equally isotropic, that is, independent of one another. The shift to the event history paradigm was mandated by the observation of survey data, which are more detailed than register data. The transition seems to generalize results to heterogeneous populations and interdependent events. Likewise, the shift to Einsteinian general relativity, made necessary by major conceptual problems posed by Newtonian theory, led to a curved space-time, determined by the physical content of the universe and, in this sense, heterogeneous and non-isotropic in its dimensions. Newton's physics, one could argue, was generalized by Einstein's: some of the axioms corresponding to both views are identical, others different (Suppes 2002a).

For a number of authors, however, this generalization is not as obvious as it seems: Thomas Kuhn (1970) asked: 'Can Newtonian dynamics really be derived from relativistic dynamics?' The detailed examination of the denomination concepts that are identical for Newton and Einstein showed him that 'the physical referents of these Einsteinian concepts are by no means identical with those of the Newtonian concepts that bear the same name. (Newtonian mass is conserved; Einsteinian is convertible with energy. Only at low relative velocities may the two be measured in the same way, and even then they must not be conceived to be the same.) Let us, therefore, now take it for granted that the differences between successive paradigms are both necessary and irreconcilable.' Similarly, whereas Newton's theory was implied by observations, Einstein's theory was, on the contrary, implied by the conceptual problems in Newtonian physics: only later was Einstein able to verify his theory through observations, such as the precession of

Mercury's perihelion over a century. Under the Kuhnian definition of paradigms, there can be no cumulativeness here.

We have already been able to show, in a comparable way, that the event history approach was very different from the cohort approach, and Blayo's attack (1995) clearly illustrates this (see Sect. 5.4.1). In fact the two approaches do not use similar definitions for the populations on which they work, and the links that they study between phenomena are very different, if not incompatible. Could we therefore take this as proving the absence of cumulativeness here?

At this point in the discussion we need to distinguish between: (a) cumulativeness of paradigms, and (b) cumulativeness of knowledge acquired in different paradigms. In theory, the non-cumulativeness of paradigms does not exclude the cumulativeness of knowledge obtained through different paradigms. Thus, as noted earlier, when Thomas Kuhn emphasizes the conceptual heterogeneity of Newtonian and Einsteinian paradigms, we cannot conclude that the knowledge of classical mechanics was nullified by the theory of relativity. Let us begin by analyzing the second point in greater detail before drawing conclusions on the first.

Cumulativeness of knowledge seems self-evident throughout the history of population sciences and our presentations in this article show this perfectly: the shift from regularity of rates to their variation; the shift from independent phenomena and homogeneous populations to interdependent phenomena and heterogeneous populations; the shift from dependence on society to dependence on the individual, ending in a fully multilevel approach. Each new stage incorporates some elements of the previous one and rejects others. The discipline has thus effectively advanced thanks to the introduction of successive paradigms. Each takes the shortcomings of its predecessor as a starting point and offers a method for surmounting them—without, however, erasing all of the knowledge attained through the earlier paradigm. Indeed, for some questions that are asked to the population scientist, cross-sectional analysis can suffice, just as any other form of analysis may be sufficient for other issues. The same is true for some questions asked to the physician, that may perfectly been answered by Newtonian's physic, without taking account of Einsteinian's physics.

We can therefore say that each new paradigm comes as a complement to the preceding one for the purpose of treating cases that lie outside of the latter's scope, while partly preserving some of the results obtained with its predecessor. However, this preservation is far from being consistently guaranteed, for the new paradigm allows more accurate and detailed reasoning than the previous paradigm. In some cases this can yield very different results, as we have shown for event history analysis compared with cohort analysis and, especially, for multilevel analysis compared with cross-sectional analysis and event history analysis.

This shows a certain cumulativeness of paradigms that is far from linear. The reason is that the objects treated by population sciences are selected by the different paradigms and are therefore specific to each of them. For instance, cross-sectional analysis ignores the time lived by an individual and connects the phenomena observed at a given moment with the characteristics of populations observed in that same instant. Its objects are therefore those observed in a census, for example. Cohort analysis introduces time lived by a generation or cohort, but fails to incorporate individual characteristics or the links between known events, for it assumes a homogeneous population and mutually independent events. Its objects are those observed with the aid of vital statistics. Event history analysis finally allows population scientists to incorporate the various individual characteristics and the links between events. Its objects are those observed with the aid of event history surveys. Lastly, multilevel analysis allows a synthesis between cross-sectional analysis and event history analysis by introducing various levels of aggregation. Its objects are those observed with the aid of surveys even more complex than event history surveys, simultaneously involving events observed at different aggregation levels.

Each of these paradigms is confined in analytical scope to its own objects, but has been proved to be totally consistent with them, as a number of works on population sciences have shown since the seventeenth century. Yet the reason why we cannot demonstrate a perfectly linear relationship between them is that these objects are different. On the other hand, there exists a non-linear relationship and a very strong continuity between them, and these features can be interpreted as a form of cumulativeness, provided we do not forget that each paradigm can be verified only by its own objects (Agazzi 1985). Going beyond Thomas Kuhn's position, Agazzi shows very clearly for the natural sciences—but this could be transposed to social science—that 'scientific progress does not consist in a purely logical relationship between theories, and moreover it is not linear. Yet it exists and may even be interpreted as an accumulation of truth, provided we do not forget that every scientific theory is true only about its own specific objects.'

Without repeating his demonstration in detail here, we shall summarize the points of use to our discussion. Agazzi notes that researchers actually choose a small number of objects for inclusion in their theories. Some of these objects depend very strongly on the theory's context (contextual part); others, on the contrary, do not depend on it (referential part) and so allow a comparison between different theories. For instance, in population sciences, the events analyzed do not depend on the type of analysis conducted. Mortality, fertility, migration, and so on are objects independent of the theory used to treat them. By contrast, the relationships assumed to exist between these objects are strongly dependent on the theory: independence between objects in cohort analysis, heavy dependence in event history analysis. This makes it possible to compare theories. Most important, it ensures that the results of one theory can never be destroyed by another theory with different objects of research. To the contrary, 'the new truths remain *together with* the old ones and *complement* them.' Classical mechanics is still demonstrated by its objects, as is cohort analysis. But relativity theory complements classical mechanics, just as event history analysis complements cohort analysis. As Granger very rightly points out (1994):

> True, the human fact can indeed be scientifically understood only through multiple angles of vision, but on condition that we discover the controllable operation that uses these angles to recreate the fact stereoscopically.

We believe that the multiplicity of paradigms observed in population sciences effectively corresponds to this multiplicity of angles of vision, and that the relationships we have been able to demonstrate between paradigms enable us to obtain a stereoscopic reproduction of them that is highly promising for the future. The resulting notion of cumulativeness is, of course, far removed from basic additivity but—in our opinion—allows an advance that is entirely relevant to demography and even to the social sciences as a whole.

General Conclusion

A number of social sciences, as we have seen, were born at the same time as probability and now routinely use its concepts. These play an essential role in population sciences and in fields such as epidemiology and economics. However, the connection is not always as close in other social sciences.

The first part of this conclusion will describe the current situation more specifically in sociology and in artificial-intelligence, a science using mainly nonprobabilistic methods in the past.

This last theory using causal diagrams, the notions of counterfactual causality and of structural equations, will lead us to examine in broader terms how different causality theories fit into the social sciences.

We shall then return to the notions of individual and levels before discussing how probabilistic reasoning is incorporated into the forecasting of individual and collective behavior.

In this General Conclusion, we shall therefore need to address these topics in greater detail. Although the scope of our book precludes an exhaustive treatment, we offer some suggestions for more clearly assessing the situation in a larger number of social sciences.

Our epilogue summarizes the main findings of our study, the issues that still need to be addressed, and the pathways toward a fuller analysis of societies.

Generality of the Use of Probability and Statistics in Social Science

In our detailed examination of the history of population sciences over three and a half centuries, we have seen how strongly their concepts and methods depended on the notions of probability and statistics, which emerged almost simultaneously. Although the links may have seemed looser at certain moments, population scientists, probabilists, and statisticians cooperated closely most of the time. Often, it was the same scientist who, like Laplace, designed the probabilistic methods, developed the appropriate statistics, and applied them to population issues (see Chaps. 3 and 4).

In Chaps. 1, 2, and 3, we saw how other social sciences, as well, relied heavily on probability and statistics for tackling certain problems. Those disciplines include, together with population sciences, economics, epidemiology, jurisprudence, education sciences, and sociology. Admittedly, we have not examined them in depth, and it is possible that they may not always need probability in their work.

For instance, we have shown (Chaps. 1 and 4) that Durkheim's sociology required the concomitant-variation method, i.e., linear regressions, to establish causality relationships (Durkheim 1895):

> We have only one means of demonstrating that a phenomenon is the cause of another: it is to compare the cases where they are present or absent simultaneously and to determine if the variations that they display in these different combinations of circumstances are evidence that one depends on the other.

In his study on suicide (Durkheim 1897), for example, he observed that suicide rates varied with the local percentage of Protestants, and he deduced the more general conclusion that:

> [s]uicide varies in inverse proportion to the degree of integration of religious society.

He showed that the same reasoning applied to domestic and political society. To explain suicide, he therefore sought a cause common to all these societies:

> Now the only one that meets this condition is that these are all strongly integrated social groups. We therefore arrive at this general conclusion: suicide varies in inverse proportion to the degree of integration of the social groups to which the individual belongs.

In other words, his demonstration, while based on probability, transcends the probabilistic approach in order to identify the more general causes of a specific sociological phenomenon: suicide.

The same is likely true in other social sciences, but we can also assume that while many use probability calculus, some do not make it their prime method. We have seen this assumption confirmed in sociology; below, we shall examine whether it also applies to artificial intelligence.

Another point is that some approaches used in population sciences are common to other social sciences as well.

For instance, the event-history approach, whose probabilistic bases we have shown to be essential, is used not only in many social sciences, but in mechanics and physics, as it applies to the more general study of phenomena occurring over time. Examples for which it is perfectly suited include: measuring task performance in psychological experiments; medical and epidemiological studies on the development of diseases; studies on the durability of manufactured parts and machines; studies on the length of strikes and unemployment spells in economics; and studies on the length of traces left on a photographic plate in particle physics.

Likewise, the multilevel approach—which studies data that are ranked hierarchically or belong to different levels—is widely used in education sciences, medical sciences, organization sciences, economic, epidemiology, biology, sociology, and other fields. Here as well, scientists use characteristics measured at different aggregation levels in their search for an overall treatment of a more general problem posed by the existence of levels in all sciences. These methods, too, are based on probability, and in particular the crucial notion of exchangeability.

However—like Durkheim, who sought to generalize the results obtained with the aid of regression methods—most social sciences aim beyond the mere observation of statistical regularities, identified with the aid of probabilistic and statistical models. Hence the importance of intensifying the search for whatever tools can supplement the use of probability in the social sciences.

Shafer (1990a) clearly frames the problem of the limits of the application of probability to certain social sciences:

> An understanding of the intellectual content of applied probability and applied statistics must therefore include an understanding of their limits. What are the characteristics of problems in which statistical logic is not helpful? What are the alternatives that scientists, engineers, and others use? What for example are the characteristics of problems for which expert systems should use nonprobabilistic tools of inference?

He suggests that we should seek the reasons for the use of these non-probabilistic methods in certain sciences: 'We must, for example, understand the nonprobabilistic methods of inference for artificial intelligence [...].' Accordingly, we shall review the situation in artificial intelligence, but not in the same detail as we have analyzed population sciences.

While the origins of artificial intelligence go back to Antiquity, it is once again Pascal (1645) who, with his arithmetic machine, stands out as one of the true forerunners of the science[1]:

> [T]he instrument compensates the failings due to ignorance or lack of habit, and, by performing the required movements, it executes alone, without even requiring the user's intention to do so, all the shortcuts of which nature is capable, and every time that the numbers are arranged on it.

Although he does not actually claim that the machine can think, he does note that it can perform operations without memory errors, particularly all arithmetical calculations regardless of complexity.

However, it was not until the twentieth century that ways were found to formalize arithmetical reasoning, then set theory, by means of Gödel's incompleteness theorems (1931), Turing's machine (1936), and Church's Lambda calculus (1932). First, Gödel's two incompleteness theorems showed that those axiomatized theories contain true but unprovable expressions. Second, Turing's machine, similar to a computer but with no limitations on its memory space, made it possible to analyze a problem's effective computability. Lastly, Church's Lambda calculus provided a formal system for defining a function, applying it, and repeating it recursively.

This sequence paved the way for artificial intelligence with Turing's article (1950) envisaging the creation of machines endowed with true intelligence. In its most outspoken form, artificial intelligence refers to a machine capable not only of producing intelligent behavior, but also of experiencing true self-consciousness and

---

[1] Guillaume Schickart reportedly built a similar machine in 1624, but it was destroyed in a fire. Pascal was clearly unaware of it.

of understanding its own logic. Let us now examine some stages in the development of the science and their connections to probability.

Solomonoff had elaborated a general theory of inductive inference. Taking a long sequence of symbols that contained all the information to be used in an induction, he sought to design the best prior distribution of the following symbol (Solomonoff 1964a, b).[2] He relied especially on Turing's work. Interestingly, many probabilists largely overlooked this theory of algorithmic probability for a very long time: as we shall see later, symbolic logic was the main qualitative tool for representing intelligence before 1980.

Solomonoff's method is based on the following principle. Let us take, for instance, the sequence of numbers 2, 4, 6, 8 and try to determine the probability distribution of the following number. It should be noted that very often—for example, in IQ tests—the respondent is asked to give the following number directly, not the distribution. Indeed, when we examine the sequence, we immediately assume that the nth term should be $2n$. In principle, therefore, the answer for the fifth term is 10. But in fact there are many sequences that begin with the same four terms. For example, the sequence expressed by the formula $2n^4 - 20n^3 + 70n^2 - 90n + 48$ also begins with the first four numbers and yields another solution to our problem: 98. Why, then, do we regard the first formula as the most likely? No doubt because we unconsciously apply the principle of Occam's razor: 'entities must not be multiplied beyond necessity'.[3] To solve this problem, we thus need to consider all possible solutions and give their distribution. More specifically, it is preferable to weight each of these answers using a function reflecting the complexity of each. The function may consist of Kolmogorov's complexity,[4] $K(s)$, defined as the length of the shortest description of the sequence s in a universal description language such as Church's Lambda calculus, used by a Turing machine. Solomonoff defines a prior *algorithmic probability*, on the space of all possible binary sequences, equal to $P(x) = \sum 2^{-K(s)}$, where the sum applies to all descriptions of infinite sequences starting with the string x. Of this probability's many properties, the most interesting is that the sum of quadratic errors in the set of sequences is limited by a constant term, which implies that the algorithmic probability tends toward the true probability when $n \to \infty$ faster than $\frac{1}{n}$.

Unfortunately, the method's main drawback is that the model is generally incomputable—or rather is calculable only asymptotically—because Kolmogorov's

---

[2] Back in 1960, Solomonoff had already presented a preliminary report on this theory. He noted that at the Summer Study Group in Artificial Intelligence at Dartmouth (1956), McCarthy, who coined the term 'artificial intelligence,' asked him the following question: 'Suppose we were wandering about in an old house, and we suddenly opened a door to a room and in that room was a computer that was printing out your sequence. Eventually it came to the end of the sequence and was about to print the next symbol. Wouldn't you bet that it would be correct?' (Solomonoff 1997). Solomonoff later succeeded in answering the question with his theory of algorithmic probability.

[3] Entia non sunt multiplicanda praeter necessitatem.

[4] In fact, this concept was introduced by Solomonoff in 1960, and Kolmogorov presented it later.

complexity is incomputable as well. However, there are proxy solutions that make allowance for the calculation time and, under these assumptions, offer a partial solution to the problem.

This theory is applicable to many problems in artificial intelligence, using probability distributions to represent all the relevant information for solving them. Solomonoff (1986) applies the theory to passive-learning problems, where the fact that a current prediction by the agent is correct or not has no impact on the future series. But we need to go one step further and examine the general case of an agent capable of performing actions that will affect his or her future behavior. Hutter (2001) extended Solomonoff's model to active learning, combining it with sequential decision theory. This allowed the development of a very general theory applicable to a large class of interactive environments.

However, the forecasts based on this broader theory are limited not only by the fact that the model is usually incomputable, but also by the fact that the convergence for the algorithmic probability may not be possible in certain environments (Legg 1997). It therefore remains an ideal but unattainable model for inductive inference in artificial intelligence.

In fact, most artificial-intelligence specialists have long viewed symbolic logic as the ideal tool for representing intelligent knowledge and solving problems. For this purpose, symbolic logic relied on essentially qualitative methods. Shafer and Pearl (1990) described this period as follows:

> Ray Solomonoff, for example, has long argued that AI should be based on the use of algorithmic probability to learn from experience (Solomonoff 1986). Most of the formal work in AI before 1980s, however, was based on symbolic logic rather than probability theory.

At the beginning of the 1980s, however, many artificial-intelligence specialists came to realize that symbolic logic would never be able to describe all human processes, such as perception, learning, planning, and form recognition. By the mid-1980s, researchers were developing truly probabilistic methods to address these issues (Pearl 1985).

Pearl's theories initially focused on *Bayesian networks*. He introduced the term, and the networks themselves, in an article published in 1985:

> Bayesian networks are directed acyclic graphs in which the nodes represent proportions (or variables), the arcs signify the existence of direct causal dependencies between the linked propositions, and the strengths of these dependencies are quantified by conditional probabilities. A network of this sort can be used to represent the deep causal knowledge of an agent or a domain expert and turns into a computational architecture if the links are used not merely for storing factual knowledge but also for directing and activating the data flow in the computations which manipulate this knowledge.

Pearl elaborated the theory in a book (Pearl 1988) that used the graphs to represent the dependency structures occurring in a number of multivariate probability distributions. Let us see in greater detail how this matching is achieved.

When we analyze human reasoning, we aim to identify the mechanism whereby people integrate data from different sources in order to arrive at a coherent interpretation of them. We can always plot a graph showing these data—or, rather, these propositions—and the links between them. We can then

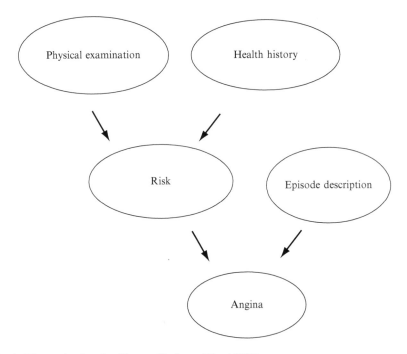

Fig. 1 Diagnostic of angina (Source: Shafer and Pearl 1990)

observe that the dependency graph forms a tree structure with nodes representing the propositions and links, arrowed or not, between the propositions that we regard as directly connected. For example, Fig. 1 (Shafer and Pearl 1990) shows how a doctor:

> combines evidence from a physical examination and a health history to get a judgement about how much at risk of heart disease the patient is, and then he or she combines this with the patient's description of an apparent angina episode to get a judgement about whether the patient really has angina.

From this figure, Shafer and Pearl conclude that:

> Physical examination and Health history are conditionally independent of Episode description given Risk.

This conditional independence is the sought-for link with probability theory, where two events A and B are conditionally independent given a third event C if and only if:

$$P(A \cap B \mid C) = P(A \mid C) P(B \mid C).$$

It is thus easy to see that, in principle, conditionally independent events have no reason to be independent of one another.

For the moment, the figure does not contain any numbers. As applied here, probability theory is more fundamentally concerned with the structure of reasoning and the causality links contained therein than with the actual values of the probabilities. Pearl (1985) summed up the approach as follows:

> This suggests that the fundamental structure of human judgemental knowledge can be represented by dependency graphs and that mental tracing links in these graphs are responsible for the basic steps in querying and updating that knowledge.

We can thus link probability theory to other formalisms used in artificial intelligence—in particular, symbolic logic.

Lastly, we need to quantify the links between propositions that will indicate the strength and type of the conditional dependencies between the propositions. These weights can be regarded as conditional probabilities. Such probabilities are in fact subjective, for they represent degrees of belief in events; the data serve to strengthen, update, or reduce the degrees. That is why Pearl called these figures 'Bayesian networks.' They also enable us to identify relationships lasting over a period of time.

We shall not take our discussion of the probability-based approach in artificial intelligence further: for more details, see Pearl 1988, 2000; Shafer and Pearl 1990. Bayesian networks are now used in many other fields such as econometrics, epidemiology, speech recognition, signal processing, error-control codes, medical diagnosis, weather forecasting, and cellular networks.

Revisiting Causality in Social Science

Using Pearl's work as our starting point, we shall now examine the more general conditions for the validity of the counterfactual theory in most social sciences. We shall also investigate whether alternative theories provide a more effective approach to causality in social science as a whole.

Pearl (1995, 2001) takes the models that he initially proposed for artificial intelligence and generalizes them to other social sciences. He shows that the causal models derived from the graphic models described above are generalizations of structural analyses used in engineering (Duncan and Collar 1934), biology and genetics (Wright 1921), economics (Tinbergen 1939; Manski and McFadden 1981), epidemiology (Greenland and Poole 1988), and many other social sciences (Degenne and Forsé 1994, 1999; Sobel 1995).[5] Counterfactual analyses (Lewis 1973a, 1973b; Holland 1986; Rubin 1974, 1977)—which we outlined briefly in the Conclusion to Part II—are also intimately linked to causal models.

Most of the discussions published in conjunction with Pearl's first article (1995) on this generalization note the value of addressing causality in probabilistic models but are highly critical of the author's conclusions. David Cox, for instance, has this to say about structural analyses:

> The difficulties here are related to those of interpreting structural equations with random terms, difficulties emphasised by Haavelmo[6] many years ago: we cannot see that Pearl's discussion resolves the matter.

---

[5] These models are still referred to as 'social networks'.

[6] Nobel price in economic sciences, 1989: he wrote a paper (Haavelmo 1943) which is at the early roots of structural equations.

In the same vein, Dawid shows that counterfactual analysis is of little value:

> To build either a distributional or a counterfactual causal model, we need to assess evidence on how interventions affect the system, and what remains unchanged. This will typically require a major scientific undertaking. [...] In most branches of science such a goal is quite unattainable.

For our final quotation, we take Imbens and Rubin, but other discussants concur:

> We feel that Pearl's methods, although formidable tools for manipulating directed acyclical graphs, can easily lull the researcher into false confidence in the resulting causal conclusions. Consequently, until we see convincing applications of Pearl's approach to substantive questions, we remain somewhat sceptical about its general applicability as a conceptual framework for causal inference in practice.

However, as noted in our Conclusion to Part II, the most powerful attack on counterfactual analysis was the article by Dawid (2000), who rests his case with these words:

> I have argued that the counterfactual approach to causal inference is essentially metaphysical, and full of temptation to make "inferences" that cannot be justified on the basis of empirical data and are thus unscientific.

He clearly demonstrates the dangers of counterfactual approaches, graphic models, and structural analyses, which leave implicit too many assumptions needed for causal inference. In structural graphic models, there are no scientific grounds for using counterfactuals, which are by definition unobservable, or latent variables, which are not genuine concomitant variables, i.e., measurable variables not affected by the treatment. His position is unambiguous:

> I term such functional models *pseudodeterministic* and regard it as misleading to base analyses on them. In particular, I regard it as unscientific to impose intrinsically unverifiable assumed forms for functional relationships, in a misguided attempt to eliminate the essential ambiguity in our inferences.

Many discussants of this article, particularly Pearl, actually confined themselves to arguments on principle without truly addressing the more basic issues. Despite these reactions, Dawid confirms that there is nothing to be gained by introducing vague and unverifiable information into probabilistic reasoning in addition to basic information.

In fact, in the same article, Dawid (2000) distinguishes two types of causality already singled out by Holland (1986): the 'effects of causes' and the 'causes of effects.' The first type answers the question: 'I have a headache. Will it help if I take aspirin?'

The aim here is to compare the expected consequences of different possible interventions. Whereas this type of question is barely addressed by counterfactual analysis, it is effectively dealt with by *decision theory* (DT), which offers a clear solution.

The second type of causality answers the question: 'My headache is gone. Is it because I took aspirin?' The goal here is to understand the causal relation between a phenomenon that has already occurred and an earlier intervention. This is the effect examined by counterfactual analysis—an effect that decision theory finds problematic. As Dawid explains:

> Since, within DT, both indicative and subjunctive conditioning are affected by the same formal conditioning rule, this would require conditioning my initial uncertainty both

(indicatively) on $X=1$[7] and (subjunctively and counterfactually) on $X=0$. But the conjunction of these two conditions is the impossible event Ø—and conditioning on Ø is not meaningful within DT. (Dawid 2007)

As noted earlier, Dawid, along with many other researchers (Cartwright 2007, 2009; Lecoutre 2004), considers that neither counterfactual analysis nor structural-equation modeling nor graphic models are ultimately capable of dealing with the 'causes of effects' properly.

Moreover, we should not forget that there are many other ways of 'seeking a reason for causes'—as the title of a volume edited by Franck clearly indicates [*Faut-il chercher aux causes une raison?*] (1994). An excellent contribution by Hespel (1994) to this gathering explains that there are at least eight major contemporary theories of causation, and that counterfactual theory is just one of many: classical theory (Nagel 1961), nomologico-deductive theory (Hempel and Oppenheim 1948), functional theory (Pearson 1911), conditional theories (Mill 1843), probabilistic theories (Suppes 1970), manipulability theories (von Wright 1971), activity theory (Madden 1969), and counterfactual theory (Lewis 1973a, b; Pearl 1995). Accordingly, we would be well advised to reconsider the rejection of the notion of 'cause' on the grounds that counterfactual causality is unsuitable: we should not throw out the causality baby with the counterfactual bathwater.

The approach to causality by various philosophers of science (Railton 1978; Salmon 1984; Franck 1994, 2002; Bechtel and Richardson 1993; Craver 2007; Darden 2002, 2006; Glennan 2002, 2005; Little 2010; et al.) working in close cooperation with scientists offers another answer to the question—thanks to the notion of 'mechanism' or underlying process. First used in discussing machines, the notion was rapidly adopted in the seventeenth century for describing more complex systems such as cells and biological processes. The recent application by Illari and Williamson (2010) to natural selection and protein synthesis seems to augur well for its implementation in biology. Meanwhile, Frank has proposed its use in the social sciences (1994, 2002).

As the focus of our book is not the philosophy of science, we can provide only a brief description of the approach set out by these authors. The term 'mechanism' was introduced into the discussion and explanation of causality by Railton (1978) and Salmon (1984), who view it as a network of interactive processes (Glennan 2002). More recent writings describe it as a complex system instead. Glennan (2002), for example, defines it thus:

> A mechanism for a behavior is a complex system that produces that behavior by the interaction of a number of parts, where interaction between parts can be characterized by direct, invariant, change-relating generalizations.

This definition, initially applied to biological and neurological sciences, is also valid for many other fields such as the social sciences, where one speaks of social mechanisms. The concept is also hierarchical (Machamer et al. 2000), for the parts of a mechanism may themselves be full mechanisms and vice versa.

---

[7] $X=1$ means that I have taken aspirin, $X=0$ that I have not. [D.C. note].

Let us now see how the approach operates more specifically in the social sciences. The first step is to systematically observe the social phenomenon or phenomena that we want to explain. This is, for example, what demographers have been doing for the past 350 years by measuring mortality, fertility, nuptiality, then internal and international migrations. The second step is to infer from the observation of this phenomenon the functions of the mechanism that are needed to generate it. Unfortunately, it is a step that a number of social sciences, including demography, have not yet succeeded in taking. However, the study by Illari and Williamson (2010) shows, for example, that protein synthesis can be understood as a process whose function is to decode the information contained in DNA in order to allow the production of proteins. The third step is to use the identified functions as the basis for modeling the social or more general mechanism that produces the phenomenon studied. Some of these mechanisms have already been studied in demographic cases, but the lack of understanding of the functions makes this study incomplete and—most important—almost impossible to generalize. Lastly, the causes of the mechanism are those that perform its functions through certain operations. The causes can vary according to the societies and the phenomena studied, whereas the functions will be stabler, albeit not eternal.

The 'mechanism' approach can also satisfy the wish to influence behavior, although such a capability may, on the face of it, seem far removed from the purposes of the approach. If we understand the social mechanism that generates a behavior, it becomes possible to act on that behavior. But, if we have not identified the mechanism's functions, the action may be misdirected.

Lastly, by providing full knowledge of the social or biological mechanism studied, the approach allows a more effective use of Bayesian networks (Casini et al. 2011) without the problems posed by unobservable and latent variables, as in counterfactual and structural analysis.

For the moment, however—apart from some fields such as game and sports modeling (Parlebas 2002), archeology (Gardin 2002), and written communication (Pratt 2010)—the 'mechanism' approach has not yet managed to model broader domains in the social sciences, such as sociology, economics, and demography. In sum, this is a highly promising avenue for strengthening the validity of the social sciences; unfortunately, it has been little explored so far, owing to the complexity of social phenomena.

Revisiting the Notions of Individual and Levels

In Chap. 5, we solved the complicated problem of individual cases by introducing the notion of statistical individual, which informs all paradigms of population sciences by showing that each paradigm corresponds to a different statistical individual, and that:

> [a]dmittedly, scientific knowledge of the human fact cannot be gained except through different planes, but only if one discovers the controllable operation that reproduces the fact stereoscopically from those planes (Granger 1994).

This reasoning extends to all the social sciences, and the more general problem now is not 'how do we move from the statistical individual to the population?' but 'how do we move from the actual individual to the statistical individual?' and, in the opposite direction, from the statistical individual to different supra- or infra-individual aggregation levels.

Let us begin by examining how the probabilist can formulate the problem more precisely. Suppose that each member of a given population follows a personal process, whether demographic, economic, sociological, or other. We pick a random set of paths in this population. As any random process can be viewed as a probability distribution over a set of paths, it is ultimately as if we were making repeated observations of a particular random process. We can thus construct the process underlying the set of observed paths, whose probabilist structure is identifiable here. The process will be applicable to the statistical individual, defined by the observation set, but not to any random person selected from the total population.

How do we go from observed individuals to the statistical individual, for example in the event-history analysis of an event such as death or first childbirth?[8] Let i be an observed individual whose instantaneous rate in t is $h_i(t)$. As a specific performance of the process represents an individual picked at random, it is possible to eliminate index i.

For a particular individual, the distribution of the random instant of event T is determined by the instantaneous rate in the form:

$$S(t \mid h) = P(T > t \mid h) = \exp\left(-\int_{x=0}^{x=t} h(x)\, dx\right),$$

where $S(t \mid h)$ is the individual survivor function. To obtain the survivor function for the population of statistical individuals, we simply take the formula's mathematical mean:

$$S(t) = P(T > t) = E\left[S(t \mid h)\right] = E \exp\left(-\int_{x=0}^{x=t} h(x)\, dx\right). \tag{1}$$

In other words, the population's survivor function is the mean of the individual ones, calculated from the distribution of individuals in the initial instant.

Similarly, there is a link between each individual's instantaneous rate, $h(t)$, and that of the population of statistical individuals, $\overline{h}(t)$:

$$\overline{h}(t) = E\left[h(t) \mid T > t\right]. \tag{2}$$

The instantaneous rate for the population is a mean of the individual instantaneous rates, but the mean is calculated only for individuals at risk in instant t.

To illustrate the above, let us consider a population, P, that we assume to be either homogeneous in its entirety, or composed of two homogeneous groups,

---

[8] See Yashin and Manton (1997) and Aalen et al. (2008) for more details.

P_1 and P_2. If we say that the statistical individuals drawn in P follow an identical process, we obtain the preceding estimates of the survivor function (1) and the instantaneous rate (2). But we can also state, without the slightest contradiction, that the statistical individuals picked separately in P_1 and P_2 respectively follow the processes defined by $h_1(t)$ and $h_2(t)$. In this case, if the groups contain N_1 and N_2 individuals with $N = N_1 + N_2$, then the survivor function will be:

$$S(t) = \frac{N_1}{N} S_1(t) + \frac{N_2}{N} S_2(t),$$

and the instantaneous rate will be:

$$\bar{h}(t) = \frac{N_1}{N} \frac{S_1(t)}{S(t)} h_1(1) + \frac{N_2}{N} \frac{S_2(t)}{S(t)} h_2(t).$$

Thus, depending on whether we make our selection from P or separately from P_1 and P_2, the statistical individuals are not identical, despite the fact that they are the same people.

More generally, we can connect the risks faced by statistical individuals to those faced by the population when we introduce observed or even unobserved characteristics. For further details, we refer the interested reader to the article by Yashin and Manton (1997).

But, in Aristotle's words, 'none of the arts theorize about individual cases' (Rhetoric, I:2). Whatever characteristics we include in the analysis, it will never be possible to predict the behavior of a given individual, even if we know his or her past history in many areas of life. All we can do is estimate a probability for the behavior, whose variance will be all the smaller as the number of observed characteristics is high and well chosen. In this way, we can approach Jacob Bernoulli's wish of being able to determine, from a study of a large number of individuals, the probability that 'Titus will die before age ten.'

We can, however, explore the possibility of identifying the type of process that, independently of the characteristics that influence it, governs the occurrence of demographic, medical, economic, and social events in people's lives.

This is worth doing for the following reason: some theoretical processes have been studied for their ability to generate the standard forms of curves showing the instantaneous rates, but the efforts to prove the validity of the processes remain uncertain because of the absence of robust theories on the occurrence of these events. Two main approaches have been followed.

In this chapter, we already discussed the concept of frailty, which assumes that individual instantaneous rates are proportional to one another, each individual being characterized by his or her specific proportionality ratio. The inability to identify these individual ratios (Trussell 1992) makes the approach highly speculative and weakens its results. In some fields such as medicine, however, biological information can provide a more robust underpinning for the analysis.

Regarding testicular cancer, for instance, there is good evidence to suggest that variations in individual risk exposure are determined by events in fetal life

(Klotz 1999). This makes it possible to develop a more robust frailty model (Aalen and Tretli 1999) with a very good fit to observations, even though it cannot fully demonstrate the model's validity.

Another approach consists in introducing a frailty that develops like a stochastic process, such as a diffusion process. Aalen et al. (2008) describe it thus:

> The stochastic process would then have a deterministic component which could be controlled though observed covariates, and a random component describing a level of uncertainty regarding unobserved covariates and a fundamental level of time-changing heterogeneity.

They accordingly introduce different types of stochastic processes, including diffusion models[9] and models based on Levy processes.[10] However, it is hard to separate the mechanisms underlying these models solely on the basis of the observation of instantaneous rates measured for the total population. This leaves little hope, for the time being, of identifying the types of processes at work in the various sciences.

The introduction of multiple aggregation levels makes the analysis even more complex. In Chaps. 4 and 5, we noted that multilevel analysis can supplement an event-history analysis in demography and other social sciences. In our view, multilevel analysis is necessary in most biological and social sciences, and it is useful to generalize it.

Let us return to the example of Illari and Williamson's study (2010) on natural selection and protein synthesis envisaged in terms of 'mechanisms.' The authors have this to say:

> Once the phenomenon is identified, mechanistic explanation characteristically proceeds by decomposing the phenomenon into lower-level components. The activities of lower-level components are often regarded as further phenomena and further explanations are sought, so that decomposition moves to another level down. This may iterate many times. So mechanisms discovered are usually located in just such a nested hierarchy, with relations to both lower-level and higher-level mechanisms in the hierarchy.

While multilevel analysis allows an examination of some of these issues, it does so under assumptions that we may regard as restrictive, such as the normality of the distribution of randoms at a given aggregation level. This field of analysis is in fact far wider and deserves further exploration. As Franck very rightly observes (1995):

> The point is to determine how the different stages or levels connect, from top to bottom and from bottom to top.

He shows that, once we have accepted the concept of hierarchy, there is no longer any reason to choose between holism and individualism.[11] But he does not propose a specific method to analyze these levels.

By contrast, as noted in Chap. 5, Goldstein (2003) has developed multilevel-analysis methods applicable to many social sciences including education sciences,

---

[9] A diffusion process is a Markov process in continuous time, with continuous sample paths.

[10] A Levy process is a stochastic process whose increments are stationary and independent.

[11] This concept had already been introduced by Jacob (1970) with his notion of a hierarchy of 'integrons.'

epidemiology, demography, and geography. But, despite such successes, these models still require considerable development.

First, we need to determine more specifically which aggregation levels to use in order to ensure that such an analysis will be efficient. These levels, very often mandated by the nature of the survey (Courgeau 2003, 2007a), are not necessarily the ideal ones for analytical purposes. We must try to identify which levels are truly needed in a given analysis.

It would also be essential to round out such an analysis—which starts from a specific, often individual aggregation level—with other analyses that would improve the links between levels. For instance, at the level of a given community, isolated individual actions may address a problem concerning the entire community. At a more aggregated level, those actions may move institutions to offer proposals eventually resulting in policy measures. The latter will, of course, influence individual behavior, producing new actions in response to their unexpected but undesirable effects, and so on.

This feedback loop offers a broad research topic that remains to be investigated in many fields.

Predicting Behavior in Social Science

In the Conclusion to Part I and in this General Conclusion, we have discussed decision-making in response to uncertainty. Another, related issue arising in the social sciences is forecasting. Here as well, our aim is not to offer an in-depth discussion of so vast a subject, but to outline some recently explored paths and show the connections with probability.

One approach—the systemic approach—regards many sets, particularly social and economic, as complex systems whose functioning cannot be understood unless they are examined in their totality. Founded by Bertalanffy in 1968, the approach considers a system of time-dependent simultaneous equations that enable us to incorporate a large number of characteristics as well as their interactions. It therefore does not introduce probability, and that is why we shall not discuss it further here. However, as an example, we should mention the Club of Rome Report (Meadows et al. 1972)—the target of well-known mathematical, economic, demographic, biological, and environmental criticisms (Berlinski 1976).

The second approach—known as the agent-based model or multi-agent simulation—simulates the behaviors of a set of individuals or collective entities belonging to a complex system. Like the first approach, it is time-dependent, but it can also be spatially situated. These models are used in many social and biological sciences. They are based on the very often complex modeling of agent behavior in a wide variety of fields, simulated stochastically on computers. For this purpose, the models use the results of event-history and multilevel studies that model behavior using their full array of probabilistic tools.

As space precludes a detailed presentation here of forecasting methods using agent-based models in biological, environmental, and social sciences, we refer the interested reader to the following studies in specific fields: population sciences (Billari and Prskawetz 2003), epidemiology (Hooten et al. 2010), economics (Remenik 2009), biology (Inkelmann et al. 2010), sociology (Macy and Willer 2002), geography (Gimblett 2002), and ecology (Hooten and Wilke 2010).

Let us, however, take a closer look at the theoretical bases of these models of computer-based simulation of human behavior, and at the procedures for verifying the validity of their underlying assumptions.

The models for predicting human behavior, particularly agent-based models, are essentially theoretical models that use certain aspects of the phenomena studied in order to try to reconstruct the phenomena as fully as possible with the aid of computers. Burch (2003) describes computer-assisted modeling as follows:

> The genre of agent-based modelling will likely occupy a central place in this work. It provides a feasible approach to study interrelations between the macro- and micro-levels in demography—exploring links between individual decisions and aggregate demographic patterns, a realm that up until now has resisted analysis.

What he says about demography fully applies to all social and biological sciences.

As we can also see, this is the opposite of the empirical approach, which prevailed in the social sciences at least throughout the twentieth century. The empirical approach sought to test the validity of statistical models and theories, and rejected those that did not fit the data. By contrast, the new approach is less restrictive from a statistical standpoint, but seeks to deduce observed facts from a formal system of connections between different characteristics and multiple aggregation levels. The search for these underlying processes requires a theoretical reflection on the observed properties of the phenomenon or phenomena studied, and an abstraction of the formal model that explains the phenomena. This new approach therefore resembles the mechanist approach to causality described in Sect. 2 of this General Conclusion. Both are based on induction, in contrast to the primacy of deduction in the empirical approach.

To test the validity of such processes, we examine whether they can reconstruct the development of the phenomena studied over time. This verification is far more complex than the simple statistical test of an empirical model, for we need to evaluate both a simulation model and its underlying assumptions. Such testing, therefore, will inevitably be incomplete and will require a variety of approaches.

We can begin by testing the model in standard fashion, i.e., by comparing its results with the observed changes in the phenomena studied. Some changes may be estimated correctly, others far less so. Unlike with the empirical approach, these findings do not invalidate the model but will allow us to improve it by trying to model the improperly estimated changes differently.

Reeves (1987), for instance, developed a microsimulation model for households, families, and kinship members based on U.S. observations ranging from 1900 to 1981. Wachter et al. (1997, 1998) tested the quality of these results using the 1987–1988 round of the National Survey of Families and Households, which supplied detailed information on the numbers of persons and their ages. Here are some of the positive results of these tests (1998):

(i) Some kinship statistics—average grandchildren below age 70, average siblings below age 40—were predicted with impeccable accuracy. Where the survey results are themselves more precise, the microsimulations achieve good accuracy.
(ii) The random error in the simulations done in 1981 was as small as the sampling error in the 1987–1988 survey.

Others are negative:

(v) There are occasional substantial systematic discrepancies.
(vi) Not surprisingly, wrong guesses about future demographic rates produce wrong numbers for kinship forecasts.

The authors conclude that while some results are very accurate, the negative ones are due to assumptions on mortality at older ages and the heterogeneity of fecundability that need to be reviewed in order to improve the projections. As suggested above, however, these negative results do not call into question the theoretical model developed for simulation purposes; instead, they will allow us to improve it.

Bijak (2011) has proposed another type of model to forecast international migration flows. He notes the many problems involved, such as the diversity of definitions and measurement errors. One way to overcome these inconsistencies is to use a Bayesian approach, which Bijak regards as an 'axiomatic reduction of the notion of "uncertain" to the notion of "random" ' (Robert 2006). For this purpose, he uses observations collected in the previous 15–20 years in order to compare the forecasts prepared by means of Bayesian and frequentist methods with observations for the period 2005–2007. The forecasts concern several European countries. The results clearly show the superiority of Bayesian methods: the frequency of empirical observations lying within the confidence intervals predicted by the Bayesian projection consistently exceeds—by far—the one predicted by the frequentist projection under the same assumptions.

Reviewing many other examples of estimates of future behavior in various social sciences, Burch (2002) concludes:

> The key to all of this is that the computer and associated software has extended much more our ability to do numerical computations. It had in effect extended our powers of logical inference and reasoning. We are able to deduce the strict logical consequences or entailments of systems of propositions much more complicated than can be dealt with using logic or even analytic mathematics.

We can also conclude that, while probability has a role to play in these forecasts, it is a modest one for the time being, although a Bayesian approach seems more capable of dealing with the uncertainty of the projections.

However, while the issue has barely been addressed so far, the links between the mechanist approach and probability will need to be examined in depth, despite the fact that the two approaches seem—at least on the face of it—hard to reconcile.

Epilogue

While we have sometimes touched upon the contribution of probability to the natural and biological sciences, the main purpose of our book is to discuss their contribution to the social sciences. We use this term to designate all the sciences that study social groups (whether human or animal), their behavior, and their evolution. Throughout the volume we have offered many examples of the application of probability to sociology, demography, epidemiology, education sciences, legal sciences, actuarial sciences, economics, criminology, political sciences, communication theory, paleodemography, and artificial intelligence. There are clearly many other uses of probability in these sciences that cannot all be mentioned here, but the overriding point is that the applications concern nearly all the other social sciences, including archeology (Buck et al. 1996), anthropology (Thomas 1986), linguistics (Bod et al. 2003), ecology (Patil and Rao 1994), and history (Roehner and Syme 2002).

We can therefore conclude that probability is used throughout the social sciences, none of which seems to elude its hold, even if other approaches and theories are also used in these sciences.

This permits Lazarsfeld (1954) to note:

> There is a general awareness that probability ideas play a dominant role, explicitly or implicitly, in the study of human behaviour. [...] The predictions of the social scientist will always be probabilistic ones [...].

Indeed, since those words were written, probability has steadily extended its reach. However, in some social sciences, particularly the last-mentioned above, the applications have become less frequent and have sometimes been criticized.

The attacks often proceed from a misunderstanding of the various approaches to probability that we have described in detail: objective probability and subjective or logical epistemic probability. As we noted, social scientists barely distinguish between these approaches, often assuming that probability can only be objective, or interpreting the Bayesian approach incorrectly.

For instance, we showed in Chap. 4 the difficulties encountered by paleodemographers in estimating the age structure of past populations in the absence of civil-registration data. This difficulty is largely due to the use of frequentist methods or to an interpretation of the terms 'Bayesian' or 'epistemic' that differs from the ones offered here. Masset (1982), using the approximations method—which is frequentist—obtained disappointing results, with many null age groups. This led him to prefer the probability-vectors method, which he regarded as more rustic but truer. Similarly, Konigsberg and Frankenberg (1992) describe the IALK method as Bayesian, because it makes some use of Bayes's theorem. In fact, however, it closely resembles Masset's approximations method: the unknown parameters are always assumed to be fixed, whereas a Bayesian method will assume them to be random. As we have seen, the use of a Bayesian method overcomes all these difficulties.

Likewise, Bonneuil (2004) recognizes the usefulness of statistical methods in history, for example to study inter-minority conflicts (Gurr 1993). But he criticizes their use by Roehner and Syme (2002), who justify a frequentist approach on the grounds of similarities between different historical events. Bonneuil argues that an approach based on dynamic game theory would be better suited, as it allows decision-making in the historical domain. His suggestion seems eminently sensible, but he forgets to point out that the approach based on game theory and decision theory, although amenable to other theories, can also resort to subjective probability theory. This has been noted by a number of economists and probabilists (von Neumann and Morgenstern 1944; Robert 2006), and by us in Chap. 2.

However, although the role of probability in the social sciences has sometimes been misinterpreted, it is hardly possible to assert that its importance in all these sciences is identical. In particular, many aspects of social phenomena must use other, non-probabilistic approaches. As Bartholomew (1975) observed:

> The statistician fully recognizes that his contribution concerns only one of many aspects and that when major policy decisions are made his part must be weighed with others and, in the end, may not be decisive. Yet he insists that to ignore the quantitative dimension is as serious an error as to rely on it alone.

Bartholomew was responding to critics who argued that human values can be neither measured nor quantified, and that any method claiming that they could be is inefficient at best, and dangerous at worst.

Keeping this major restriction in mind, we can say, at the end of this work, that the notion of probability truly fostered the emergence of the main social sciences in the seventeenth century by enabling their practitioners to formalize the uncertainty that lies at the heart of all those disciplines. This formalization, of course, was steadily enhanced to the point of allowing an axiomatization of probability in the twentieth century. We have shown that, despite the resulting fragmentation into at least three broad types—objective, subjective, and logical probability—a reunification seems possible and has even already been attempted. New paths have also been opened to extend the use of probability outside the decision-making sphere and to develop intuitionist approaches.

In most social sciences, by contrast, axiomatization remains a very remote prospect. However, our discussion of the paradigms of population sciences has shed light on the topic and provided a 'stereoscopic' reproduction of the various angles from which that science has been approached. We have shown that there is room for non-additive cumulativity in these sciences. A similar exercise concerning the other social sciences would be needed in order to understand them better, and we strongly encourage it.

Lastly, we have highlighted new alternatives—such as the event-history and multilevel approaches—that can be implemented in many social sciences. While these shared methods enable us to adopt a synthetic view of the social sciences, they do not diminish the need for each of these sciences. Nevertheless, the social sciences will be able to converge toward the real individual only by integrating the statistical individual into a world whose nature is all at once social, political, economic, religious, and so on—a world ever closer to the complexity of the world inhabited by the people actually observed, a world where these diversities are experienced simultaneously and not in separate sciences. However, the observed individual will never be attained, but can only be envisaged as the unattainable asymptote of all the worlds to which the statistical individuals belong.

Glossary

Axioms Collection of formally stated assertions deduced from the properties of experimental phenomena, from which other formally stated assertions follow by the application of well-defined rules.

Coherence (probability) One should assign and manipulate probabilities so that one cannot be made a sure looser in betting based on them.

Completeness (axioms) A set of axioms is complete if, for any statement in the axiom's language, either that statement or its negation is provable from the axioms.

Consistency (axioms) A set of axioms is consistent if there is no statement such that both the statement and his negation are provable by the axioms.

Consistency (plausibility) (1) If a conclusion can be reasoned out in more than one way, then every possible way must lead to the same result. (2) All the evidence relevant to the question must be taken into account. (3) Equivalent states of knowledge are always represented by equivalent plausibility assignments.

Cumulativity A dynamic principle of consistency during the revisions of social sciences, which ensures that the results of a new theory remain together with some results of the old one and complement them.

Entropy (Shannon) A numerical measure of the information provided by a full set of propositions on a subject.

Equipossibility Concept that allows one to assign equal probabilities to outcomes when they are judged to be equally likely.

Exchangeability Random variables are exchangeable if their joint distribution is invariant under permutation of its arguments.

Paradigm Theoretical framework within which one moves from experimental phenomena to a scientific object.

Physical impossibility Random event corresponding to a measure zero set.

Population sciences Studies of populations, including: size, composition and distribution, and the causes and consequences of changes in these characteristics.

Probability A numerical measure, between 0 and 1, of the certainty of some event or some proposition.

Probability space A triplet (Ω, $\mathscr{B}$, P), consisting of a set Ω (called the sample space), a σ-algebra of sub-sets $\mathscr{B}$, (called events) and a measure P (called the probability measure).

Set theory Branch of mathematics that studies collections of objects.

Social science Study of social groups.

Statistical inference To reach the most robust conclusion possible by making the best use of the incomplete information one may have on a given phenomenon.

Statistics Study of the collection, organization, analysis and interpretation of data.

Utility A function that takes a numerical value for each possible state of a system and is intended to measure the benefice or usefulness of that state.

References

Aalen, O. O. (1975). *Statistical inference for a family of counting processes*. PhD thesis, University of California, Berkeley.
Aalen, O. O. (1978). Nonparametric inference for a family of counting processes. *The Annals of Statistics, 6*(4), 701–726.
Aalen, O. O., & Tretli, S. (1999). Analysing incidence of testis cancer by means of a frailty model. *Cancer Causes & Control, 10*, 285–292.
Aalen, O. O., Borgan, Ø., Keiding, N., & Thorman, J. (1980). Interaction between life history events. Nonparametric analysis for prospective and retrospective data in the presence of censoring. *Scandinavian Journal of Statistics, 7*, 161–171.
Aalen, O. O., Borgan, Ø., & Gjessing, H. K. (2008). *Survival and event history analysis*. New York: Springer.
Aalen, O. O., Andersen, P. K., Borgan, O., Gill, R. D., & Keiding, N. (2009). History of applications of martingales in survival analysis. *Journ@l Electronique d'Histoire des Probabilités et de la Statistique, 5*(1), 1–28.
Abel, N. H. (1826). Untersuchung der Functionen zweier unabhängig veränderlichen Gröszen x und y, wie $f(x, y)$, welche die Eigenschaft haben, dasz $f[z, f(x,y)]$ eine symmetrische Function von z, x und y ist. *Journal Reine und angewandte Mathematik (Crelle's Journal), 1*, 11–15.
Aczèl, J., Forte, B., & Ng, C. T. (1974). Why the Shannon and Hartley entropies are 'natural'. *Advanced Applied Probabilities, 6*, 131–146.
Agazzi, E. (1985). Commensurability, incommensurability and cumulativity in scientific knowledge. *Erkenntnis, 22*(1–3), 51–77.
Agliardi, E. (2004). Axiomatization and economic theories: Some remarks. *Revue Economique, 55*(1), 123–129.
Allais, M. (1953). Le comportement de l'homme rationnel devant le risque: critique des postulats et axiomes de l'école américaine. *Econometrica, 21*(4), 503–546.
Andersen, P. K., & Gill, R. D. (1982). Cox's regression model for counting processes: A large sample study. *The Annals of Statistics, 10*, 1100–1120.
Andersen, P. K., Borgan, Ø., Gill, R. D., & Keiding, N. (1993). *Statistical models based on counting processes*. New York/Berlin/Heidelberg: Springer.
Arbuthnott, J. (1710). An argument for divine providence, taken from the constant regularity observ'd in the birth of both sexes. *Philosophical Transactions of the Royal Society of London, 27*, 186–190.
Aristotle. (around 330 B.C.). *Nichomachean ethics*. The Internet Classic Archive (W. D. Ross, Trans.). http://classics.mit.edu/Aristotle/nicomachaen.html. Accessed August 30, 2011.
Aristotle. (around 330 B.C.). *Physics*. The Internet Classic Archive (R. P. Hardie & R. K. Gaye, Trans.). http://classics.mit.edu/Aristotle/physics.html. Accessed August 30, 2011.

Aristotle. (around 330 B.C.). *Politics*. The Internet Classic Archive (B. Jowett, Trans.). http://classics.mit.edu/Aristotle/politics.html. Accessed August 30, 2011.
Aristotle. (around 330 B.C.). *Rhetoric*. The Internet Classic Archive: translated by Roberts, W. R. http://classics.mit.edu/Aristotle/rethoric.html. Accessed August 30, 2011.
Aristotle. (around 330 B.C.). *Topics*. The Internet Classic Archive (W. A. Pickard-Cambridge, Trans.). http://classics.mit.edu/Aristotle/topics.html. Accessed August 30, 2011.
Armatte, M. (2004). L'axiomatisation et les théories économiques: un commentaire. *Revue Economique, 55*(1), 130–142.
Armatte, M. (2005). Lucien March (1859–1933). Une statistique mathématique sans probabilité? *Journ@l Electronique d'Histoire des Probabilités et de la Statistique, 1*(1), 1–19.
Arnauld, A., & Nicole, P. (1662). *La logique ou l'Art de penser*. Paris: Chez Charles Savreux.
Arnborg, S. (2006). Robust Bayesianism: Relation to evidence theory. *Journal of Advances in Information Fusion, 1*(1), 75–90.
Arnborg, S., & Sjödin, G. (2000). Bayes rules in finite models. *Proceedings of the European Conference on Artificial Intelligence, Berlin*, pp. 571–575.
Arnborg, S., & Sjödin, G. (2001). On the foundations of Bayesianism. In *Bayesian Inference and Maximum Entropy Methods in Science and Engineering*, 20th International Workshop, American Institute of Physics, Gif-sur-Yvette, pp. 61–71.
Ayton, P. (1997). How to be incoherent and seductive: Bookmaker's odds and support theory. *Organizational Behavior and Human Decision Processes, 72*, 99–115.
Bacon, F. (1605). *The two books of Francis Bacon, of the proficience and advancement of learning, divine and humane*. London: Henrie Tomes.
Bacon, F. (1620). *Novum Organum*. London: J. Bill.
Bacon, F. (1623). *Historia vitae et mortis*. Londini: In Officio Io. Haviland, impensis Matthei Lownes.
Barbin, E., & Lamarche, J. P. (Eds.). (2004). *Histoires de probabilités et de statistiques*. Paris: Ellipses.
Barbin, E., & Marec, Y. (1987). Les recherches sur la probabilité des jugements de Simon-Denis Poisson. *Histoire & Mesure, 11*(2), 39–58.
Barbut, M. (1968). Les treillis des partitions d'un ensemble fini et leur représentation géométrique. *Mathématiques et Sciences Humaines, 22*, 5–22.
Barbut, M. (2002). Une définition fonctionnelle de la dispersion en statistique et en calcul des probabilités: les fonctions de concentration de Paul Lévy. *Mathématiques et Sciences Humaines, 40*(158), 31–57.
Barbut, M.; & Monjardet, B. (1970). *Ordre et classification. Algèbre et combinatoire*. Tomes 1, 2. Paris: Librairie Hachette.
Bartholomew, D. J. (1975). Probability and social science. *International Social Science Journal, XXVII*(3), 421–436.
Bateman, B. W. (1987). Keynes changing conception of probability. *Economics and Philosophy, 3*(1), 97–120.
Bayes, T. R. (1763). An essay towards solving a problem in the doctrine of chances. *Philosophical Transactions of the Royal Society of London, 53*, 370–418.
Bechtel, W., & Richardson, R. C. (1993). *Discovering complexity*. Princeton: Princeton University Press.
Bellhouse, D. R. (2011) A new look at Halley's life table. *Journal of the Royal Statistical Society: Series A (Statistics in Society), 174*(3), 823–832. The Royal Statistical Society Data Set Website: http://www.blackwellpublishing.com/rss/Readmefiles/A174p3bellhouse.htm. Accessed August 1, 2011.
Bentham, J. (1823). *Traité des preuves judiciaires*. Paris: Etienne Dumont.
Benzécri, J. P., et collaborateurs. (1973). *L'analyse des données* (2 vols). Paris: Dunod.
Berger, J. O., Bernardo, J. M., & Sun, D. (2009). The formal definition of reference priors. *The Annals of Statistics, 37*(2), 905–938.
Berlinski, D. (1976). *On systems analysis: An essay concerning the limitations of some mathematical methods in the social, political and biological sciences*. Cambridge, MA: MIT Press.

Bernardo, J. M. (2011). Integrated objective Bayesian estimation and hypothesis testing (with discussion). In J. M. Bernardo, M. J. Bayarri, J. O. Berger, A. P. Dawid, D. Heckerman, A. F. M. Smith, & M. West (Eds.), *Bayesian statistics 9* (pp. 1–68). Oxford: Oxford University Press.

Bernardo, J. M., & Smith, A. F. M. (1994). *Bayesian theory*. Chichester: Wiley.

Bernoulli, N. (1709). *Dissertatio inauguralis mathematico-juridica, de usu artis conjectandi in jure*. Bâle.

Bernoulli, J. I. (1713). *Ars conjectandi*. Bâle: Impensis Thurnisiorum fratrum.

Bernoulli, D. (1738). Specimen theoriae novae de mensura sortis. *Commentarii Academiae Scientiarum Imperialis Petropolitanae, V*, 175–192.

Bernoulli, D. (1760). Essai d'une nouvelle analyse de la mortalité causée par la petite vérole, et des avantages de l'inoculation pour la prévenir. *Mémoires de l'Académie Royale des Sciences de l'Année 1760*, 1–45.

Bernoulli, D. (1777). Diiudicatio maxime probabilis plurium observationum discrepantium atque verissimillima inductio inde formanda. *Acta Academia Petropolitanae*, pp. 3–23. (English translation by M. G. Kendall, Studies on the history of probability and statistics. XI. Daniel Bernoulli on maximum likelihood, *Biometrika, 48*(1), 1–18).

Bernoulli, J. I., & Leibniz, G. W. (1692–1704). Lettres échangées (French translation by N. Meusnier (2006). Quelques échanges ? *Journ@l électronique d'Histoire des Probabilités et de la Statistique, 2*(1), 1–12).

Bernstein, F. (1917). Опыт аксиоматического обоснования теории вероятностей. *Communication de la Société Mathématique de Karkov, 15*, 209–274.

Bertalanffy, L. (1968). *General system theory. Foundations, development, applications*. New York: John Braziler.

Berthelot, J.-M. (Ed.). (2001). *Épistémologie des sciences sociales*. Paris: Presses Universitaires de France.

Bertrand, J. (1889). *Calcul des probabilités*. Paris: Gauthier-Villard et Fils.

Bhattacharya, S. K., Singh, N. K., & Tiwari, R. C. (1992). Hierarchical Bayesian survival analysis based on generalized exponential model. *Metron, 50*(3), 161–183.

Bienaymé, J. (1838). Mémoire sur la probabilité des résultats moyens des observations; démonstration directe de la règle de Laplace. *Mémoires Présentés à l'Académie Royale des Sciences de l'Institut de France, 5*, 513–558.

Bienaymé, J. (1855). Sur un principe que M. Poisson avait cru découvrir et qu'il avait appelé loi des grands nombres. *Séances et travaux de l'Académie des sciences morales et politiques, 31*, 379–389.

Bienvenu, L., Shafer, G., & Shen, A. (2009). On the history of martingales in the study of randomness. *Journ@l électronique d'Histoire des Probabilités et de la Statistique, 5*(1), 1–40.

Bijak, J. (2011). *Forecasting international migration* (Springer series on demographic methods and population analysis, Vol. 24). Dordrecht/Heidelberg/London/New York: Springer.

Billari, F., & Prskawetz, A. (Eds.). (2003). *Agent-based computational demography. Using simulation to improve our understanding of demographic behaviour*. Heidelberg/New York: Physica-Verlag.

Birkhoff, G. (1935). Abstract-linear dependence and lattices. *American Journal of Mathematics, 57*(4), 800–804.

Blayo, C. (1995). La condition d'homogénéité en analyse démographique et en analyse statistique des biographies. *Population, 50*(6), 1501–1518.

Bocquet-Appel, J.-P. (2005). La paléodémographie. In O. Dutour, J. J. Hublin, & B. Vandermeersch (Eds.), *Objets et méthodes en paléoanthropologie* (pp. 271–313). Paris: Comité des travaux historiques et scientifiques.

Bocquet-Appel, J.-P., & Bacro, J.-N. (2008). Estimation of age distribution with its confidence intervals using an iterative Bayesian procedure and a bootstrap sampling approach. In J.-P. Bocquet-Appel (Ed.), *Recent advances in paleodemography: Data, techniques, patterns* (pp. 63–82). Dordrecht: Springer.

Bocquet-Appel, J.-P., & Masset, C. (1982). Farewell to paleodemography. *Journal of Human Evolution, 11*, 321–333.

Bocquet-Appel, J.-P., & Masset, C. (1996). Paleodemography: Expectancy and false hope. *American Journal of Physical Anthropology, 99*, 571–583.
Bod, R., Hay, J., & Jannedy, S. (Eds.). (2003). *Probabilistic linguistic*. Cambridge, MA: MIT Press.
Boisguilbert, P. (1695). *Le Détail de la France, ou Traité de la cause de la cause de la diminution des biens, et des moyens d'y remédier*. Rouen.
Boltzmann, L. (1871). Einige allgemeine Sätze über wärmegleichgewicht. *Sitzungberichte, K. Akademie der Wissenshaften, Wien, Mathematisch-Naturwissenchaftlichte Klasse, 63*, 679–711.
Bonneuil, N. (2004). Analyse critique de Pattern and repertoire in history, by Roehner B. and Syme, T. *History and Theory, 43*, 117–123.
Bonvalet, C., Bry, X., & Lelièvre, E. (1997). Analyse biographique des groupes. Les avancées d'une recherche en cours. *Population, 52*(4), 803–830 (English translation (1998). Event history analysis of groups. The findings of an on going research project. *Population, 10*(1), 11–38).
Boole, G. (1854). *An investigation of the laws of thought: On which are founded the mathematical theories of logic and probability*. London: Walton and Maberly.
Bordas-Desmoulins, J.-B. (1843). *Le cartésianisme ou la véritable rénovation des sciences*. Paris: J. Hetzel.
Borel, E. (1898). *Leçons sur la théorie des fonctions*. Paris: Gauthier-Villars et fils.
Borel, E. (1909). *Eléments de la théorie des probabilités*. Paris: Librairie Hermann.
Borel, E. (1914). *Le hasard*. Paris: Librairie Félix Alcan.
Borsboom, D., Mellenberg, G. J., & van Herden, J. (2003). The theoretical status of latent variables. *Psychological Review, 110*(2), 203–219.
Bourgeois-Pichat, J. (1994). *La dynamique des populations. Populations stables, semi-stables, quasi-stables*. Travaux et Documents, Cahier no. 133, Paris: INED/PUF.
Braithwaite, R. B. (1941). Book review of Jeffreys' Theory of probability. *Mind, 50*, 198–201.
Bremaud, P. (1973). *A martingale approach to point processes*. Memorandum ERL-M345, Electronic research laboratory. Berkeley: University of California.
Bretagnolle, J., & Huber-Carol, C. (1988). Effects of omitting covariates in Cox's model for survival data. *Scandinavian Journal of Statistics, 15*, 125–138.
Brian, E. (2001). Nouvel essai pour connaître la population du royaume. Histoire des sciences, calcul des probabilités et population de la France vers 1780. *Annales de Démographie Historique, 102*(2), 173–222.
Brian, E. (2006). Les phénomènes sociaux que saisissait Jakob Bernoulli, aperçus de Condorcet à Auguste Comte. *Journ@l Electronique d'Histoire des Probabilités et de la Statistique, 2*(1b), 1–15.
Brian, E., & Jaisson, M. (2007). *The descent of human sex ratio at birth* (Methodos series, Vol. 4). Dordrecht: Springer.
Broggi, U. (1907). *Die Axiome der Wahrscheinlichkeitsrechnung*. PhD thesis, Universität Göttingen, Göttingen.
Browne, W. J. (1998). *Applying MCMC methods to multi-level models*. PhD thesis, University of Bath, Bath, citeseerx.ist.edu: http://www.ams.ucsc.edu/~draper/browne-PhD-dissert. Accessed July 10, 2011.
Buck, C. E., Cavanagh, W. G., & Litton, W. G. (1996). *Bayesian approach to interpreting archaeological data*. Chichester: Wiley.
Burch, T. (2002). Computer modelling of theory: Explanation for the 21st century. In R. Franck (Ed.), *The explanatory power of models. Bridging the gap between empirical and theoretical models in the social sciences* (Methodos series, Vol. 1, pp. 245–266). Boston/Dordrecht/London: Kluwer Academic Publishers.
Burch, T. (2003). Data, models and theory: The structure of demographic knowledge. In F. C. Billari & A. Prskawetz (Eds.), *Agent-based computational demography. Using simulation to improve our understanding of demographic behaviour* (pp. 19–40). Heidelberg/New York: Physica-Verlag.

Cantelli, F. P. (1905). Sui fondamenti del calcolo delle probabilità. *Il Pitagora. Giornale di matematica per gli alunni delle scuole secondarie, 12*, 21–25.
Cantelli, F. P. (1932). Una teoria astratta del calcolo delle probabilità. *Giornale dell'Istituto Italiano degli Attuari, 8*, 257–265.
Cantelli, F. P. (1935). Considérations sur la convergence dans le calcul des probabilités. *Annales de l'I.H.P., 5*(1), 3–50.
Cantillon, R. (1755). *Essai sur la nature du commerce en général*. London: Fletcher Gyles.
Cantor, G. (1873). Notes historiques sur le calcul des probabilités. In *Comptes-rendus de la session de l'association de recherche scientifique*, Halle, pp. 34–42.
Cantor, G. (1874). Über eine Eigenschaft des Inbegriffes aller reellen algebraischen Zahlen. *Journal de Crelle, 77*, 258–262.
Cardano, J. (1663). Liber De Ludo Aleae. In *Opera Omnia,Tomus I*, Lugduni.
Carnap, R. (1928). *Der logische Aufbau der Welt*. Leipzig: Felix Meiner Verlag (English translation 1967. *The logical structure of the world. Pseudo problems in philosophy*. Berkeley: University of California Press).
Carnap, R. (1933). *L'ancienne et la nouvelle logique*. Paris: Hermann.
Carnap, R. (1950). *The logical foundations of probability*. Chicago: University of Chicago Press.
Carnap, R. (1952). *The continuum of inductive methods*. Chicago: University of Chicago Press.
Carnot, S. (1824). *Réflexions sur la puissance motrice du feu et les machines propres à développer cette puissance*. Paris: Bachelier.
Cartwright, N. (2006). Counterfactuals in economics: A commentary. In M. O'Rourke, J. K. Campbell, & H. Silverstein (Eds.), *Explanation and causation: Topics in contemporary philosophy* (pp. 191–221). Cambridge, MA: MIT Press.
Cartwright, N. (2007). *Hunting causes and using them: Approaches in philosophy and economics*. Cambridge, MA: Cambridge University Press.
Cartwright, N. (2009). Causality, invariance and policy. In H. Kincaid & D. Ross (Eds.), *The Oxford handbook of philosophy of economics* (pp. 410–421). New York: Oxford University Press.
Casini, L., Illari, P. M., Russo, F., & Williamson, J. (2011). Models for prediction, explanation and control: Recursive Bayesian networks. *Theoria, 70*, 5–33.
Caticha, A. (2004). Relative entropy and inductive inference. In G. Erickson & Y. Zhai (Eds.), *Bayesian inference and maximum entropy methods in science and engineering* (Vol. 107, pp. 75–96). Melville: AIP.
Caussinus, H., & Courgeau, D. (2010). Estimating age without measuring it: A new method in paleodemography. *Population-E, 65*(1), 117–144 (Estimer l'âge sans le mesurer en paléodémographie. *Population, 65*(1), 117–145).
Caussinus, H., & Courgeau, D. (2011). Une nouvelle méthode d'estimation de la structure par âge des décès des adultes. In I. Séguy & L. Buchet (Eds.), *Manuel de paléodémographie* (pp. 291–325). Paris: INED.
Charbit, Y. (2010). *The classical foundations of population thought from Plato to Marx*. Heidelberg/London: Springer.
Chikuni, S. (1975). Biological study on the population of the Pacific Ocean perch in the North Pacific. *Bulletin of the Far Seas Fisheries Research Laboratory, 12*, 1–19.
Choquet, G. (1953). Theory of capacities. *Annales de l'Institut Fourier, 5*, 131–296.
Chung, K.-L. (1942). On mutually favourable events. *The Annals of Statistics, 13*, 338–349.
Church, A. (1932). An unsolvable problem of elementary number theory. *American Journal of Mathematics, 58*(2), 345–363.
Church, A. (1940). On the concept of a random sequence. *Bulletin of the American Mathematical Society, 46*(2), 130–135.
Clausius, R. (1865). Ueber verschiedene für die Anwendung bequeme Formen der Hauptgleichungen der mechanischen Wärmetheorie. *Annalen der Physik und Chemie, CXXV*, 353–400.
Clayton, D. (1991). A Monte-Carlo method for Bayesian inference in frailty models. *Biometrics, 47*, 467–485.

Cohen, M. R., & Nagel, E. (1934). *An introduction to logic and scientific method*. New York: Harcourt Brace.
Colom, R., Rebollo, I., Palacios, A., Juan-Espinosa, M., & Kyllonen, P. (2004). Working memory is (almost) perfectly predicted by g. *Intelligence, 32*, 277–296.
Colyvan, M. (2004). The philosophical significance of Cox's theorem. *International Journal of Approximate Reasoning, 37*(1), 71–85.
Colyvan, M. (2008). Is probability the only coherent approach to uncertainty? *Risk Analysis, 28*(3), 645–652.
Comte, A. (1830–1842). *Cours de philosophie positive*. Paris: Bachelier.
Condorcet. (1778). Sur les probabilités. *Histoire de l'Académie Royale de Sciences, 1781*, 43–46.
Condorcet. (1785). *Essai sur l'application de l'analyse à la probabilité des décisions rendues à la pluralité des voix*. Paris: Imprimerie Nationale.
Condorcet. (1786). Mémoire sur le calcul des probabilités. Cinquième partie. Sur la probabilité des faits extraordinaires. *Mémoire pour l'Académie Royale des Sciences pour, 1783*, pp. 553–559 (In *Arithmétique politique. Textes rares ou inédits (1767–1789)*. Paris: Institut National d'Etudes Démographiques, pp. 431–436).
Condorcet. (2004). In J.-P. Schandeler & P. Crépel (Eds.), *Tableau historique des progrès de l'esprit humain. Projets, esquisse, fragments et notes (1772–1794)*. Paris: Institut National d'Etudes Démographiques.
Copeland, A. (1928). Admissible numbers in the theory of probability. *American Journal of Mathematics, 50*, 535–552.
Copeland, A. (1936). Point set theory applied to the random selection of the digits of an admissible number. *American Journal of Mathematics, 58*, 181–192.
Copernic, N. (1543). *De revolutionibus orbium coelestum*. Nuremberg: Johanes Petreius ed.
Coumet, E. (1970). La théorie du hasard est-elle née par hasard? *Annales: Economies, Sociétés Civilisations, 25*(3), 574–598.
Coumet, E. (2003). Auguste Comte, le calcul des chances. Aberration radicale de l'esprit mathématique. *Mathématiques et Sciences Humaines, 41*(162), 9–17.
Council of the Statistical Society of London. (1838). Introduction. *Journal of the Statistical Society of London, 1*(1), 1–5.
Courgeau, D. (1982). Proposed analysis of the French Migration, family and occupation history survey. Multistate Life-History Analysis Task Force Meeting, Laxenburg: IIASA, pp. 1–14.
Courgeau, D. (1985). Interaction between spatial mobility, family and career life-cycle: A French survey. *European Sociological Review, 1*(2), 139–162.
Courgeau, D. (1991). Analyse de données biographiques erronées. *Population, 46*(1), 89–104.
Courgeau, D. (1992). Impact of response error on event history analysis. *Population: An English Selection, 4*, 97–110.
Courgeau, D. (2002). Evolution ou révolutions dans la pensée démographique? *Mathématiques et Sciences Humaines, 40*(160), 49–76.
Courgeau, D. (Ed.). (2003). *Methodology and epistemology of multilevel analysis. Approaches from different social sciences* (Methodos series, Vol. 2). Dordrecht/Boston/London: Kluwer Academic Publishers.
Courgeau, D. (2004a). *Du groupe à l'individu. Synthèse multiniveau*. Paris: INED.
Courgeau, D. (2004b). Probabilités, démographie et sciences sociales. *Mathématiques et Sciences Humaines, 42*(3), 5–19.
Courgeau, D. (2007a). *Multilevel synthesis. From the group to the individual*. Dordrecht: Springer.
Courgeau, D. (2007b). Inférence statistique, échangeabilité et approche multiniveau. *Mathématiques et Sciences Humaines, 45*(179), 5–19.
Courgeau, D. (2009). Paradigmes démographiques et cumulativité. In B. Walliser (Ed.), *La cumulativité du savoir en sciences sociales* (pp. 243–276). Lassay-les-Châteaux: Éditions de l'École des Hautes Études en Sciences Sociales.
Courgeau, D. (2010). Dispersion of measurements in demography: A historical view. *Electronic Journal for History of Probability and Statistics, 6*(1), 1–19 (French version: La dispersion des

mesures démographiques: vue historique. *Journ@l Electronique d'Histoire des Probabilités et de la Statistique, 6*(1), 1–20).
Courgeau, D. (2011). Critiques des méthodes actuellement utilisés. In I. Séguy & L. Buchet (Eds.), *Manuel de paléodémographie* (pp. 255–290). Paris: INED.
Courgeau, D., & Franck, R. (2007). Demography, a fully formed science or a science in the making. *Population-E, 62*(1), 39–45 (La démographie, science constituée ou en voie de constitution? Esquisse d'un programme. *Population, 62*(1), 39–46).
Courgeau, D., & Lelièvre, E. (1986). Nuptialité et agriculture. *Population, 41*(2), 303–326.
Courgeau, D., & Lelièvre, E. (1989). *Analyse démographique des biographies*. Paris: INED (English translation: (1992). *Event history analysis in demography*. Oxford: Clarendon Press. Spanish translation: (2001). *Análisis demográfico de las biografías*. México: El Colegio de México).
Courgeau, D., & Lelièvre, E. (1996). Changement de paradigme en démographie. *Population, 51*(2), 645–654 (English translation: (1997) Changing paradigm in demography. *Population. An English Selection, 9,* 1–10).
Courgeau, D., & Pumain, D. (1993). Spatial population issues. 6. France. In N. van Nimwegen, J.-C. Chesnais, & P. Dykstra (Eds.), *Coping with sustained low fertility in France and the Netherlands* (pp. 127–160). Amsterdam: Swetz & Zeitlinger Publishers.
Cournot, A.-A. (1843). *Exposition de la théorie des chances et des probabilités*. Paris: Hachette.
Cox, R. (1946). Probability, frequency, and reasonable expectation. *American Journal of Physics, 14,* 1–13.
Cox, R. (1961). *The algebra of probable inference*. Baltimore: The John Hopkins Press.
Cox, D. R. (1972). Regression models and life-tables (with discussion). *Journal of the Royal Statistical Society, 34*(2), 187–220.
Cox, D. R. (1975). Partial likelihood. *Biometrika, 62,* 269–276.
Cox, R. (1979). On inference and inquiry. In R. D. Levine & M. Tribus (Eds.), *The maximum entropy formalism* (pp. 119–167). Cambridge, MA: MIT Press.
Cox, D. R. (2006). *Principles of statistical inference*. Cambridge, UK: Cambridge University Press.
Cox, D. R., & Hinkley, D. V. (1974). *Theoretical statistics*. London: Chapman & Hall.
Cox, D. R., & Oakes, D. (1984). *Analysis of survival data*. London/New York: Chapman & Hall.
Craver, C. F. (2007). *Explaining the brain: Mechanisms and the mosaic unity of neurosciences*. Oxford: Clarendon Press.
Cribari-Neto, F., & Zarkos, S. G. (1999). Yet another econometric programming environment. *Journal of Applied Econometrics, 14,* 319–329.
D'Alembert, J. l. R. (1761a). Réflexions sur le calcul des Probabilités. In *Opuscules Mathématiques* (Tome II, pp. 1–25). Dixième Mémoire, Paris: David.
D'Alembert, J. l. R. (1761b). Sur l'application du calcul des probabilités à l'inoculation de la petite vérole. In *Opuscules Mathématiques* (Tome II, pp. 26–46). Onzième mémoire, Paris : David.
D'Alembert, J. l. R. (1768a). Extrait de plusieurs lettres de l'auteur sur différents sujets, écrites dans le courant de l'année 1767. In *Opuscules Mathématiques* (Tome IV, pp. 61–105). Vingt-troisième mémoire, Paris: David.
D'Alembert, J. l. R. (1768b). Extraits de lettres sur le calcul des probabilités et sur les calculs relatifs à l'inoculation. In *Opuscules Mathématiques* (Tome IV, pp. 283–341). Vingt-septième mémoire, Paris: David.
Darden, L. (2002). Strategies for discovering mechanisms: Schema instantiation, modular subassembly, forward/backward chaining. *Philosophy of Science (Supplement), 69,* S354–S365.
Darden, L. (Ed.). (2006). *Reasoning in biological discoveries*. Cambridge, MA: Cambridge University Press.
Daston, L. (1988). *Classical probabilities in the enlightenment*. Princeton: Princeton University.
Daston, L. (1989). L'interprétation classique du calcul des probabilités. *Annales: Economies, Sociétés, Civilisations, 44*(3), 715–731.
David, F. N. (1955). Studies in the history of probability and statistics. I. Dicing and gaming (A note on the history of probability). *Biometrika, 42*(1–2), 1–15.

Davis, J. (2003). The relationship between Keynes's early and later philosophical thinking. In J. Runde & S. Mizuhara (Eds.), *The philosophy of Keynes's economics: Probability, uncertainty, and convention* (pp. 100–110). London/New York: Routledge.
Dawid, A. P. (2000). Causal inference without counterfactuals. *Journal of the American Statistical Association, 95*(450), 407–448.
Dawid, A. P. (2007). Counterfactuals, hypothetical and potential responses: A philosophical examination of statistical causality. In F. Russo & J. Williamson (Eds.), *Causality and probability in the sciences* (Texts in philosophy series, Vol. 5, pp. 503–532). London: College Publications.
Dawid, A. P., & Mortera, J. (1996). Coherent analysis of forensic identification evidence. *Journal of the Royal Statistical Society, 58*(2), 425–453.
de Fermat, P. (1679). *Varia opera mathemetica Petri de Fermat, Senatoris Tolosianis*. Toulouse: Joannen Pech.
de Finetti, B. (1931a). Sul significato soggettivo della probabilità. *Fundamenta Mathematicae, 17*, 298–329.
de Finetti, B. (1931b). Sul concetto di media. *Giornale dell'Instituto Italiano degli Attuari, 2*, 367–396.
de Finetti, B. (1937). La prévision: ses lois logiques, ses sources subjectives. *Annales de l'Institut Henri Poincaré, 7*(Paris), 1–68.
de Finetti, B. (1951). Recent suggestions for the reconciliation of theories of probability. In J. Neyman (Ed.), *Proceedings of the second Berkeley symposium on mathematical statistics and probability* (pp. 217–225). Berkeley: University of California Press.
de Finetti, B. (1952). Sulla preferibilità. *Giornale degli Economisti e Annali de Economia, 11*, 685–709.
de Finetti, B. (1964). Foresight: Its logical laws, its subjective sources. In H. E. Kyburg & H. E. Smokler (Eds.), *Studies in subjective probability* (pp. 95–158). New York: Wiley.
de Finetti, B. (1974). *Theory of probability* (2 vols). London/New York: Wiley.
de Finetti, B. (1985). Cambridge probability theorists. *The Manchester School, 53*, 348–363.
de Moivre, A. (1711). De mensura sortis, seu, de probabilitate eventuum in ludis a casu fortuito pendentibus. *Philosophical Transactions, 27*(329), 213–264.
de Moivre, A. (1718). *The doctrine of chances: Or a METHOD of calculating the probabilities of events in PLAY*. London: Millar (Third edition: 1756).
de Montessus, R. (1908). *Leçons élémentaires sur le calcul des probabilités*. Paris: Gauthier Villars.
de Montmort, P. R. (1713). *Essay d'analyse sur les jeux de hazard* (2nd ed.). Paris: Jacques Quillau.
Degenne, A., & Forsé, M. (1994). *Les réseaux sociaux. Une approche structurale en sociologie*. Paris: Armand Colin.
Degenne, A., & Forsé, M. (1999). *Introducing social networks*. London: Sage.
DeGroot, M. H. (1970). *Optimal statistical decision*. New York: McGraw-Hill.
Delannoy, M. (1895). Sur une question de probabilités traitée par d'Alembert. *Bulletin de la S.M.F., Tome 23*, 262–265.
Delaporte, P. (1941). *Evolution de la mortalité en Europe depuis les origines des statistiques de l'état civil*. Paris: Imprimerie Nationale.
Dellacherie, C. (1978). Nombres au hasard de Borel à Martin Löf, *Gazette des Mathématiques du Québec, 11*, 1978. (Version remaniée de l'Institut de Mathématiques, Université Louis-Pasteur de Strasbourg (1978), 30 p).
Demming, W. E. (1940). On a least squares adjustment of a sample frequency table when the expected marginal totals are known. *Annals of Mathematical Statistics, 11*, 427–444.
Dempster, A. P. (1967). Upper and lower probabilities induced by a multilevel mapping. *Annals of Mathematical Statistics, 38*, 325–339.
Dempster, A. P. (1968). A generalization of Bayesian inference. *Journal of the Royal Statistical Society, Series B, 30*, 205–245.
Deparcieux, A. (1746). *Essai sur les probabilités de la durée de la vie humaine*. Paris: Frères Guerin.
Descartes, R. (1647). Méditations, objections et réponses. In *Œuvres et lettres*. Paris: Gallimard.
Desrosières, A. (1993). *La politique des grands nombres: histoire de la raison statistique*. Paris: La Découverte.

Destutt de Tracy, A. L. C. (1801). *Elémens d'idéologie* (Vol. 4). Paris: Pierre Firmin Didot, An IX (Seconde édition (1804–1818)).
Doob, J. L. (1940). Regularity properties of certain families of chance variables. *Transactions of the American Mathematical Society, 44*(1), 455–486.
Doob, J. L. (1949). Application of the theory of martingales. In *Actes du Colloque International: le Calcul des Probabilités et ses Applications* (pp. 23–27). Paris: CNRS.
Doob, J. L. (1953). *Stochastic processes*. New York: Wiley.
Dormoy, E. (1874). Théorie mathématique des assurances sur la vie. *Journal des Actuaires Français, 3*, 283–299, 432–461.
Draper, D. (1995). Inference and hierarchical modelling in the social sciences (with discussion). *Journal of Educational and Behavioural Statistics, 20*, 115–147, 233–239.
Draper, D. (2008). Bayesian multilevel analysis and MCMC. In J. de Leeuw & E. Meyer (Eds.), *Handbook of multilevel models* (pp. 77–140). New York: Springer.
Dubois, D., & Prade, H. (1988). An introduction to possibilistic and fuzzy logics. In P. Smets, A. Mandani, D. Dubois, & H. Prade (Eds.), *Non standard logics for automated reasoning* (pp. 287–326). New York: Academic.
Duncan, W. J., & Collar, A. R. (1934). A method for the solution of oscillations problems by matrices. *Philosophical Magazine, 17*(Series 7), 865.
Dupâquier, J. (1996). *L'invention de la table de mortalité*. Paris: Presses Universitaires de France.
Durkheim, E. (1895). *Les règles de la méthode sociologique*. Paris: Alcan.
Durkheim, E. (1897). *Le suicide*. Paris: Alcan.
Dussause, H., & Pasquier, M. (1905). *Les Œuvres Économiques de Sir William Petty* (2 vols). Paris: Giard & Brière (French translation of *The Economic Writings of Sir William Petty*, edited by C. H. Hull (2 vols). Cambridge, MA: Cambridge University Press, 1899).
Edgeworth, F. Y. (1883). The method of least squares. *Philosophical Magazine, 5th Series, 16*, 360–375.
Edgeworth, F. Y. (1885a). Observation and statistics. An essay on the theory of errors of observation and the first principles of statistics. *Transactions of the Cambridge Philosophical Society, 14*, 138–169.
Edgeworth, F. Y. (1885b). On methods of ascertaining variations in the rate of births, deaths and marriage. *Journal of the Royal Statistical Society of London, 48*, 628–649.
Edgeworth, F. Y. (1892). Correlated averages. *Philosophical Magazine, 5th Series, 34*, 190–204.
Edgeworth, F. Y. (1893a). Note on the correlation between organs. *Philosophical Magazine, 5th Series, 36*, 350–351.
Edgeworth, F. Y. (1893b). Statistical correlations between social phenomena. *Journal of the Royal Statistical Society, 56*, 852–853.
Edgeworth, F. Y. (1895). On some recent contributions to the theory of statistics. *Journal of the Royal Statistical Society, 58*, 506–515.
Einstein, A. (1905). Zur Elektrodynamik bewegter Körper. *Annalen der Physik, 332*(10), 891–921.
Ellis, R. L. (1849). On the Foundations of the theory of probability. *Transactions of the Cambridge Philosophical Society, VIII*, 1–16.
Ellsberg, D. (1961). Risk, ambiguity and the Savage axioms. *Quarterly Journal of Economics, 75*(4), 643–669.
Ellsberg, D. (2001). *Risk, ambiguity and decision*. New York/London: Garland Publishing Inc.
Eriksson, L., & Hájek, A. (2007). What are degrees of belief? *Studia Logica, 86*, 185–215.
Euler, L. (1760). Recherches générales sur la mortalité et la multiplication du genre humain. *Histoire de l'Académie Royale des Sciences et des Belles Lettres de Berlin, 16*, 144–164.
Feller, W. (1934). Review of Kolmogorov (1933). *Zentralblatt für Mathematik und ihre Grenzegebiete, 7*, 216.
Feller, W. (1950). *An introduction to the theory of probability and its applications* (Vol. 1). New York: Wiley.
Feller, W. (1961). *An introduction to the theory of probability and its applications* (Vol. 2). New York: Wiley.

Fergusson, T. S. (1973). A Bayesian analysis of some parametric problems. *The Annals of Statistics, 1*, 209–230.
Fishburn, P. C. (1964). *Decision and value theory*. New York: Wiley.
Fishburn, P. C. (1975). A theory of subjective probability and expected utilities. *Theory and Decision, 6*, 287–310.
Fishburn, P. C. (1986). The axioms of subjective probability. *Statistical Science, 1*(3), 335–345.
Fisher, R. A. (1922a). *The mathematical theory of probability*. London: Macmillan.
Fisher, R. A. (1922b). On the mathematical foundations of theoretical statistics. *Philosophical Transactions of the Royal Society, Series A, 222*, 309–368.
Fisher, R. A. (1923). Statistical tests of agreement between observation and hypothesis. *Economica, 3*, 139–147.
Fisher, R. A. (1925a). Theory of statistical estimation. *Proceedings of the Cambridge Philosophical Society, 22*, 700–725.
Fisher, R. A. (1925b). *Statistical methods for research workers*. Edinburgh: Olivier and Boyd.
Fisher, R. A. (1933). The concepts of inverse probability and fiducial probability referring to unknown parameters. *Proceedings of the Royal Society, Series A, 139*, 343–348.
Fisher, R. A. (1934). Probability likelihood and quantity of information in the logic of uncertain inference. *Proceedings of the Royal Society, Series A, 140*, 1–8.
Fisher, R. A. (1935). The logic of inductive inference. *Journal of the Royal Statistical Society, 98*, 39–82.
Fisher, R. A. (1956). *Statistical methods and scientific inference*. Edinburgh: Oliver and Boyd.
Fisher, R. A. (1958). The nature of probability. *Centennial Review, 2*, 261–274.
Fisher, R. A. (1960). Scientific thought and the refinement of human reasoning. *Journal of the Operations Research Society of Japan, 3*, 1–10.
Florens, J.-P. (2002). Modèles de durée. In J.-J. Droesbeke, J. Fine, & G. Saporta (Eds.), *Méthodes bayésiennes en statistique* (pp. 315–330). Paris: Editions Technip.
Florens, J. P., Mouchart, M., & Rolin, J.-M. (1999). Semi and non-parametric Bayesian analysis of duration models. *International Statistical Review, 67*(2), 187–210.
Franck, R. (Ed.). (1994). *Faut-il chercher aux causes une raison? L'explication causale dans les sciences humaines*. Paris: Librairie Philosophique Vrin.
Franck, R. (1995). Mosaïques, machines, organismes et sociétés. Examen métadisciplinaire du réductionnisme. *Revue Philosophique de Louvain, 93*, 67–81.
Franck, R. (Ed.). (2002). *The explanatory power of models. Bridging the gap between empirical and theoretical research in the social sciences*. Boston/Dordrecht/London: Kluwer Academic Publishers.
Franck, R. (2007). Peut-on accroître le pouvoir explicatif des modèles en économie? In A. Leroux, & P. Livet (dir.), *Leçons de philosophie économique* (Tome III, pp. 303–354). Paris: Economica.
Franck, R. (2009). Allier l'investigation empirique et la recherché théorique: une priorité. In B. Walliser (Ed.), *La cumulativité du savoir en sciences sociales* (pp. 57–84). Lassay-les-Châteaux: Éditions de l'École des Hautes Études en Sciences Sociales.
Franklin, J. (2001). Resurrecting logical probability. *Erkenntnis, 55*, 277–305.
Fréchet, M. (1915). Sur l'intégrale d'une fonctionnelle étendue à un ensemble abstrait. *Bulletin de la S.M.F., Tome 43*, 248–265.
Fréchet, M. (1937). *Généralités sur le calcul des probabilités. Variables aléatoires*. Paris: Gauthier Villars.
Fréchet, M. (1938). Exposé et discussion de quelques recherches récentes sur les fondements du calcul des probabilités. In R. Wavre (Ed.), *Les fondements du calcul des probabilités* (Vol. II, pp. 23–55). Paris: Hermann.
Fréchet, M. (1951). Rapport général sur les travaux du calcul des probabilités. In R. Bayer (Ed.), *Congrès International de Philosophie des Sciences, Paris, 1949; IV: Calcul des probabilités* (pp. 3–21). Paris: Hermann.
Freund, J. E. (1965). Puzzle or paradox? *The American Statistician, 19*(4), 29–44.

Fridriksson, A. (1934). On the calculation of age distribution within a stock of cods by means of relatively few age-determinations as a key to measurements on a large scale. *Rapports et Procès-verbaux des Réunions du Conseil Permanent International pour l'Exploration des Mers, 86*, 1–14.

Friedman, M., & Savage, L. J. (1948). The utility analysis of choices involving risk. *The Journal of Political Economy, LVI*(4), 279–304.

Frischhoff, B., Slovic, B., & Lichteinstein, S. (1978). Fault trees: Sensitivity of estimated failure probabilities to problem representation. *Journal of Experimental Psychology. Human Perception and Performance, 4*, 330–344.

Gacôgne, L. (1993). *About a foundation of the Dempster's rule*, Rapport 93/27 Laforia.

Gail, M., Wieand, S., & Piantadosi, S. (1984). Biased estimates of treatment effect in randomized experiments with nonlinear regressions and omitted covariates. *Biometrika, 71*, 431–444.

Galileo, G. (1613). *Istoria e dimostrazioni intorno alle macchie solari e loro accidenti*. Roma: Giacomo Mascardi.

Galileo, G. (1898). Sopra le scoperte de i dadi. In *Opere* (Vol. VIII, pp. 591–594). Firenze: Barbera Editore.

Galton, F. (1875). Statistics by intercomparison, with remarks on the law of frequency of error. *Philosophical Magazine, 4th Series, 49*, 33–46.

Galton, F. (1886a). Family likeness in stature. *Proceedings of the Royal Society of London, 40*, 42–72 (Appendix by Hamilton Dickson, J. D., pp. 63–72).

Galton, F. (1886b). Regression towards mediocrity in hereditary stature. *Journal of the Anthropological Institute, 15*, 246–263.

Galton, F. (1888). Co-relations and their measurement, chiefly from anthropometric data. *Proceedings of the Royal Society of London, 45*, 135–145.

Gärdenfors, P., Hansson, B., & Sahlin, N.-E. (Eds.). (1983). *Evidential value: Philosophical, judicial and psychological aspects of a theory*. Lund: Gleerups.

Gardin, J.-C. (2002). The logicist analysis of explanatory theories in archæology. In R. Franck (Ed.), *The explanatory power of models. Bridging the gap between empirical sciences and theoretical research in the social sciences* (pp. 267–284). Boston/Dordrecht/London: Kluwer Academic Publishers.

Garnett, J. C. M. (1919). On certain independent factors in mental measurements. *Proceedings of the Royal Society London, Series A, 96*, 91–111.

Gauss, C. F. (1809). *Theoria motus corporum celestium*. Hamburg: Perthes et Besser.

Gauss, C. F. (1816). Bestimmung der Genauigkeit der boebechtungen. *Zeitshrifte für Astronomie und Verwandte Wissenschaften, 1*, 185–216.

Gauss, C. F. (1823). *Theoria combinationis observationum erroribus minimis obnoxiae*. Göttingen: Dieterich.

Gavrilova, N. S., & Gavrilov, L. A. (2001). Mortality measurement and modeling beyond age 100. *Living to 100 Symposium*, Orlando, Florida. Website: http://www.soa.org/library/monographs/life/living-to-100/2011/mono-li11-5b-gavrilova.pdf. Accessed September 20, 2011.

Gelfand, A. E., & Solomon, H. (1973). A study of Poisson's model for jury verdicts in criminal and civil trials. *Journal of the American Statistical Association, 68*(342), 271–278.

Gelman, A., Karlin, J. B., Stern, H. S., & Rubin, D. B. (1995). *Bayesian data analysis*. New York: Chapman & Hall.

Gergonne, J.-D. (1818–1819). Examen critique de quelques dispositions de notre code d'instruction criminelle. *Annales de Mathématiques Pures et Appliquées (Annales de Gergonne), 9*, 306–319.

Gerrard, B. (2003). Keynesian uncertainty: What do we know? In J. Runde & S. Mizuhara (Eds.), *The philosophy of Keynes's economics: Probability, uncertainty, and convention* (pp. 239–251). London/New York: Routledge.

Ghosal, S. (1996). A review of consistency and convergence rates of posterior distributions. *Proceedings of Varanashi Symposium in Bayesian Inference*. India: Banaras Hindu University, pp. 1–10.

Gibbs, J. W. (1902). *Elementary principles in statistical mechanics*. New Haven: Yale University Press.

Gignac, G. E. (2007). Working memory and fluid intelligence are both identical to g?! Reanalyses and critical evaluation. *Psychological Science, 49*(3), 187–207.
Gignac, G. E. (2008). Higher-order models versus direct hierarchical models: g as superordinate or breadth factor? *Psychology Science Quarterly, 50*(1), 21–43.
Gill, J. (2008). *Bayesian methods. A social and behavioral sciences approach.* Boca Raton: Chapman & Hall.
Gillies, D. (2000). *Philosophical theories of probability.* London/New York: Routledge.
Gillies, D. (2003). Probability and uncertainty in Keynes's economics. In J. Runde & S. Mizuhara (Eds.), *The philosophy of Keynes's economics: Probability, uncertainty, and convention* (pp. 111–129). London/New York: Routledge.
Gimblett, R. (Ed.). (2002). *Integrating geographic information systems and agent-based modelling techniques for simulating social and ecological processes.* New York: Oxford University Press.
Gingerenzer, G., Swijtink, Z., Daston, L. J., Beatty, L., & Krüger, L. (Eds.). (1989). *The empire of chance: How probability changed science and everyday life.* Cambridge, MA: Cambridge University Press.
Glennan, S. (2002). Rethinking mechanical explanations. *Philosophy of Science, 69*(Proceedings), S342–S353.
Glennan, S. (2005). Modeling mechanisms. *Studies in History and Philosophy of Science Part C: Studies in History and Philosophy of Biological and Biomedical Sciences, 36*(2), 375–388.
Gödel, K. (1931). Über formal unentscheidbare Sätze der Principia Mathematica und verwandter Systeme, I. *Monatshefte für Mathematik und Physik, 38*, 173–198.
Goldstein, H. (1986). Multilevel mixed linear model analysis using iterative generalized least-squares. *Biometrika, 73*, 43–56.
Goldstein, H. (1987). Multilevel covariance component models. *Biometrika, 74*, 430–431.
Goldstein, H. (1991). Nonlinear multilevel models, with an application to discrete response data. *Biometrika, 78*, 45–51.
Goldstein, H. (2003). *Multilevel statistical models.* London: Edward Arnold.
Gompertz, B. (1825). On the nature of the function expressive of the law of human mortality. *Philosophical Transactions, 115*, 513.
Gonseth, F. (1975). *Le référentiel, univers obligé de médiatisation.* Lausanne: Editions l'Age d'Homme.
Good, I. J. (1952). Rational decisions. *Journal of the Royal Statistical Society, Series B, 14*, 107–114.
Good, I. J. (1956). Which comes first, probability or statistics? *Journal of the Institute of Actuaries, 42*, 249–255.
Good, I. J. (1962). Subjective probability as the measure of a non-measurable set. In E. Nagel, P. Suppes, & A. Tarski (Eds.), *Logic, methodology and philosophy of science* (pp. 319–329). Stanford: Stanford University Press.
Good, I. J. (1971). 46656 varieties of Bayesians. *The American Statistician, 25*, 62–63.
Good, I. J. (1980). Some history of the hierarchical Bayesian methodology. In J. M. Bernardo et al. (Eds.), *Bayesian statistics* (pp. 489–519). Valencia: University of Valencia Press.
Good, I. J. (1983). *Good thinking.* Minneapolis: University of Minnesota Press.
Gosset, W. S. (Student) (1908a). The probable error of a mean. *Biometrika, 6*(1), 1–25.
Gosset, W. S. (Student) (1908b). Probable error of a correlation coefficient. *Biometrika, 6*(2–3), 302–310.
Gould, S. J. (1981). *The mismeasure of man.* New York: W.W. Norton & Co.
Gouraud, C. (1848). *Histoire du calcul des probabilités depuis ses origines jusqu'à nos jours.* Paris: Auguste Durand.
Graetzer, J. (1883). *Edmund Halley und Caspar Neumann: Ein Beitrag zur Geschichte der Bevölkerungsstatistik.* Breslau: Schottlaender.
Granger, G.-G. (1967). Épistémologie économique. In J. Piaget (Ed.), *Logique et connaissance scientifique* (pp. 1019–1055). Paris: Editions Gallimard.
Granger, G.-G. (1976). *La théorie aristotélicienne de la science.* Paris: Éditions Aubier Montaigne.
Granger, G.-G. (1988). *Essai d'une philosophie du style.* Paris: Editions Odile Jacob.

Granger, G.-G. (1992). A quoi sert l'épistémologie? *Droit et Société, 20/21*, 35–42.
Granger, G.-G. (1994). *Formes, opérations, objets*. Paris: Librairie Philosophique Vrin.
Graunt, J. (1662). *Natural and political observations mentioned in a following index, and made upon the bills of mortality*. London: Tho: Roycroft, for John Martin, James Allestry, and Tho: Dicas (French translation by Vilquin, E. (1977). *Observations Naturelles et Politiques répertoriées dans l'index ci-après et faites sur les bulletins de mortalité*. Paris: INED).
Greenland, S. (1998a). Probability logic and probabilistic induction. *Epidemiology, 9*, 322–332.
Greenland, S. (1998b). Induction versus Popper: Substance versus semantics. *International Journal of Epidemiology, 27*, 543–548.
Greenland, S. (2000). Principles of multilevel modelling. *International Journal of Epidemiology, 29*, 158–167.
Greenland, S., & Poole, C. (1988). Invariants and noninvariants in the concept of interdependent effects. *Scandinavian Journal of Work, Environment and Health, 14*, 125–129.
Grether, D. M., & Plott, C. R. (1979). Economic theory of choice and the preference reversal phenomenon. *The American Economic Review, 69*(4), 623–638.
Guillard, A. (1855). *Eléments de statistique humaine ou démographie comparée*. Paris: Guillaumin.
Gurr, T. R. (1993). *Minorities at risk: A global view of ethnopolitical conflicts*. Washington, DC: United States Institute of Peace Progress.
Gustafson, P. (1998). Flexible Bayesian modelling for survival data. *Lifetime Data Analysis, 4*, 281–299.
Gustafsson, J.-E. (1984). A unifying model of the structure of intellectual abilities. *Intelligence, 8*, 179–203.
Haavelmo, T. (1943). The statistical implications of a system of simultaneous equations. *Econometrica, 11*, 1–12.
Hacking, I. (1965). *Logic of statistical inference*. Cambridge, UK: Cambridge University Press.
Hacking, I. (1975). *The emergence of probability*. Cambridge, UK: Cambridge University Press.
Hacking, I. (1990). *The taming of science*. Cambridge, UK: Cambridge University Press.
Hacking, I. (2001). *An introduction to probability and inductive logic*. Cambridge, UK: Cambridge University Press.
Hadamard, J. (1922). Les principes du calcul des probabilités. *Revue de Métaphysique et de Morale, 29*(3), 289–293.
Hájek, A. (2008a). Dutch book arguments. In P. Anand, P. Pattanaik, & C. Pup (Eds.), *The Oxford handbook of corporate social responsibility*, Phil papers: http://philrsss.anu.edu.au/people-defaults/alanh/papers/DBA.pdf. Accessed July 10, 2011.
Hájek, A. (2008b). Probability. In B. Gold (Ed.), *Current issues in the philosophy of mathematics: From the perspective of mathematicians*, Mathematical Association of America, Phil papers. http://philrsss.anu.edu.au/people-defaults/alanh/papers/overview.pdf. Accessed July 10, 2011.
Hald, A. (1990). *A history of probability and statistics and their applications before 1750*. New York: Wiley.
Hald, A. (1998). *A history of mathematical statistics from 1750 to 1930*. New York: Wiley.
Hald, A. (2007). *A history of parametric statistical inference from Bernoulli to Fisher, 1713–1935*. New York: Springer.
Halley, E. (1693). An estimate of the degrees of the mortality of mankind, drawn from curious tables of the births and funeral's at the City of Breslau; with an attempt to ascertain the price of the annuities upon lives. *Philosophical Transactions Giving some Accounts of the Present Undertaking, Studies and Labour of the Ingenious in many Considerable Parts of the World, XVII*(196), 596–610.
Halpern, J. Y. (1999a). A counterexample of to theorems of Cox and Fine. *The Journal of Artificial Intelligence Research, 10*, 67–85.
Halpern, J. Y. (1999b). Technical addendum, Cox's theorem revisited. *The Journal of Artificial Intelligence Research, 11*, 429–435.
Halpern, J. Y., & Koller, D. (1995). Representation dependence in probabilistic inference. *Proceedings of the Fourteenth International Joint Conference on Artificial Intelligence (IJCAI'95)*, Montreal, Quebec, Canada, pp. 1853–1860.

Halpern, J. Y., & Koller, D. (2004). Representation dependence in probabilistic inference. *The Journal of Artificial Intelligence Research, 21*, 319–356.

Hanson, T. E. (2006). Modeling censored lifetime data using a mixture of Gamma baseline. *Bayesian Analysis, 1*(3), 575–594.

Hanson, T. E., & Johnson, W. E. (2002). Modeling regression error with a mixture of Polya trees. *Journal of the American Statistical Association, 97*, 1020–1033.

Harr, A. (1933). Der massbegrieff in der Theorie der kontinuierlichen Gruppen. *Annals of Mathematical Statistics, 34*, 147–169.

Hasofer, A. M. (1967). Studies in the history of probability and statistics. XVI. Random mechanisms in Talmudic literature. *Biometrika, 54*(1/2), 316–321.

Hasselblad, V. (1966). Estimation of parameters for a mixture of normal distribution. *Technometrics, 8*(3), 431–444.

Hecht, J. (1977). L'idée de dénombrement jusqu'à la révolution. In *Pour une histoire de la statistique* (Vol. Tome 1/Contributions, pp. 21–81). Paris: INSEE.

Heckman, J., & Singer, B. (1982). Population heterogeneity in demographic models. In K. Land & A. Rogers (Eds.), *Multidimensional mathematical demography* (pp. 567–599). New York: Academic.

Heckman, J., & Singer, B. (1984a). Econometric duration analysis. *Journal of Econometrics, 24*, 63–132.

Heckman, J., & Singer, B. (1984b). A method for minimizing distributional assumptions in econometric models for duration data. *Econometrica, 52*(2), 271–320.

Hempel, C. G., & Oppenheim, P. (1948). Studies in the logic explanation. *Philosophy of Science, 15*, 567–579.

Henry, L. (1957). Un exemple de surestimation de la mortalité par la méthode de Halley. *Population, 12*(1), 141–142.

Henry, L. (1959). D'un problème fondamental de l'analyse démographique. *Population, 14*(1), 9–32.

Henry, L. (1966). Analyse et mesure des phénomènes démographiques par cohorte. *Population, 13*(1), 465–482.

Henry, L. (1972). *Démographie. Analyse et modèles*. Paris: Larousse.

Henry, L. (1981). Dictionnaire démographique multilingue. Liège: UIESP, Ordina éditions (English translation: Adapted by Van de Valle, E. (1982). Multilingual demographic dictionary. Liège: IUSSP, Ordina éditions).

Henry, L., & Blayo, Y. (1975). La population de la France de 1740 à 1860. *Population, 30*, 71–122.

Hespel, B. (1994). Revue sommaire des principales théories contemporaines de la causation. In R. Franck (Ed.), *Faut-il chercher aux causes une raison ? L'explication causale dans les sciences humaines* (pp. 223–231). Paris: Librairie Philosophique J. Vrin.

Hilbert, D. (1902). Mathematical problems. *Bulletin of the American Mathematical Society, 8*(2), 437–439.

Hoem, J. (1983). Multistate mathematical demography should adopt the notions of event history analysis. *Stockholm Research Reports in Demography, 10*, 1–17.

Hofacker, J. D. (1829). Extrait d'une lettre du professeur Hofacker au rédacteur de la Gazette médico-chirurgicale d'Innsbruck. *Annales d'hygiène publique et de médecine légale, 1*(01), 557–558.

Holland, P. (1986). Statistics and causal inference (with comments). *Journal of the American Statistical Association, 81*, 945–970.

Hooper, G. (1699). A calculation of the credibility of human testimony. Anonymous translation in *Philosophical Transactions of the Royal Society, 21*, 359–365.

Hooten, M. B., & Wilke, C. K. (2010). Statistical agent-based models for discrete spatio-temporal systems. *Journal of the American Statistical Association, 105*(489), 236–248.

Hooten, M. B., Anderson, J., & Waller, L. A. (2010). Assessing North American influenza dynamics with a statistical SIRS model. *Spatial and Spatio-temporal Epidemiology, 1*, 177–185.

Hoppa, R. D., & Vaupel, J. W. (Eds.). (2002). *Paleodemography. Age distribution from skeletons samples*. Cambridge, UK: Cambridge University Press.

Horn, J. L., & Catell, R. B. (1966). Refinement and test of the theory of fluid and crystallised general intelligence. *Journal of Educational Psychology, 57*, 253–270.

Hunt, G. A. (1966). *Martingales et processus de Markov*. Paris: Dunod.

Hunter, D. (1989). Causality and maximum entropy updating. *International Journal of Approximate Reasoning, 3*, 87–114.

Hutter, M. (2001). Towards a universal theory of artificial intelligence based on an algorithmic probability and sequential decisions. In L. De Raedt & P. Flash (Eds.), *Proceedings of the 12th European conference on machine learning* (Lecture notes on artificial intelligence series). New York/Berlin/Heidelberg: Springer.

Huygens, C. (1657). *De ratiociniis in ludo aleae*. Leyde: Elzevier.

Huygens, C. (1895). Correspondance 1666–1669. In *Œuvres complètes* (Vol. Tome Sixième). La Haye: Martinus Nijhoff.

Ibrahim, J. G., Chen, M.-H., & Sinha, D. (2001). *Bayesian survival analysis*. New York/Berlin/Heidelberg: Springer.

Illari, P. M., & Williamson, J. (2010). Function and organization: Comparing the mechanisms of protein synthesis and natural selection. *Studies in History and Philosophy of Biological and Biomedical Sciences, 41*, 279–291.

Illari, P. M., Russo, F., & Williamson, J. (Eds.). (2011). *Causality in the sciences*. Oxford: Oxford University Press.

Inkelmann, F., Murrugarra, D., Jarrah, A. S., & Laubenbacher, R. (2010). *A mathematical framework for agent based models of complex biological networks*. ArXiv: 1006.0408v5 [q-bio.QM], 23 p.

Irwin, J. O. (1941). Book review of Jeffreys' Theory of probability. *Journal of the Royal Statistical Society, 104*, 59–64.

Jacob, F. (1970). *La logique du vivant*. Paris: Gallimard.

Jacob, P. (1980). *L'empirisme logique: ses antécédents, ses critiques*. Paris: Editions de Minuit.

Jaynes, E. T. (1956). *Probability theory in science and engineering*. Dallas: Socony-Mobil Oil Co.

Jaynes, E. T. (1957). *How does brain do plausible reasoning?* (Report 421), Microwave Laboratory, Stanford University (Published 1988, In G. J. Erickson, & C. R. Smith (Eds.), *Maximum entropy and Bayesian methods in science and engineering* (Vol. 1, pp. 1–24). Dordrecht: Kluwer Academic Publishers).

Jaynes, E. T. (1963). Information theory and statistical mechanics. In K. Ford (Ed.), *Statistical physics* (pp. 181–218). New York: Benjamin.

Jaynes, E. T. (1968). Prior probabilities. *IEEE Transactions on Systems Science and Cybernetics, SSC-4*, 227–241.

Jaynes, E. T. (1976). Confidence intervals vs. Bayesian intervals. In R. G. Harper & G. Hooker (Eds.), *Foundations of probability theory, statistical inference, and statistical theories of science* (Vol. II, pp. 175–257). Dordrecht-Holland: D. Reidel Publishing Company.

Jaynes, E. T. (1979). Where do we stand on maximum entropy? In R. D. Levine & M. Tribus (Eds.), *The maximum entropy formalism* (pp. 15–118). Cambridge, MA: MIT Press.

Jaynes, E. T. (1980). Marginalization and prior probabilities. In A. Zellner (Ed.), *Bayesian analysis in econometrics and statistics* (pp. 43–87). Amsterdam: North Holland.

Jaynes, E. T. (1988). How does the brain do plausible reasoning. In G. J. Erickson & C. R. Smith (Eds.), *Maximum-entropy and Bayesian methods in science and engineering* (Vol. 1, pp. 1–24). Dordrecht: Kluwer.

Jaynes, E. T. (1990). Probability theory as logic. In P. F. Fougère (Ed.), *Maximum entropy and Bayesian methods* (pp. 1–16). Dordrecht: Kluwer Academic Publishers.

Jaynes, E. T. (1991). *How should we use entropy in economics?* Unpublished works by Edwin Jaynes. http://bayes.wustl.edu/etj/articles/entropy.in.economics.pdf. Accessed July 10, 2011.

Jaynes, E. T. (2003). *Probability theory: The logic of science*. Cambridge, UK: Cambridge University Press.

Jeffreys, H. (1931). *Scientific inference*. Cambridge, UK: Cambridge University Press.
Jeffreys, H. (1932). On the theory of errors and least squares. *Proceedings of the Royal Society, Series A, 138*, 48–55.
Jeffreys, H. (1933a). Probability, statistics, and the theory of errors. *Proceedings of the Royal Society, Series A, 140*, 523–535.
Jeffreys, H. (1933b). On the prior probability in the theory of sampling. *Proceedings of the Cambridge Philosophical Society, 29*, 83–87.
Jeffreys, H. (1934). Probability and scientific method. *Proceedings of the Royal Society, Series A, 146*, 9–16.
Jeffreys, H. (1937). On the relation between direct and inverse methods in statistics. *Proceedings of the Royal Society, Series A, 160*, 325–348.
Jeffreys, H. (1939). *Theory of probability*. New York: Clarendon Press.
Jeffreys, H. (1946). An invariant form for the prior probability in estimation problems. *Proceedings of the Royal Society of London, Series A, 186*, 453–461.
Jeffreys, H. (1955). The present position in probability theory. *The British Journal for the Philosophy of Science, 5*, 275–289.
Johnson, W. E. (1932). Probability: The deductive and inductive problem. *Mind, 41*, 409–423.
Jones, K. (1993). *'Everywhere is nowhere': Multilevel perspectives on the importance of place*. Portsmouth: The University of Portsmouth Inaugural Lectures.
Jöreskog, K. G., & van Thillo, M. (1972). *LISREL: A general computer program for estimating a linear structural equations system involving indicators of unmeasured variables*. Princeton: Educational Testing Service. Educational Resources Information Center (ERIC). http://www.eric.ed.gov/ERICWebPortal/search/detailmini.jsp?_nfpb=true&_&ERICExtSearch_SearchValue_0=ED073122&ERICExtSearch_SearchType_0=no&accno=ED073122. Accessed August 19, 2011.
Kahneman, D., & Tversky, A. (1972). Subjective probability: A judgement of representativeness. *Cognitive Psychology, 3*, 430–454.
Kahneman, D., & Tversky, A. (1979). Prospect theory: An analysis of decision under risk. *Econometrica, 47*(2), 263–291.
Kalbfleisch, J. D. (1978). Non-parametric Bayesian analysis of survival time data. *Journal of the Royal Statistical Society, Series B, 40*, 214–221.
Kalbfleisch, J. D., & Prentice, R. L. (1980). *The statistical analysis of failure time data*. New York/Chichester/Brisbane/Toronto: Wiley.
Kamke, E. (1932). Über neure Begründungen der Wahrscheinlichkeitsrechnung. *Jahresbericht der Deutschen Mathematiker-Vereinigung, 42*, 14–27.
Kaplan, M. (1996). *Decision theory as philosophy*. Cambridge, MA: Cambridge University Press.
Kaplan, E. L., & Meier, P. (1958). Non-parametric estimation from incomplete observations. *Journal of the American Statistical Association, 53*, 457–481.
Kardaun, O. J. W. F., Salomé, D., Schaafsma, W., Steerneman, A. G. M., Willems, J. C., & Cox, D. R. (2003). Reflections on fourteen cryptic issues concerning the nature of statistical inference. *International Statistical Review, 71*(2), 277–318.
Karlis, D., & Patilea, V. (2007). Confidence hazard rate functions for discrete distributions using mixtures. *Computational Statistics & Data Analysis, 51*(11), 5388–5401.
Kass, R. E., & Wassetman, L. (1996). The selection of prior distribution by formal rules. *Journal of the American Statistical Association, 91*, 1343–1369.
Kaufmann, W. (1906). Über die Konstitution des Elektrons. *Annalen der Physik, 324*(3), 487–553.
Keiding, N. (1990). Statistical inference in the Lexis diagram. *Philosophical Transactions of the Royal Society of London, 332*, 487–509.
Kendall, M. G. (1956). Studies in the history of probability and statistics. II. The beginnings of a probability calculus. *Biometrika, 43*(1/2), 1–14.
Kendall, M. G. (1960). Studies on the history of probability and statistics. X. Where shall the history of statistics begin? *Biometrika, 47*(3/4), 447–449.

Kendall, M. G. (1963). Ronald Aylmer Fisher, 1890–1962. *Biometrika, 50*(1/2), 1–15.
Kersseboom, W. (1742). Troisième traité sur la grandeur probable de la population de Hollande et de Frise occidentale. In *Essais d'Arithmétique politique contenant trois traités sur la population de la province de Hollande et de Frise occidentale*, Paris: Editions de l'Ined, 1970.
Keynes, J. M. (1921). *A treatise on probability*. London: Macmillan.
Keynes, J. M. (1971). Essay in biography. In *The collected writings of John Maynard Keynes* (Vol. X). London: Macmillan.
Kimura, D. K. (1977). Statistical assessment of the age-length key. *Journal of the Fisheries Research Board of Canada, 34*, 317–324.
Kimura, D. K., & Chikuni, S. (1987). Mixture of empirical distributions: An iterative application of the age-length key. *Biometrics, 43*, 23–35.
Klotz, L. H. (1999). Is the rate of testicular cancer increasing? *Canadian Medical Association Journal, 160*, 213–214.
Knuth, K. H. (2002). What is a question? In C. Williams (Ed.), *Bayesian inference and maximum entropy methods in science and engineering, Moscow ID, 2002* (AIP conference proceedings, Vol. 659, pp. 227–242). Melville: American Institute of Physics.
Knuth, K. H. (2003a). Intelligent machines in the twenty-first century: Foundations of inference and inquiry. *Philosophical transactions of the Royal Society of London, Series A, 361*, 2859–2873.
Knuth, K. H. (2003b). Deriving laws from ordering relations. In G. J. Erickson & Y. Zhai (Eds.), *Bayesian inference and maximum entropy methods in science and engineering, Jackson Hole WY, USA, 2003* (AIP conference proceedings, Vol. 707, pp. 204–235). Melville: American Institute of Physics.
Knuth, K. H. (2005). Lattice duality: The origin of probability and entropy. *Neurocomputing, 67*, 245–274.
Knuth, K. H. (2007). Lattice theory, measures, and probability. In K. H. Knuth, A. Caticha, J. L. Center, A. Giffin, & C. C. Rodríguez (Eds.), *Bayesian inference and maximum entropy methods in science and engineering, Saratoga Springs, NY, USA, 2007* (AIP conference proceedings, Vol. 954, pp. 23–36). Melville: American Institute of Physics.
Knuth, K. H. (2008). The origin of probability and entropy. In M. de Souza Lauretto, C. A. I. de Bragança Pereira, & J. M. Stern (Eds.), *Bayesian inference and maximum entropy methods in science and engineering, São Paulo, Brazil 2008* (AIP conference proceedings, Vol. 1073, pp. 35–48). Melville: American Institute of Physics.
Knuth, K. H. (2009). *Measuring on lattices*. Arxiv preprint: arXiv0909.3684v1 (math..GM). Accessed July 11, 2011.
Knuth, K. H. (2010a). *Foundations of inference*. Arxiv preprint: arXiv1008.4831v1 (math..PR). Accessed July 11, 2011.
Knuth, K. H. (2010b). *Information physics: The new frontier*. Arxiv preprint: arXiv1009. 5161v1 (math.ph). Accessed July 11, 2011.
Kolmogorov, A. (1933). Grundbegriffe der wahrscheinlichkeitsrenung. In *Ergebisne der Mathematik* (Vol. 2). Berlin: Springer (English translation, Morrison, N. (1950). *Foundations of the theory of probability*. New York: Chelsea).
Kolmogorov, A. (1951). Вероятность (Probability). In *Great Soviet encyclopedia* (Vol. 7, pp. 508–510). Moscow: Soviet Encyclopedia Publishing House.
Kolmogorov, A. (1965). Three approaches to the quantitative definition of information. *Problems of Information Transmission, 1*(1), 1–7.
Konigsberg, L. W., & Frankenberg, S. R. (1992). Estimation of age structure in anthropological demography. *American Journal of Physical Anthropology, 89*, 235–256.
Konigsberg, L. W., & Frankenberg, S. R. (2002). Deconstructing death in paleodemography. *American Journal of Physical Anthropology, 117*, 297–309.
Konigsberg, L. W., & Herrmann, N. P. (2002). Markov Chain Monte Carlo estimation of hazard models parameters in paleodemography. In R. D. Hoppa & J. W. Vaupel (Eds.), *Paleodemography. Age distribution from skeletons samples* (pp. 222–242). Cambridge, MA: Cambridge University Press.

Koopman, B. O. (1940). The axioms and algebra of intuitive probability. *Annals of Mathematics, 41*(2), 269–292.
Koopman, B. O. (1941). Intuitive probabilities and sequences. *Annals of Mathematics, 42*(1), 169–187.
Kraft, C. H., Pratt, J. W., & Seidenberg, A. (1959). Intuitive probability on finite sets. *Annals of Mathematical Statistics, 30*, 408–419.
Krüger, L., Daston, L. J., & Heidelberg, M. (Eds.). (1986). *The probabilistic revolution*. Cambridge, UK: Cambridge University Press.
Kruithof, R. (1937). Telefoonverkeersreking. *De Ingenieur, 52*(8), E15–E25.
Kuhn, T. (1962). *The structure of scientific revolutions*. Chicago/London: The University of Chicago Press.
Kuhn, T. (1970). Postscript-1969. In *The structure of scientific revolutions* (2nd ed., pp. 174–210). Kuhn/Chicago/London: The University of Chicago Press.
Kuhn, R., Everett, B., & Silvey, R. (2011). The effects of children's migration on elderly kin's health: A counterfactual approach. *Demography, 48*(1), 183–209.
Kullbach, S., & Leiber, R. A. (1951). On information and sufficiency. *Annals of Mathematical Statistics, 22*(1), 79–86.
Kumar, D. (2010). Bayesian and hierarchical analysis of response-time data with concomitant variables. *Journal of Biomedical Science and Engineering, 3*, 711–718.
Kyburg, H. (1978). Subjective probability: Criticisms, reflections and problems. *Journal of Philosophical Logic, 7*, 157–180.
La Harpe, J.-F. (1799). *Lycée, ou cours de littérature ancienne et moderne* (Vol. 14). Paris: Chez H. Agasse (An VII).
Laemmel, R. (1904). *Untersuchungen über die Ermittlung von Wahrscheinlichkeiten*. PhD thesis, Universität Zürich, Zurich.
Lambert, J. H. (1764). *Neues Organon oder gedanken über die Erforschung und Bezeichnung des wahren und dessen underscheidung von irrthum und schein*. Leipzig: Johan Wendler.
Lambert, J. H. (1772). *Beyträge zum Gebrauche der Mathematik und deren Anwendung* (Vol. III). Berlin: Verlag der Buchhandlung der Realschule.
Landry, A. (1909). Les trois théories principales de la population. *Scientia, 6*, 1–29.
Landry, A. (1945). *Traité de démographie*. Paris: Payot.
Laplace, P. S. (1774). Mémoire sur la probabilité des causes par les événements. *Mémoires de l'Académie Royale des Sciences de Paris, Tome VI*, 621–656.
Laplace, P. S. (1778). Mémoire sur les probabilités. *Mémoires de l'Académie Royale des sciences de Paris, 1781*, 227–332.
Laplace, P. S. (1782). Mémoire sur les approximations des formules qui sont fonction de très grands nombres. *Mémoires de l'Académie Royale des sciences de Paris, 1785*, 1–88.
Laplace, P. S. (1783a). Mémoire sur les approximations des formules qui sont fonctions de très-grands nombres (suite). *Mémoires de l'Académie Royale des sciences de Paris, 1786*, 423–467.
Laplace, P. S. (1783b). Sur les naissances, les mariages et les morts à Paris, depuis 1771 jusqu'à 1784, et dans toute l'étendue de la France, pendant les années 1781 et 1782. *Mémoires de l'Académie Royale des sciences de Paris, 1786*, 693–702.
Laplace, P. S. (1809a). Mémoire sur les approximations des formules qui sont fonction de très grands nombres et sur leur application aux probabilités. *Mémoires de l'Académie Royale des sciences de Paris, 1810*, 353–415.
Laplace, P. S. (1809b). Supplément au mémoire sur les approximations des formules qui sont fonction de très grands nombres. *Mémoires de l'Académie Royale des sciences de Paris, 1810*, 559–565.
Laplace, P. S. (1812). *Théorie analytique des probabilités* (Vols. 2). Paris: Courcier Imprimeur.
Laplace, P. S. (1814). *Essai philosophique sur les probabilités*. Paris: Courcier Imprimeur.
Laplace, P. S. (1816). Premier supplément sur l'application du calcul des probabilités à la philosophie naturelle. In *Œuvres complètes* (Vol. 13, pp. 497–530). Paris: Gauthier-Villars.
Laplace, P. S. (1827). Mémoire sur le flux et le reflux lunaire atmosphérique. In *Connaissance des Temps pour l'an 1830* (pp. 3–18). Paris: Veuve Coursier Imprimeur.

Lazarsfeld, P. (Ed.). (1954). *Mathematical thinking in the social science.* Glencoe: The Free Press.
Lazarsfeld, P. F., & Henry, N. W. (1968). *Latent structure analysis.* Boston: Hougthton Mifflin.
Lazarsfeld, P. F., & Menzel, H. (1961). On the relation between individual and collective properties. In A. Etzioni (Ed.), *Complex organizations* (pp. 422–440). New York: Holt, Reinhart and Winston.
Le Bras, H. (1971). Géographie de la fécondité française depuis 1921. *Population, 26*(6), 1093–1124.
Le Bras, H. (2000). *Naissance de la mortalité. L'origine politique de la statistique et de la démographie.* Paris: Seuil/Gallimard.
Lebesgue, H. (1901). Sur une généralisation de l'intégrale définie. *Comptes Rendus de l'Académie des Sciences, 132,* 1025–1028.
Lecoutre, B. (2004). Expérimentation, inférence statistique et analyse causale. *Intellectica, 38*(1), 193–245.
Lee, P. M. (1989). *Bayesian statistics.* London: Arnold.
Legendre, A. M. (1805). *Nouvelles méthodes pour la détermination des orbites des comètes.* Paris: Coursier.
Legg, S. (1997). *Solomonoff induction* (Technical Report CDMTCS-030), Centre for Discrete Mathematics and Theoretical Computer Science, University of Auckland. http://www.vetta.org/documents/disSol.pdf. Accessed August 20, 2011.
Leibniz, G. W. (1666). *Dissertatio de arte combinatoria.* Leipzig (French translation: Peyroux, J. (1986). *Dissertation sur l'art combinatoire.* Paris: Blanchard).
Leibniz, G. W. (1675). *De problemata mortalitatis propositum per ducem de Roannez.* Partie A du manuscrit traduite en français par M. Parmentier (1995). Paris: Librairie Philosophique Vrin.
Leibniz, G. W. (1765). *Nouveaux essais sur l'entendement humain.* Paris: GF-Flammarion.
Leibniz, G. W. (1995). L'estime des apparences: *21 manuscrits de Leibniz sur les probabilités, la théorie des jeux, l'espérance de vie,* Texte établi, traduit, introduit et annoté par M. Parmentier. Paris: Librairie Philosophique Vrin.
Leonard, T., & Hsu, S. J. (1999). *Bayesian methods. An analysis for statisticians and interdisciplinary researchers.* Cambridge, UK: Cambridge University Press.
Lévy, P. (1925). *Calcul des probabilités.* Paris: Gauthier-Villars.
Lévy, P. (1936). Sur quelques points de la théorie des probabilités dénombrables. *Annales de l'Institut Henri Poincaré, 6*(2), 153–184.
Lévy, P. (1937). *Théorie de l'addition des variables aléatoires.* Paris: Gauthier-Villars.
Lewis, D. K. (1973a). Causation. *The Journal of Philosophy, 70,* 556–567.
Lewis, D. K. (1973b). *Counterfactuals.* Oxford: Basil Blackwell.
Lexis, W. (1877). *Zur Theorie der Massenerscheinungen in der menschlichen Gesellschaft.* Freiburg: Wagner.
Lexis, W. (1879). Über die Theorie der Stabilität statistischer Reihen. *Jahrbücher für Nationalökonomie und Statistik, 32,* 60–98.
Lexis, W. (1880). Sur les moyennes normales appliquées aux mouvements de la population et sur la vie normale. *Annales de Démographie internationale, 4,* 481–497.
Lillard, L. A. (1993). Simultaneous equations fir hazards: Marriage duration and fertility timing. *Journal of Econometrics, 56,* 189–217.
Lindley, D. V. (1956). On a measure of information provided by an experiment. *Annals of Mathematical Statistics, 27*(4), 986–1005.
Lindley, D. V. (1962). Book review of the third edition of Jeffreys' Theory of probability. *Journal of the American Statistical Association, 57,* 922–924.
Lindley, D. V. (1977). A problem in forensic science. *Biometrika, 64*(2), 207–213.
Lindley, D. V. (2000). The philosophy of statistics. *Journal of the Royal Statistical Society, Series D (The Statistician), 49*(3), 293–337.
Lindley, D. V., & Novick, M. R. (1981). The role of exchangeability in inference. *The Annals of Statistics, 9,* 45–58.
Lindley, D. V., & Smith, A. F. M. (1972). Bayes estimates for the linear model. *Journal of the Royal Statistical Society, Series B (Methodological), 34,* 1–41.

Lindsay, B. G. (1995). *Mixture models: Theory, geometry and applications*. Hayward: Institute of Mathematical Statistics.
Little, D. (2010). *New contributions to the philosophy of history* (Methodos series, Vol. 6). Dordrecht/Heidelberg/London/New York: Springer.
Łomnicki, A. (1923). Nouveaux fondements du calcul des probabilités. *Fundamenta Mathematicae, 4*, 34–71.
Loomes, G., & Sugden, R. (1982). Regret theory: An alternative theory of rational choice under uncertainty. *The Economic Journal, 92*, 805–824.
Lotka, A. J. (1939). *Théorie analytique des associations biologiques; Deuxième partie. Analyse démographique avec application particulière à l'espèce humaine*. Paris: Herman.
Louçã, F. (2007). *The years of high econometrics: A short history of the generation that reinvented economics*. London/New York: Routledge.
Luce, R. D., & Krantz, D. H. (1971). Conditional expected utility. *Econometrica, 39*(2), 253–271.
Luce, R. D., & Suppes, P. (1965). Preference, utility and subjective probability. In R. D. Luce, R. R. Bush, & E. H. Galanter (Eds.), *Handbook of mathematical psychology* (Vol. 3, pp. 249–410). New York: Wiley.
Machamer, P., Darden, L., & Craver, C. (2000). Thinking about mechanisms. *Philosophy of Science, 67*, 1–25.
Machina, M. J. (1982). "Expected utility" analysis without the independence axiom. *Econometrica, 50*(2), 277–323.
Macy, M. W., & Willer, R. (2002). From factors to actors: Computational sociology and agent based modelling. *Annual Review of Sociology, 28*, 143–166.
Madden, E. H. (1969). A third view of causality. *The Review of Metaphysics, XXIII*, 67–84.
Mandel, D. R. (2005). Are risk assessments of a terrorist attack coherent? *Journal of Experimental Psychology Applied, 11*, 277–288.
Mandel, D. R. (2008). Violations of coherence in subjective probability: A representational and assessment processes account. *Cognition, 106*, 130–156.
Manski, C. F., & McFadden, D. L. (1981). *Structural analysis of discrete data and econometric applications*. Cambridge, UK: The MIT Press.
Manton, K. G., Singer, B., & Woodbury, M. A. (1992). Some issues in the quantitative characterization of heterogeneous populations. In J. Trussel, R. Hankinson, & J. Tilton (Eds.), *Demographic applications of event history analysis* (pp. 9–37). Oxford: Clarendon Press.
March, L. (1908). Remarques sur la terminologie en statistique. In *Congrès de mathématiques de Rome*, JSSP, pp. 290–296.
Martin, T. (Ed.). (2003). *Arithmétique politique dans la France du XVIIIe siècle*. Classiques de l'économie et de la population. Paris: INED.
Martin-Löf, P. (1966). The definition of random sequences. *Information and Control, 7*, 602–619.
Masset, C. (1971). Erreurs systématiques dans la détermination de l'âge par les sutures crâniennes. *Bulletins et Mémoires de la Société d'Anthropologie de Paris, 12*(7), 85–105.
Masset, C. (1982). Estimation de l'âge au décès par les sutures crâniennes. PhD thesis, University Paris VII, Paris.
Masset, C. (1995). Paléodémographie: problèmes méthodologiques. *Cahiers d'Anthropologie et Biométrie Humaine, XIII*(1–2), 27–38.
Masterman, M. (1970). The nature of a paradigm. In I. Lakatos & A. Musgrave (Eds.), *Criticism and the growth of knowledge* (pp. 59–89). Cambridge, MA: Cambridge University Press.
Matalon, B. (1967). Epistémologie des probabilités. In J. Piaget (Ed.), *Logique et connaissance scientifique* (pp. 526–553). Paris: Gallimard.
Maupin, M. G. (1895). Note sur une question de probabilités traitée par d'Alembert dans l'encyclopédie. *Bulletin de la S.M.S., Tome 23*, 185–190.
Maxwell, J. C. (1860). Illustration of the dynamical theory of gases. *The London, Edimburg, and Dubling Philosophical Magazine and Journal of Science, XIX*, 19–32.

McCrimmon, K., & Larson, S. (1979). Utility theory: Axioms versus "paradoxes". In M. Allais & O. Hagen (Eds.), *Expected utility hypotheses and the Allais paradox* (pp. 333–409). Dordrecht: D. Reidel.
McCulloch, W. S., & Pitts, W. (1943). A logical calculus of the ideas immanent in nervous activity. *The Bulletin of Mathematical Biophysics, 5*, 115–133.
McKinsey, J. C. C., Sugar, A., & Suppes, P. (1953). Axiomatic foundations of classical particle mechanics. *Journal of Rational Mechanics and Analysis, 2*, 273–289.
McLachlan, G. J., & Peel, D. (2000). *Finite mixture models*. New York: Wiley.
Meadows, D. H., Meadows, D. L., Randers, J., & Behrens, W. W., III. (1972). *The limits of growth*. New York: Universe Books.
Menken, J., & Trussel, J. (1981). Proportional hazards life table models: An illustrative analysis of socio-demographic influences on marriages and dissolution in the United States. *Demography, 18*(2), 181–200.
Meusnier, N. (2004). Le problème des partis avant Pacioli. In E. Barbin & J.-P. Lamarche (Eds.), *Histoires de probabilités et de statistiques* (pp. 3–23). Paris: Ellipses.
Meyer, P.-A. (1972). *Martingales and stochastic integrals*. Berlin/Heidelberg/New York: Springer.
Mill, J. S. (1843). *A system of logic, ratiocinate and inductive, being a connected view of the principles of evidence, and the methods of scientific investigation* (Vol. II). London: Harrison and co.
Missiakoulis, S. (2010). Cecrops, King of Athens: The first (?) recorded population census in history. *International Statistical Review, 78*(3), 413–418.
Moheau, M. (1778). *Recherches et considérations sur la population de la France*. Edition annotée par Vilquin E., Paris: INED PUF.
Mongin, P. (2003). L'axiomatisation et les théories économiques. *Revue Economique, 54*(1), 99–138.
Morrison, D. (1967). On the consistency of preferences in Allais's paradox. *Behavioral Science, 12*, 373–383.
Mosteller, F., & Wallace, D. L. (1964). *Applied Bayesian and classical inference: The case of the federalist papers*. New York: Springer.
Muliere, P., & Parmigiani, G. (1993). Utility and means in the 1930s. *Statistical Science, 8*(4), 421–432.
Müller, H.-G., Love, B., & Hoppa, R. D. (2002). Semiparametric methods for estimating paleodemographic profiles from age indicator data. *American Journal of Physical Anthropology, 117*, 1–14.
Nadeau, R. (1999). *Vocabulaire technique et analytique de l'épistémologie*. Paris: Presses Universitaires de France.
Nagel, E. (1940). Book review of Jeffreys' Theory of probability. *The Journal of Philosophy, 37*, 524–528.
Nagel, E. (1961). *The structure of science*. London: Routledge and Kegan Paul.
Narens, L. (1976). Utility, uncertainty and trade-off structures. *Journal of Mathematical Psychology, 13*, 296–332.
Neuhaus, J. M., & Jewell, N. P. (1993). A geometric approach to assess bias due to omitted covariates in generalized linear models. *Biometrika, 80*(4), 807–815.
Neveu, J. (1972). *Martingales à temps discret*. Paris: Masson.
Newton, I. (1687). *Philosophia naturalis principia mathematica*. Londini: S. Pepys.
Neyman, J. (1937). Outline of the theory of statistical estimation based on the classical theory of probability. *Philosophical Transactions of the Royal Society of London Series A, 236*, 333–380.
Neyman, J. (1940). Book review of Jeffreys' Theory of probability. *Journal of the American Statistical Association, 35*, 558–559.
Neyman, J., & Pearson, E. S. (1928). On the use and interpretation of certain test criteria for purposes of statistical inference. Part I. *Biometrika, 20A*, 175–240.
Neyman, J., & Pearson, E. S. (1933a). On the problem of the most efficient tests of statistical hypotheses. *Philosophical Transactions of the Royal Society of London, Series A, 231*, 289–337.

Neyman, J., & Pearson, E. S. (1933b). The testing of statistical hypotheses in relation to probabilities *a priori*. *Proceedings of the Cambridge Philosophical Society, 26*, 492–510.

O'Donnel, R. (2003). The thick and the think of controversy. In J. Runde & S. Mizuhara (Eds.), *The philosophy of Keynes's economics: Probability, uncertainty, and convention* (pp. 85–99). London/New York: Routledge.

Orchard, T., & Woodbury, M. A. (1972). A missing information principle: Theory and applications. *Proceedings of the VIth Berkeley Symposium on Mathematical Statistical Probability, 1*, 697–715.

Paris, J. B. (1994). *The uncertain reasoner's companion. A mathematical perspective*. Cambridge, UK: Cambridge University Press.

Paris, J., & Vencovská, A. (1997). In defence of maximum entropy inference process. *International Journal of Approximate Reasoning, 17*(1), 77–103.

Parlebas, P. (2002). Elementary mathematic modelization of games and sports. In R. Franck (Ed.), *The explanatory power of models. Bridging the gap between empirical sciences and theoretical research in the social sciences* (pp. 197–228). Boston/Dordrecht/London: Kluwer Academic Publishers.

Pascal, B. (1640). *Essay sur les coniques*. B.N. Imp. Res. V 859.

Pascal, B. (1645). *Lettre dédicatoire à Monseigneur le Chancelier sur le sujet de la machine nouvellement inventée par le sieur B.P. pour faire toutes sortes d'opérations d'arithmétique par un mouvement réglé sans plume ni jetons avec un avis nécessaire à ceux qui auront la curiosité de voir ladite machine et s'en servir*. In 1er manuscrit gros in 4° de Guerrier (archives de la famille Bellaigues de Bughas), pp. 721 et suiv.

Pascal, B. (1648). *Récit de l'expérience de l'équilibre des liqueurs*. Paris: C. Savreux.

Pascal, B. (1654a). *Traité du triangle arithmétique, avec quelques autres traités sur le même sujet*. Paris: Guillaume Desprez.

Pascal, B. (1654b). *Celeberrimæ mathesos academiæ Pariensi*. Paris: Académie Parisienne.

Pascal, B. (1670). *Pensées*. Paris: Édition de Port-Royal.

Pascal, B. (1922). *Les lettres de Blaise Pascal accompagnées de lettres de ses correspondants*. Paris: Les Éditions G. Grès & Cie (Voir le courrier échangé avec Pierre de Fermat en 1654, pp. 188–229).

Pasch, M. (1882). *Vorlesungen über neure Geometrie*. Leipzig: Verlag from Julius Springer.

Patil, G. P., & Rao, C. R. (1994). *Environmental statistics*. Amsterdam: Elsevier Science B.V.

Pearl, J. (1985). *Bayesian networks: A model of self activated memory for evidential reasoning*. Paper submitted to the Seventh Annual Conference of the Cognitive Science Society, Irvine, CA, 20 p.

Pearl, J. (1988). *Probabilistic reasoning and intelligent systems: Networks of plausible inference*. San Mateo: Morgan Kaufmann.

Pearl, J. (1995). Causal diagrams for empirical research (with discussion). *Biometrika, 82*(4), 669–710.

Pearl, J. (2000). *Causality, reasoning and inference*. Cambridge, UK: Cambridge University Press.

Pearl, J. (2001). Bayesianism and causality, or, why I am only a half –Bayesian. In D. Corfield & J. Williamson (Eds.), *Foundations of Bayesianism* (Kluwer applied logic series, Vol. 24, pp. 19–36). Dordrecht: Kluwer Academic Publishers.

Pearson, K. (1894). Mathematical contributions to the theory of evolution. *Philosophical Transactions of the Royal Society of London, Series A, 185*, 71–110.

Pearson, K. (1896). Mathematical contributions to the theory of evolution, III: Regression, heredity and panmixia. *Philosophical Transactions of the Royal Society of London, Series A, 187*, 253–318.

Pearson, K. (1900). On the criterion that a given system of deviations from the probable in the case of a correlated system of variables is such that it can be reasonably supposed to have arisen from random sampling. *Philosophical Magazine, 50*(5th series), 157–175.

Pearson, K. (1911). *The grammar of science*. London: Adam and Charles Black.

Pearson, K. (1920). The fundamental problem of practical statistics. *Biometrika, 13*(1), 1–16.

Pearson, K. (1925). Bayes' theorem, examined in the light of experimental sampling. *Biometrika, 17*(3/4), 388–442.
Pearson, K., & Filon, L. N. G. (1898). Mathematical contributions to the theory of evolution, IV: On the probable errors of frequency constants and on the influence of random selection on variation and correlation. *Philosophical Transactions of the Royal Society of London, Series A, 191*, 229–311.
Peirce, C. S. (1883). A theory of probable inference. In *Studies in logic: Members of the John Hopkins University* (pp. 126–203). Boston: Little, Brown and Company.
Petty, W. (1690). *Political arithmetick*. London: Robert Clavel & Hen. Mortlock.
Piaget, J. (1967). Les deux problèmes principaux de l'épistémologie des sciences de l'homme. In J. Piaget (Ed.), *Logique et connaissance Scientifique* (pp. 1114–1146). Paris: Gallimard.
Plato. (around 360 B.C.). *Laws*. The Internet Classic Archive (B. Jowett, Trans.). Website: http://classics.mit.edu/Plato/laws.html. Accessed August 30, 2011.
Plato. (around 360 B.C.). *Republic*. The Internet Classic Archive (B. Jowett, Trans.). Website: http://classics.mit.edu/Plato/republic.html. Accessed August 30, 2011.
Plato. (around 360 B.C.). *Stateman*. The Internet Classic Archive (B. Jowett, Trans.). Website: http://classics.mit.edu/Plato/stateman.html. Accessed August 30, 2011.
Poincaré, H. (1912). *Calcul des probabilités*. Paris: Gauthier-Villars.
Poinsot, L., & Dupin, C. (1836). Discussion de la « Note sur le calcul des probabilités » de Poisson. *Comptes Rendus Hebdomadaires des Séances de l'Académie des Sciences, 2*, 398–399.
Poisson, S. D. (1835). Recherches sur la probabilité des jugements. *Comptes Rendus Hebdomadaires des Séances de l'Académie des Sciences, 1*, 473–474.
Poisson, S. D. (1836a). Note sur la loi des grands nombres. *Comptes Rendus Hebdomadaires des Séances de l'Académie des Sciences, 2*, 377–382.
Poisson, S. D. (1836b). Note sur le calcul des probabilités. *Comptes Rendus Hebdomadaires des Séances de l'Académie des Sciences, 2*, 395–399.
Poisson, S.-D. (1837). *Recherches sur la probabilité des jugements en matière criminelle et en matière civile*. Paris: Bachelier.
Polya, G. (1954). *Mathematics and plausible reasoning* (2 vols). Princeton: Princeton University Press.
Popper, K. (1934). *Logik der forshung*. Vienna: Springer.
Popper, K. (1956). Adequacy and consistency: A second reply to Dr. Bar Illel. *The British Journal for the Philosophy of Science, 7*, 249–256.
Popper, K. (1959). The propensity interpretation of probability. *Philosophy of Science, 10*, 25–42.
Popper, K. (1982). *The postscript to The logic of scientific discovery III. Quantum theory and the schism in physics*. London: Hutchinson.
Popper, K. (1983). *The postscript of the logic of scientific discovery. I. Realism and the aim of science*. London: Hutchinson.
Porter, T. M. (1986). *The rise of statistical thinking 1820–1900*. Princeton: Princeton University Press.
Poulain, M., Riandey, B., & Firdion, J. M. (1991). Enquête biographique et registre belge de population: une confrontation des données. *Population, 46*(1), 65–88.
Poulain, M., Riandey, B., & Firdion, J. M. (1992). Data from a life history survey and the Belgian population register: A comparison. *Population: An English Selection, 4*, 77–96.
Pratt, D. (2010). *Modeling written communication. A new systems approach to modelling in the social sciences*. Dordrecht/Heidelberg/London/New York: Springer.
Prentice, R. L. (1978). Covariate measurement errors and parameter estimation in a failure time regression model. *Biometrika, 69*, 167–179.
Pressat, R. (1966). *Principes d'analyse*. Paris: INED.
Preston, M. G., & Baratta, P. (1948). An experimental study of the auction value of an uncertain outcome. *The American Journal of Psychology, 61*, 183–193.
Preston, S. H., & Coale, A. J. (1982). Age/structure growth, attrition and accession: A new synthesis. *Population Index, 48*(2), 217–259.

Quesnay, F. (1758). *Tableau oeconomique* (Document M 784 no. 71-1). Paris: Archives Nationales (Published in (2005) C. Théré, L. Charles, J.-C. Perrot (Eds.), Œuvres économiques complètes et autres textes de François Quesnay. Paris: INED).

Quetelet, A. (1827). Recherches sur la population, les naissances, les décès, les prisons, les dépôts de mendicité, etc., dans le royaume des Pays-Bas. *Nouveaux mémoires de l'académie royale des sciences et des belles-lettres de Bruxelles, 4*, 117–192.

Quetelet, A. (1835). *Sur l'homme et le développement de ses facultés, ou Essai de physique sociale*. Tome premier et Tome second. Paris: Bachelier.

Rabinovitch, N. L. (1969). Studies in the history of probability and statistics. XXII. Probability in the Talmud. *Biometrika, 56*(2), 437–441.

Rabinovitch, N. L. (1970). Studies on the history of probability and statistics. XXIV. Combinations and probability in rabbinic literature. *Biometrika, 57*(1), 203–205.

Radon, J. (1913). Theorie und Anwendungen der absolut Additiven Mengenfunktionen. *Sitzungsberichte der kaiserlichen Akademie der Wissenschaften, Mathematisch-Naturwissenschaftliche Klasse* 122IIa, pp. 1295–1438.

Railton, P. (1978). A deductive-nomological model of probabilistic explanation. *Philosophy of Science, 45*, 206–226.

Ramsey, F. P. (1922). Mr. Keynes and probability. *The Cambridge Magazine, 11*(1), 3–5.

Ramsey, F. P. (1926). Truth and probability. In F. P. Ramsey. (1931). *The foundations of mathematics and other logical essays*, R. B. Braithwaite (Ed.), Chapter VII, pp. 156–198. London: Kegan, Trubner & Co., New York: Harcourt, Brace and Company.

Ramsey, F. P. (1931). In R. B. Braithwaite (Ed.), *The foundations of mathematics and other logical essays*. London/New York: Kegan, Trubner & Co/Harcourt, Brace and Company.

Reeves, J. (1987). Projection of number of kin. In J. Bongaarts, T. Burch, & K. Wachter (Eds.), *Family demography* (pp. 228–248). Oxford: The Clarendon Press.

Reichenbach, H. (1935). *Wahrscheinlichkeitslehre : eine Untersuchung über die logischen und mathematischen Grundlagen der Wahrscheinlichkeitsrechnung*. Leiden: Sijthoff (English translation: 2nd edition (1949) *The theory of probability, an inquiry into the logical and mathematical foundations of the calculus of probability*, Berkeley-Los Angeles: University of California Press).

Reichenbach, H. (1937). Les fondements logiques du calcul des probabilités. *Annales de l'Institut Henri Poincaré, 7*, 267–348.

Remenik, D. (2009). *Limit theorems for individual-based models in economics and finance*. arXiv: 0812813v4 [math.PR], 38 p.

Reungoat, S. (2004). *William Petty. Observateur des Îles Britanniques*. Classiques de l'économie et de la population. Paris: INED.

Richardson, S., & Green, P. J. (1997). On Bayesian analysis of mixtures with an unknown number of components. *Journal of the Royal Statistical Society, Series B, 59*(4), 731–792.

Ripley, B. D. (1994). Neural networks and related methods for classification. *Journal of the Royal Statistical Society, Series B, 56*(3), 409–456.

Ripley, R. M. (1998). *Neural networks models for breast cancer prognoses*. Doctor of Philosophy thesis. Oxford: Oxford University, http://portal.stats.ox.ac.uk/userdata/ruth/thesis.pdf. Accessed July 11, 2011.

Ripley, B. D., & Ripley, R. M. (1998). Neural networks as statistical methods in survival analysis. In R. Dybrowski & V. Gant (Eds.), *Artificial neural networks: Prospects for medicine*. Austin: Landes Biosciences Publishers.

Robert, C. P. (2006). *Le choix bayésien*. Paris: Springer.

Robert, C. P., Chopin, N., & Rousseau, J. (2009). Harold Jeffreys's theory of probability revisited. *Statistical Science, 24*(2), 141–172.

Roberts, H. V. (1974). Reporting of Bayesian studies. In S. E. Fienberg & A. Zellner (Eds.), *Studies in Bayesian econometrics and statistics: In honor of Leonard J. Savage* (pp. 465–483). Amsterdam: North Holland.

Robertson, B., & Vignaux, G. A. (1991). Inferring beyond reasonable doubt. *Oxford Journal of Legal Studies, 11*(3), 431–438.

Robertson, B., & Vignaux, G. A. (1993). Probability – The logic of the law. *Oxford Journal of Legal Studies, 13*(4), 457–478.

Robertson, B., & Vignaux, G. A. (1995). *Interpreting evidence: Evaluation forensic science in the courtroom*. New York/Chichester/Brisbane/Toronto: Wiley.

Robinson, W. S. (1950). Ecological correlations and the behavior of individuals. *American Sociological Review, 15*, 351–357.

Roehner, B., & Syme, T. (2002). *Pattern and repertoire in history*. Cambridge, MA: Harvard University Press.

Rohrbasser, J.-M. (2002). Qui a peur de l'arithmétique? Les premiers essais de calcul sur les populations dans la seconde moitié du XVIIe siècle. *Mathématiques et Sciences Humaines, 159*, 7–41.

Rohrbasser, J.-M., & Véron, J. (2001). *Leibniz et les raisonnements sur la vie humaine. Classiques de l'économie et de la population*. Paris: INED.

Rouanet, H., Bernard, J.-M., Bert, M.-C., Lecoutre, B., Lecoutre, M.-P., & Le Roux, B. (1998). *New ways in statistical methodology. From significance tests to Bayesian inference*. Bern: Peter Lang.

Rouanet, H., Lebaron, F., Le Hay, V., Ackermann, W., & Le Roux, B. (2002). Régression et analyse géométrique des données: réflexions et suggestions. *Mathématiques et Sciences Humaines, 160*, 13–45.

Royall, R. M. (1970). On finite population theory under certain linear regression models. *Biometrika, 57*(2), 377–387.

Rubin, D. B. (1974). Estimating the causal effects of treatments in randomized and non randomized studies. *Journal of Educational Psychology, 66*, 688–701.

Rubin, D. B. (1977). Assignment to treatment group on the basis of a covariate. *Journal of Educational Psychology, 2*, 1–26.

Rubin, D. B. (1978). Bayesian inference for causal effects: The role of randomization. *The Annals of Statistics, 6*, 34–58.

Russo, F. (2009). *Causality and causal modelling in the social sciences. Measuring variations* (Methodos series, Vol. 5). Dordrecht: Springer.

Ryder, N. B. (1951). *The cohort approach. Essays in the measurement of temporal variations in demographic behaviour*. PhD dissertation. New York: Princeton University.

Ryder, N. B. (1954). La mesure des variations de la fécondité au cours du temps. *Population, 11*(1), 29–46.

Ryder, N. B. (1964). Notes on the concept of population. *The American Journal of Sociology, 69*(5), 447–463.

Sadler, M. T. (1830). *The law of population. A treatise, in six books, in disproof of the superfecondity of human beings, and developing the real principle of their increase*. London: Murray.

Salmon, W. C. (1961). Vindication of induction. In H. Feigl & G. Maxwell (Eds.), *Current issues in the philosophy of science* (pp. 245–264). New York: Holt, Reinhart, and Winston.

Salmon, W. C. (1984). *Scientific explanation and the causal structure of the world*. Princeton: Princeton University Press.

Salmon, W. C. (1991). Hans Reichenbach's vindication of induction. *Erkenntnis, 35*(1–3), 99–122.

Sarkar, S. (Ed.). (1996). *Decline and obsolescence of logical empirism: Carnap vs. Quine and the critics*. New York/London: Garland Publishing, Inc.

Savage, L. J. (1954). *The foundations of statistics*. New York: Wiley.

Savage, L. J. (1962). *The foundations of statistical inference*. New York: Wiley.

Savage, L. J. (1967). Implications of personal probability for induction. *The Journal of Philosophy, 64*(19), 593–607.

Savage, L. J. (1976). On reading R.A. Fisher. *The Annals of Statistics, 4*(3), 441–500.

Schmid, J., & Leiman, J. M. (1957). The development of hierarchical factor solutions. *Psychometrika, 22*, 83–90.

Schmitt, R. C., & Crosetti, A. H. (1954). Accuracy of the ratio-correlation method for estimating postcensal population. *Land Economics, 30*, 279–281.

Scott, D. (1964). Measurement structures and linear inequalities. *Journal of Mathematical Psychology, 1*, 233–247.
Séguy, I., & Buchet, L. (Eds.). (2011). *Manuel de paléodémographie*. Paris: INED.
Seidenfeld, T. (1987). Entropy and uncertainty. In I. B. MacNeill & G. J. Umphrey (Eds.), *Foundations of statistical inference* (pp. 259–287). Boston: Reidel.
Seidenfeld, T., Schervish, M. J., & Kadane, J. B. (1990). When fair betting odds are not degrees of belief. *Philosophical Science Association, 1*, 517–524.
Shafer, G. (1976). *A mathematical theory of evidence*. Princeton: Princeton University Press.
Shafer, G. (1979). Allocations of probability. *Annals of Probability, 7*(5), 827–839.
Shafer, G. (1982). Thomas Bayes's Bayesian inference. *Journal of the Royal Statistical Society, Series A, 145*(2), 250–258.
Shafer, G. (1985). Conditional probability. *International Statistical Review, 53*(3), 261–277.
Shafer, G. (1986). Savage revisited. *Statistical Science, 1*(4), 463–501.
Shafer, G. (1990a). The unity and diversity of probability (with comments). *Statistical Science, 5*(4), 435–462.
Shafer, G. (1990b). The unity of probability. In G. M. von Furstenberg (Ed.), *Acting under uncertainty: Multidisciplinary conceptions* (pp. 95–126). New York: Kluwer.
Shafer, G. (1992). What is probability? In D. C. Hoaglin & D. S. Moore (Eds.), *Perspectives on contemporary statistics* (pp. 93–106). New York: Mathematical Association of America.
Shafer, G. (1996). *The art of causal conjecture*. Cambridge, MA: MIT Press.
Shafer, G. (2001). *The notion of event in probability and causality. Situating myself relative to Bruno de Finetti*. Unpublished paper, presented in Pisa and in Bologna march 2001, 14 p.
Shafer, G. (2004). Comments on "Constructing a logic of plausible inference: A guide to Cox's theorem", by Kevin S. Van Horn. *International Journal of Approximate Reasoning, 35*, 97–105.
Shafer, G. (2010). A betting interpretation for probabilities and Dempster-Shafer degrees of belief (Working paper 31), Project website: http://probabilityandfinance.com, 18 p. Accessed July 11, 2011.
Shafer, G., & Pearl, J. (Eds.). (1990). *Readings in uncertainty reasoning*. San Mateo: Morgan Kaufman Publishers.
Shafer, G., & Volk, V. (2006). The sources of Kolmogorov Grundbegriffe. *Statistical Science, 21*(1), 70–98.
Shafer, G., & Vovk, V. (2001). *Probability and finance: It's only a game!* New York: Wiley.
Shafer, G., & Vovk, V. (2005). *The origins and legacy of Kolmogorov's Grundbegriffe* (Working paper 4), Project web site: http://probabilityandfinance.com, 104 p. Accessed July 11, 2011.
Shafer, G., Gilett, P. R., & Scherl, R. B. (2000). The logic of events. *Annals of Mathematics and Artificial Intelligence, 28*, 315–390.
Shalizi, C. R. (2009). Dynamics of Bayesian updating with dependent data and misspecified models. *Electronic Journal of Statistics, 3*, 1039–1074.
Shannon, C. E. (1948). A mathematical theory of communication. *Bell System Technical Journal, 27*, 379–423, 623–656.
Shore, J. E., & Johnson, R. W. (1980). Axiomatic derivation of the principle of maximum entropy and the principle of minimum cross-entropy. *IEEE Transactions on Information Theory, IT-26*, 26.
Simpson, E. H. (1951). The interpretation of interaction in contingency tables. *Journal of the Royal Statistical Society, Series B, 13*, 238–241.
Sinha, D. (1993). Semiparametric Bayesian analysis of multiple even time data. *Journal of the American Statistical Association, 88*(423), 979–983.
Skilling, J. (1988). The axioms of maximum entropy. In G. J. Erickson & C. R. Smith (Eds.), *Maximum entropy and Bayesian methods in science and engineering* (Vol. 1, pp. 173–188). Boston/Dordrecht/London: Kluwer.
Skilling, J. (1998). Probabilistic data analysis: An introductory guide. *Journal of Microscopy, 190*(1–2), 28–36.
Slovic, P., & Tversky, A. (1974). Who accepts Savage's axioms? *Behavioral Science, 19*, 368–373.

Slutsky, E. (1922). К вопросу о логических основах теории вероятности (Sur la question des fondations logiques du calcul des probabilité). *Bulletin de Statistique, 12*, 13–21.
Smets, P. (1988). Belief functions. In P. Smets, A. Mandani, P. Dubois, & H. Prade (Eds.), *Non standard logics for automated reasoning* (pp. 253–286). London: Academic.
Smets, P. (1990). Constructing pignistic probability function in a context of uncertainty. In M. Henrion, R. D. Shachter, L. N. Kanal, & J. F. Lemmer (Eds.), *Uncertainty in artificial intelligence 5* (pp. 29–40). Amsterdam: North Holland.
Smets, P. (1991). Probability of provability and belief functions. *Logique et Analyse, 133–134*, 177–195.
Smets, P. (1994). What is Dempster-Shafer's model? In R. R. Yager, M. Fedrizzi, & J. Kacprzyk (Eds.), *Advances in the Dempster-Shafer theory of evidence* (pp. 5–34). New York/Chichester/Brisbane/Toronto: Wiley.
Smets, P. (1997). The normative representation of quantified beliefs by belief functions. *Artificial Intelligence, 92*, 229–242.
Smets, P. (1998). The transferable belief model for quantified belief representation. In P. Smets (Ed.), *Handbook of defeasible reasoning and uncertainty management systems, Vol. 1: Quantified representation of uncertainty & imprecision* (pp. 267–301). Dordrecht: Kluwer.
Smets, P., & Kennes, R. (1994). The transferable belief model. *Artificial Intelligence, 66*(2), 191–234.
Smith, A. (1776). *An inquiry into the nature and causes of the wealth of nations*. London: W Strahan and T. Cadell.
Smith, C. A. B. (1961). Consistency in statistical inference and decision (with discussion). *Journal of the Royal Statistical Society, Series B, 23*, 1–25.
Smith, C. A. B. (1965). Personal probability and statistical analysis (with discussion). *Journal of the Royal Statistical Society, Series A, 128*, 469–499.
Smith, H. L. (1990). Specification problems in experimental and nonexperimental social research. *Sociological Methodology, 20*, 59–91.
Smith, H. L. (1997). Matching with multiple controls to estimate treatments effects in observational studies. In A. E. Raftery (Ed.), *Sociological methodology 1997* (pp. 325–353). Oxford: Basil Blackwell.
Smith, H. L. (2003). Some thoughts on causation as it relates to demography and population studies. *Population and Development Review, 29*(3), 459–469.
Smith, H. L. (2009). Causation and its discontents. In H. Engelhardt, H.-P. Kohler, & A. Fürnkranz-Prskawetz (Eds.), *Causal analysis in population studies* (The Springer series on demographic methods and population analysis). Dordrecht/Heidelberg/London/New York: Springer.
Smith, R. C., & Crosetti, A. H. (1954). Accuracy of ratio-correlation method for estimating postcensal population. *Land Economics, 30*(3), 279–280.
Snow, P. (1998). On the correctness and reasonableness of Cox's theorem for finite domains. *Computational Intelligence, 14*(3), 452–459.
Sobel, M. E. (1995). Causal inference in the social and behavioural sciences. In M. E. Sobel (Ed.), *Handbook of statistical modelling for the social and behavioural sciences, Arminger, Clogg* (pp. 1–38). New York: Plenum.
Solomonoff, R. J. (1960). *A preliminary report on a general theory of inductive inference* (Report ZTB-138). Cambridge, MA: Zator CO.
Solomonoff, R. J. (1964a). A formal theory of inductive inference, Part 1. *Information and Control, 7*(1), 1–22.
Solomonoff, R. J. (1964b). A formal theory of inductive inference, Part 2. *Information and Control, 7*(2), 224–254.
Solomonoff, R. J. (1986). The applicability of algorithmic probability to problems in artificial intelligence. In L. N. Kanal & J. F. Lemmer (Eds.), *Uncertainty in artificial intelligence* (pp. 473–491). North-Holland: Elsevier Science Publishers B.V.
Solomonoff, R. J. (1997). The discovery of algorithmic probability. *Journal of Computer and System Science, 55*(1), 73–88.
Spearman, C. (1904). "General intelligence", objectively determined and measured. *The American Journal of Psychology, 15*, 201–293.

Starmer, C. (1992). Testing new theories of choice under uncertainty using the common consequence effect. *Review of Economic Studies, 59*(4), 813–830.
Starmer, C. (2000). Developments in non expected-utility theory: The hunt for a descriptive theory of choice under risk. *Journal of Economic Literature, 38*(2), 332–382.
Steinhaus, H. (1923). Les probabilités dénombrables et leur rapport à la théorie de la mesure. *Fundamenta Mathematicae, 4*, 286–310.
Stephan, F. F. (1942). An iterative method of adjusting sample frequency tables when expected marginal totals are known. *Annals of Mathematical Statistics, 13*(2), 166–178.
Stigler, S. M. (1973). Studies on the history of probability and statistics. XXXII. Laplace, Fisher, and the discovery of the concept of sufficiency. *Biometrika, 60*(3), 439–445.
Stigler, S. M. (1974). Studies on the history of probability and statistics. XXXIII. Cauchy and the witch of Agnesi: An historical note on the Cauchy distribution. *Biometrika, 61*(2), 375–380.
Stigler, S. M. (1975). Studies on the history of probability and statistics. XXXIV. Napoleonic statistics: The work of Laplace. *Biometrika, 62*(2), 503–517.
Stigler, S. M. (1982). Thomas Bayes Bayesian inference. *Journal of the Royal Statistical Society, Series A, 145*, 250–258.
Stigler, S. M. (1986). *The history of statistics: the measurement of uncertainty before 1900.* Cambridge, MA: Belknap Press of Harvard University Press.
Stigum, B. P. (1972). Finite state space and expected utility maximization. *Econometrica, 40*, 253–259.
Stuart Mill, J. (1843). *A system of logic, ratiocinative and inductive, being a connected view of the principles of evidence, and the methods of scientific investigation* (2 vols). London: John W. Parker.
Suppe, F. (1989). *The semantic conception of theories and scientific realism.* Urbana/Chicago: University of Illinois Press.
Suppes, P. (1956). The role of subjective probability and utility in decision-making. *Proceedings of the Third Berkeley Symposium on Mathematical Statistics and Probability, 1954–1955*(5), 61–73.
Suppes, P. (1960). Some open problems in the foundations of subjective probability. In R. E. Machol (Ed.), *Information and decision processes* (pp. 129–143). New York: McGraw-Hill.
Suppes, P. (1970). *A probabilistic theory of causality.* Amsterdam: North Holland.
Suppes, P. (1974). The measurement of belief. *Journal of the Royal Statistical Society, Series B (Methodological), 36*(2), 160–191.
Suppes, P. (1976). Testing theories and the foundations of statistics. In W. L. Harper & C. A. Hooker (Eds.), *Foundations of probability theory, statistical inference, and statistical theories of science, II* (pp. 437–457). Dordrecht: Reidel.
Suppes, P. (2002a). *Representation and invariance of scientific structures.* Stanford: CSLI Publications.
Suppes, P. (2002b). Representation of probability. In P. Suppes (Ed.), *Representation and invariance of scientific structures* (pp. 129–264). Stanford: CSLI Publications.
Suppes, P., & Zanotti, M. (1975). Necessary and sufficient conditions for existence of a unique measure strictly agreeing with a qualitative probability ordering. *Journal of Philosophical Logic, 5*, 431–438.
Suppes, P., & Zanotti, M. (1982). Necessary and sufficient qualitative axioms for conditional probability. *Zeitschrift für Wahrscheinlichkeitstheorie und Werwande Gebiete, 60*, 163–169.
Susarla, V., & van Rysin, J. (1976). Nonparametric Bayesian estimation of survival curves from incomplete observations. *Journal of the American Statistical Association, 71*, 897–902.
Susser, M. (1996). Choosing a future for epidemiology: I. Eras and paradigms. *American Journal of Public Health, 86*(5), 668–673.
Süssmilch, J. P. (1741). *Die göttliche Ordnung in den Veränderungen des menschlichen Geschlechts, aus der Geburt, Tod, und Fortpflanzung desselben erwiesen.* Berlin: zu finden bei J. C. Spener.
Süssmilch, J. P. (1761–1762). *Die göttliche Ordnung in den Veränderungen des menschlichen Geschlechts, aus der Geburt, Tod, und Fortpflanzung desselben erwiesen.* Berlin: Realschule.
Sylla, E. D. (1998). The emergence of mathematical probability from the perspective of the Leibniz-Jacob Bernoulli correspondence. *Perspectives on Science, 6*(1&2), 41–76.

Tabutin, D. (2007). Vers quelle(s) démographie(s)? Atouts, faiblesses et évolutions de la discipline depuis 50 ans. *Population, 62*(1), 15–32 (Wither demography? Strengths and weaknesses of the discipline over fifty years of change. *Population-E, 62*(1), 13–32).

Thomas, D. H. (1986). *Refiguring anthropology: First principles of probability & statistics*. Long Grove: Waveland Press, Inc.

Thurstone, L. L. (1927). A law of comparative judgment. *Psychological Review, 34*, 273–286.

Thurstone, L. L. (1938). *Primary mental abilities*. Chicago: University of Chicago Press.

Thurstone, L. L. (1947). *Multiple factor analysis*. Chicago: University of Chicago Press.

Tinbergen, J. (1939). *Vérification statistique des théories de cycles économiques. Une méthode d'application au mouvement des investissements*. Genève: SDN.

Titterington, D. M., Smith, A. F. M., & Makov, U. E. (1985). *Statistical analysis of finite mixtures distributions*. New York: Wiley.

Todhunter, I. (1865). *A history of the theory of probability from the time of Pascal to that of Laplace*. Cambridge/London: Macmillan and Co.

Torche, F. (2011). The effect of maternal stress on birth outcomes: Exploiting a natural experiment. *Demography, 48*(11), 1473–1491.

Tornier, E. (1929). Wahrscheinlichkeisrechnunug und zalhlentheorie. *Journal für die Teine und Angewandte Mathematik, 60*, 177–198.

Trussell, J. (1992). Introduction. In J. Trussell, R. Hankinson, & J. Tilton (Eds.), *Demographic applications of event history analysis* (pp. 1–7). Oxford: Clarendon Press.

Trussell, J., & Richards, T. (1985). Correcting for unmeasured heterogeneity in hazard models using the Heckman-Singer procedure. In N. Tuma (Ed.), *Sociological methodology* (pp. 242–249). San-Francisco: Jossey-Bass.

Trussell, J., & Rodriguez, G. (1990). Heterogeneity in demographic research. In J. Adams, D. A. Lam, A. I. Hermalin, & P. E. Smouse (Eds.), *Convergent questions in genetics and demography* (pp. 111–132). New York: Oxford University Press.

Tuma, N. B., & Hannan, M. (1984). *Social dynamics*. London: Academic.

Turing, A. M. (1936). On computable numbers, with an application to Endscheidungsproblem. *Proceedings of the London Mathematical Society, 42*(2), 230–265.

Turing, A. M. (1950). Computing machinery and intelligence. *Mind, 59*, 433–460.

Tversky, A. (1974). Assessing uncertainty. *Journal of the Royal Statistical Society, Series B (Methodological), 36*(2), 148–159.

Tversky, A., & Kahneman, D. (1981). The framing of decisions and the psychology of choice. *Science, 211*(30), 453–457.

Tversky, A., & Kahneman, D. (1992). Advances in prospect theory: Cumulative representation of uncertainty. *Journal of Risk and Uncertainty, 5*, 297–323.

Tversky, A., & Koeler, D. J. (1994). Support theory: A nonextensional representation of subjective probability. *Psychological Review, 101*, 547–567.

Ulam, S. (1932). Zum Massbegriffe in Produkträumen. In *Verhandlung des Internationalen Mathematiker-Kongress Zürich* (Vol. II, pp. 118–119), Zurich: Orell Fiisli Verlag.

Van Fraassen, B. (1980). *The scientific image*. Oxford: Clarendon Press.

Van Horn, K. S. (2003). Constructing a logic of plausible inference: A guide to Cox's theorem. *International Journal of Approximate Reasoning, 34*(1), 3–24.

Van Imhoff, E., & Post, W. (1997). Méthodes de micro-simulation pour des projections de population, *Population* (D. Courgeau (Ed.)), 52(4), pp. 889–932 ((1998). Microsimulation methods for population projections. *Population. An English Selection* (D. Courgeau (Ed.)), 10(1), pp. 97–138).

van Lambalgen, M. (1987). Von Mises' definition of random sequences reconsidered. *The Journal of Symbolic Logic, 32*(3), 725–755.

Vaupel, J. W., & Yashin, A. I. (1985). Heterogeneity's ruses: Some surprising effects of selection on population dynamics. *The American Statistician, 39*, 176–185.

Vaupel, J. W., Manton, K. G., & Stallard, E. (1979). The impact of heterogeneity in individual frailty data on the dynamics of mortality. *Demography, 16*(3), 439–454.

Venn, J. (1866). *The logic of chance*. London: Macmillan.

Vernon, P. E. (1950). *The structure of human abilities*. London: Methuen.
Véron, J., & Rohrbasser, J.-M. (2000). Lodewijck et Christian Huygens: la distinction entre vie moyenne et vie probable. *Mathématiques et sciences humaines, 149*, 7–22.
Véron, J., & Rohrbasser, J.-M. (2003). Wilhlem Lexis: la durée normale de la vie comme expression d'une « nature des choses ». *Population, 58*(3), 343–363.
Vetta, A., & Courgeau, D. (2003). Demographic behaviour and behaviour genetics. *Population-E, 58*(4–5), 401–428 (French edition: (2003). Comportements démographiques et génétique du comportement, *Population, 58*(4–5), 457–488).
Vidal, A. (1994). *La pensée démographique. Doctrines, théories et politiques de population*. Grenoble: Presses Universitaires de Grenoble.
Vignaux, G. A., & Robertson, B. (1996). Lessons for the new evidence scholarship. In G. R. Heidbreder (Ed.), *Maximum entropy and Bayesian methods*, Proceedings of the 13th International Workshop, Santa Barbara, California, August 1–5, 1993 (pp. 391–401). Dordrecht: Kluwer Academic Publishers.
Ville, J. A. (1939). *Étude critique de la notion de collectif*. Paris: Gauthier-Villars.
Vilquin, E. (1977). Introduction. In J. Graunt (Ed.), *Observations naturelles et politiques* (pp. 7–31). Paris: INED.
Voltaire. (1734). Lettre XI. Sur l'insertion de la petite vérole. In *Lettres Philosophiques, par M. de V...* (pp. 92–149). Amsterdam: Chez E. Lucas, au Livre d'or.
von Mises, R. (1919). Grundlagen der wahrscheinlichkeitesrechnung. *Mathematische Zeitschrift, 5*, 52–99.
von Mises, R. (1928). *Wahrscheinlichkeit, Statistik und Wahrheit*. Wien: Springer (English translation: (1957). *Probability, statistics and truth*. London: George Allen & Unwin Ltd.).
von Mises, R. (1932). Théorie des probabilités. Fondements et applications. *Annales de l'Institut Henri Poincaré, 3*(2), 137–190.
von Mises, R. (1942). On the correct use of Bayes's formula. *Annals of Mathematical Statistics, 13*, 156–165.
von Neumann, J., & Morgenstern, O. (1944). *Theory of games and economic behaviour*. Princeton: Princeton University Press.
Von Wright, G. H. (1971). *Explanation and understanding*. London: Routledge and Kegan Paul.
Wachter, K., Blackwell, D., & Hammel, E. A. (1997). Testing the validity of kinship microsimulation. *Journal of Mathematical and Computer Modeling, 26*, 89–104.
Wachter, K., Blackwell, D., & Hammel, E. A. (1998). Testing the validity of kinship microsimulation: An update. Website: http://citeseerx.ist.psu.edu/viewdoc/summary?doi=10.1.1.25.2243, 36 p. Accessed August 5, 2005.
Waismann, F. (1930). Logische Analyse des wahrscheinlichkeisbegriffs. *Erkenntnis, 1*, 228–248.
Wald, A. (1936). Sur la notion de collectif dans le calcul des probabilités. *Comptes Rendus des Séances de l'Académie des Sciences, 202*, 180–183.
Wald, A. (1947a). Foundations of a general theory of sequential decision functions. *Econometrica, 15*(4), 279–313.
Wald, A. (1947b). *Sequential analysis*. New York: Wiley.
Wald, A. (1949). Statistical decision functions. *Annals of Mathematical Statistics, 20*(2), 165–205.
Wald, A. (1950). *Statistical decision functions*. New York: Wiley.
Walliser, B. (Ed.). (2009). *La cumulativité du savoir en sciences sociales*. Lassay-les-Châteaux: Éditions de l'École des Hautes Études en Sciences Sociales.
Wargentin, P. W. (1766). Mortaliteten i Sverige, i anledning af Tabell-Verket. *Kongl. Svenska Vetenskap Academiens Handlingar, XXVII*, 1–25.
Wavre, R. (1938–1939). *Colloque consacré à la théorie des probabilités*, Fascicules 734–740; 766. Paris: Hermann.
Weber, M. (1998). The resilience of the Allais paradox. *Ethics, 109*(1), 94–118.
Weber, B. (2007). The effects of losses and event splitting on the Allais paradox. *Judgment and Decision Making, 2*, 115–125.

References

Whelpton, P. (1946). Reproduction rates adjusted for age, parity, fecundity and marriage. *Journal of the American Statistical Association, 41*, 501–516.
Whelpton, P. (1949). Cohort analysis of fertility. *American Sociological Review, 14*(6), 735–749.
West, M., Mueller, P., & Escobar, M.D. (1994). Hierarchical priors and mixture models, with applications in regression and density estimation. In P.R. Freeman and A.F.M. Smith, *Aspects of uncertainty: A tribute to D.V. Lindley* (pp. 363–386), London: Wiley Series in Probability and Statistics.
Wilks, S. S. (1941). Book review of Jeffreys' Theory of probability. *Biometrika, 32*, 192–194.
Williamson, J. (2005). *Bayesian nets and causality: Philosophical and computational foundations*. Oxford: Oxford University Press.
Williamson, J. (2009). Philosophies of probability. In A. Irvine (Ed.), *Handbook of the philosophy of mathematics* (Handbook of the philosophy of science, Vol. 4, pp. 1–40). Amsterdam: Elsevier/North-Holland.
Wilson, M. C. (2007). Uncertainty and probability in institutional economics. *Journal of Economic Issues, 41*(4), 1087–1108.
Wolfe, J. H. (1965). *A computer program for the maximum-likelihood analysis of types* (Technical Bulletin 65–15). U. S. Naval Personnel Research Activity, San Diego (Defense Documentation Center AD 620 026).
Wright, S. (1921). Correlation and causation. *Journal of Agricultural Research, 20*, 557–585.
Wrinch, D., & Jeffreys, H. (1919). On some aspects of the theory of probability. *Philosophical Magazine, 38*, 715–731.
Wrinch, D., & Jeffreys, H. (1921). On certain fundamental principles of scientific inquiry. *Philosophical Magazine, 42*, 369–390.
Wrinch, D., & Jeffreys, H. (1923). On certain fundamental principles of scientific inquiry. *Philosophical Magazine, 45*, 368–374.
Wunsch, G. (1994). L'analyse causale en démographie. In R. Franck (Ed.), *Faut-il chercher aux causes une raison ? L'explication causale dans les sciences humaines* (pp. 24–40). Paris: Librairie Philosophique J. Vrin.
Yager, R. R., & Liu, L. (Eds.). (2007). *Classic works on the Dempster-Shafer theory of belief functions*. Heidelberg: Springer.
Yashin, A. I., & Manton, K. G. (1997). Effects of unobserved and partially observed covariate process on system failure: A review of models and estimation strategies. *Statistical Science, 12*, 20–34.
Younes, H., Delampady, M., MacGibbon, B., & Cherkaoui, O. (2007). A hierarchical Bayesian approach to the estimation of monotone hazard rates in the random right censoring model. *Journal of Statistical Research, 41*(2), 35–62.
Yule, U. (1895). On the correlation of total pauperism with proportion of out-relief, I: All ages. *The Economic Journal, 5*, 603–611.
Yule, U. (1897). On the theory of correlation. *Journal of the Royal Statistical Society, 60*, 812–854.
Yule, U. (1899). An investigation into the causes of changes in pauperism in England, chiefly during the last two intercensal decades, I. *Journal of the Royal Statistical Society, 62*, 249–295.
Zadeh, L. A. (1965). Fuzzy sets. *Information and Control, 8*, 338–353.
Zadeh, L. A. (1978). Fuzzy sets as a basis for a theory of possibility. *Fuzzy Sets and Systems, 1*, 3–28.

Author Index

A
Aalen, O.O., xxx, xxxii, 171, 215–217, 224, 253, 255
Abel, N.H., 104
Ackermann, W., 69
Aczél, J., 111
Agazzi, E., 240
Agliardi, E., 4
Allais, M., xxvii, 61–63, 67, 79–82
Andersen, P.K., 69, 153, 172, 215–217
Anderson, J., 257
Arbuthnott, J., xxv, xxix, 34, 35, 39, 158, 204
Aristotle, xv, xvi, xxv, xxx, xxxi, 3, 85, 92, 94, 134, 193, 194, 196, 254
Armatte, M., 4, 234
Arnauld, A., 44, 45, 47, 48, 56, 133, 134, 196
Arnborg, S., 104, 130
Ayton, P., 80

B
Bacon, F., 13, 18, 112, 138, 194, 195
Bacro, J.-N., 182, 188
Baratta, P., 63
Barbin, E., xiv, 122
Barbut, M., 110, 155
Bartholomew, D.J., 260
Bateman, B.W., 88
Bayes, T.R., xx, xxi, xxvi, 48–51, 85, 98, 102, 103, 105, 116, 135, 143, 147, 159, 168, 259
Beatty, L., xiv
Bechtel, W., 251
Behrens, W.W. III., 256
Bellhouse, D.R., 32, 33
Bentham, J., 120

Benzécri, J.P., 221
Berlinski, D., 256
Bernard, J.-M., 69
Bernardo, J.M., 63, 137, 139, 143, 144, 226
Bernoulli, D., xxvi, 35, 40, 48, 49, 56, 61, 141, 142, 144
Bernoulli, J.I., xx, xxiii, 14, 15, 46, 48, 75, 135, 158, 160
Bernstein, F., xxiv, 19, 24
Bertalanffy, L., 256
Berthelot, J.-M., xiv
Bert, M.-C., 69
Bertrand, J., 16, 38, 96
Bhattacharya, S.K., 225
Bienaymé, J., 51, 166
Bienvenu, L., 215
Bijak, J., 258
Billari, F., 257
Birkhoff, G., 110
Blackwell, D., 257
Blayo, C., 163, 169, 213, 239
Blayo, Y., 163
Bocquet-Appel, J.-P., 174, 176, 177, 182, 188
Bod, R., 259
Boisguilbert, P., xix
Boltzmann, L., 91
Bonneuil, N., 259, 260
Bonvalet, C., 231
Boole, G., xxvii, 51, 92, 100, 167
Bordas-Desmoulins, J.-B., 151
Borel, E., 18, 20, 21, 24
Borgan, O., 69, 153, 171, 172, 216, 217, 224, 253, 255
Borsboom, D., 221
Bourgeois-Pichat, J., 202, 203
Braithwaite, R.B., 103

Bremaud, P., 216
Bretagnolle, J., 220
Brian, E., 128, 163, 204, 205
Broggi, U., xxiv, 19
Browne, W.J., 71
Bry, X., 231
Buchet, L., 174, 187
Buck, C.E., 259
Burch, T., 257, 258

C

Cantelli, F.P., xxiv, 21, 22, 24
Cantillon, R., xix
Cantor, G., 17
Cardano, J., xiii, xvi, xxii, 8, 14
Carnap, R., 52, 97, 130, 137
Carnot, S., 91, 136
Cartwright, N., 251
Casini, L., 252
Catell, R.B., 223
Caticha, A., 109, 111
Caussinus, H., xxx, 174, 182, 188
Cavanagh, W.G., 259
Charbit, Y., 191, 193
Chen, M.-H., 69, 172, 218, 219, 227, 235
Cherkaoui, O., 225
Chikuni, S., 177
Chopin, N., 90, 103
Choquet, G., 54
Chung, K.-L., 69
Church, A., 20, 245
Clausius, R., 91
Clayton, D., 172
Coale, A.J., 203
Cohen, M.R., 69
Collar, A.R., 249
Colom, R., 223
Colyvan, M., 104, 130
Comte, A., 128, 151
Condorcet, xx, xxviii, 48, 50, 116–120, 122, 135, 136, 151
Copeland, A., 20
Copernic, N., 194
Coumet, E., 128
Courgeau, D., xiv, xxx, 13, 37, 40, 69, 71, 73, 152, 159, 169–174, 180–182, 188, 190–192, 198, 200, 208, 213, 214, 222, 228, 229, 234, 237, 256
Cournot, A.-A., xxiii, 10, 16, 23, 133, 135, 136, 166, 167
Cox, D.R., 69, 138, 139, 143, 172, 214

Cox, R., 43, 88, 91, 93, 94, 104–107, 109–111, 137, 144, 146, 172
Craver, C.F., 251
Cribari-Neto, F., 178
Crosetti, A.H., 169

D

D'Alembert, J.l.R., 16, 36–38, 40
Darden, L., 251
Daston, L.J., xiv, xvi
David, F.N., xiii
Davis, J., 88
Dawid, A.P., 128, 237, 238, 250, 251
Degenne, A., 249
DeGroot, M.H., 84
Delampady, M., 225
Delannoy, M., 16, 38
Delaporte, P., 210
Dellacherie, C., 21
Demming, W.E., 176
de Moivre, A., 14
de Montessus, R., 15
Dempster, A.P., 54, 55, 63, 69, 87, 137
Deparcieux, A., xix, 159
Descartes, R., 13, 112
Desrosiéres, A., xiv, 120
Destutt de Tracy, A.L.C., 119, 120
Doob, J. L., xxxii, 215, 216, 218
Dormoy, E., 205
Draper, D., 173
Dubois, D., 55, 130
Duncan, W.J., 249
Dupâquier, J., 166
Dupin, C., 127, 135, 151
Durkheim, E., xiv, 37, 152, 168, 169, 205, 206, 244

E

Edgeworth, F.Y., 51, 137, 205, 206
Einstein, A., 13, 140, 192, 238, 239
Ellis, R.L., 51, 167
Ellsberg, D., 87
Eriksson, L., 84
Escobar, M.D., 226
Euler, L., xxxi, 31, 200–202, 218
Everett, B., 236

F

Feller, W., 24
Fergusson, T.S., 218

Fermat, xiii, xviii, xxi, 3, 9, 28, 133, 140, 145, 151, 156
Filon, L.N.G., 207
Firdion, J.M., 234
Fishburn, P.C., 54, 63, 66, 81
Fisher, R.A., xiv, 11, 25, 27, 28, 39, 44, 50, 51, 69, 88, 90, 102–104, 109, 137, 207
Florens, J.-P., 152, 218
Forsé, M., 249
Forte, B., 111
Franck, R., xiv, 13, 112, 191, 192, 195, 198, 202, 235, 251, 255
Frankenberg, S.R., 176–179, 259
Franklin, J., 113
Fréchet, M., 16, 18, 21, 23, 24, 38
Freund, J.E., 82
Friedman, M., 61
Frischhoff, B., 80

G
Gacôgne, L., 55
Gail, M., 206
Galileo, G., xiii, 13
Galton, F., 206, 208
Gärdenfors, P., 54
Gardin, J.-C., vii, 252
Garnett, J.C.M., 221
Gauss, C.F., xxx, 14, 20, 163–165, 169, 189, 203
Gavrilova, N.S., 210
Gavrilov, L.A., 210
Gelfand, A.E., 128
Gelman, A., 69
Gergonne, J.-D., 122
Gerrard, B., 88
Ghosal, S., 216
Gibbs, J.W., 91, 173
Gigerenzer, G., xiv
Gignac, G.E., 223
Gilett, P.R., 65
Gillies, D., 84, 88
Gill, J., 69
Gill, R.D., 69, 153, 172, 215–217
Gimblett, R., 257
Gjessing, H.K., 217, 224, 253, 255
Glennan, S., 251
Gödel, K., 139, 245
Goldstein, H., vii, 69, 71, 153, 173, 228, 255
Gompertz, B., 181
Gonseth, F., 192
Good, I.J., 3, 4, 54, 63, 137, 220, 225
Gosset, W.S., 25

Gould, S.J., 223
Gouraud, C., xiv
Graetzer, J., 32, 33
Granger, G.-G., xiv, xv, xxiv, 13, 192, 194, 199, 240, 252
Graunt, J., xiv, xviii–xxi, xxv, xxix, 3, 28–30, 34, 39, 151, 156–160, 168, 174, 195, 196, 198, 199, 233
Greenland, S., 152, 249
Green, P.J., 226
Grether, D.M., 62
Guillard, A., xix
Gurr, T.R., 259
Gustafson, P., 227
Gustafsson, J.-E., 223

H
Haavelmo, T., 249
Hacking, I., xiv, 4, 28, 29, 50, 144, 145
Hadamard, J., 16
Hájek, A., 84
Hald, A., 165, 206
Halley, E., xix, 30–32, 34–36, 158, 174
Halpern, J.Y., 104–106, 130, 131
Hammel, E.A., 257
Hannan, M., 153
Hanson, T.E., 226
Hansson, B., 54
Harr, A., 131
Hasofer, A.M., xiii
Hasselblad, V., 177
Hay, J., 259
Hecht, J., xiv
Heckman, J., 152, 220, 225
Heidelberg, M., xiv
Hempel, C.G., 251
Henry, L., vii, xxii, xxiii, 32, 163, 166, 168, 198, 208–212, 221, 234
Henry, N.W., 221
Herrmann, N.P., 178, 181
Hespel, B., 251
Hilbert, D., 17, 20
Hinkley, D.V., 172
Hoem, J., 36, 168, 216
Hofacker, J.D., 204
Holland, P., 235, 236, 249, 250
Hooper, G., xxvii, 45, 53, 64, 69, 75, 76, 78
Hooten, M.B., 257
Hoppa, R.D., 176, 181
Horman, J., 171
Horn, J.L., 223

Hsu, S.J., 69
Huber-Carol, C., 220
Hunter, D., 132
Hunt, G.A., 215
Hutter, M., 247
Huygens, C., xiv, xxii, 9, 30, 85

I

Ibrahim, J.G., 69, 172, 218, 219, 227, 235
Illari, P.M., 235, 251, 252, 255
Inkelmann, F., 257
Irwin, J.O., 102

J

Jacob, F., 255
Jacob, P., 97
Jaisson, M., 204, 205
Jannedy, S., 259
Jarrah, A.S., 257
Jaynes, E.T., xxvi, xxviii, 32, 52, 59, 84, 91–94, 96–99, 104, 107, 108, 111–113, 128, 131, 134, 137, 142–144, 146, 172, 216
Jeffreys, H., xxvi, xxvii, xxviii, 43, 52, 88–92, 94, 97–99, 101–103, 108, 112, 113, 130, 137, 140, 143, 144, 172
Jewell, N.P., 206
Johnson, R.W., 109
Johnson, W.E., 67, 226
Jones, K., 153
Jöreskog, K.G., 223
Juan-Espinosa, M., 223

K

Kadane, J.B., 84
Kahneman, D., 62, 63, 81
Kalbfleisch, J.D., 69, 172, 215, 218, 219
Kamke, E., 20
Kaplan, E.L., 214
Kaplan, M., 84
Kardaun, O.J.W.F., 138, 139, 143
Karlin, J.B., 69
Karlis, D., 226
Kass, R.E., 109
Kaufmann, W., 140
Keiding, N., 69, 153, 171, 172, 216, 217
Kendall, M.G., xiii, 39
Kersseboom, W., 159
Keynes, J.M., xiv, 43, 47, 52, 86–90, 93, 101, 137, 145

Kimura, D.K., 176, 177
Klotz, L.H., 255
Knuth, K.H., 97, 110, 111, 146, 147
Koeler, D.J., 80
Koller, D, 131
Kolmogorov, A., xv, xxiv, xxv, 7, 12, 15, 22–24, 90, 98, 137, 144, 146, 201, 246
Konigsberg, L.W., 176–179, 181, 259
Koopman, B.O., 87
Kraft, C.H., 59
Krantz, D.H., 63
Krüger, L., xiv
Kruithof, R., 175
Kuhn, R., 236
Kuhn, T., 13, 144, 191, 192, 238–240
Kullbach, S., 147
Kumar, D., 225
Kyburg, H., 84
Kyllonen, P., 223

L

Laemmel, R., xxiv, 19
La Harpe, J.-F., 119
Lamarche, J.P., xiv, 122
Lambert, J.H., 54, 76–78
Landry, A., xiv, 168, 169, 191, 206
Laplace, P.S., xx, xxviii, xxxiii, 4, 9, 48, 50, 51, 76, 77, 85, 86, 90, 94, 98, 102, 113–116, 120, 122, 123, 126–128, 135–137, 141, 142, 144, 151, 159–169, 189, 199, 203, 204, 234, 244
Larson, S., 81
Laubenbacher, R., 257
Lazarsfeld, P.F., 220, 221, 259
Lebaron, F., 69
Lebesgue, H., 18, 20
Le Bras, H., 28, 221, 222
Lecoutre, B., 69, 251
Lecoutre, M.-P., 69
Lee, P.M., 69
Legendre, A.M., 163, 203
Legg, S., 247
Le Hay, V., 69
Leiber, R.A., 147
Leibniz, G.W., xiv, 29, 43, 48, 85, 86, 89, 128, 134, 135
Leiman, J.M., 222
Leliévre, E., 170–172, 181, 214, 231
Leonard, T., 69
Le Roux, B., 69
Lévy, P., 15, 16, 18, 24
Lewis, D.K., 235, 249, 251

Lexis, W., 205
Lichteinstein, S., 80
Lillard, L.A., 225
Lindley, D.V., 67, 69, 80, 103, 128, 172, 207
Lindsay, B.G., 224
Little, D., 251
Litton, W.G., 259
Liu, L., 54
Łomnicki, A., xxiv
Loomes, G., 62, 81
Lotka, A.J., 198, 202
Louçã, F., 191
Love, B., 176
Luce, R.D., 63

M
MacGibbon, B., 225
Machamer, P., 251
Machina, M.J., 81
Macy, M.W., 257
Madden, E.H., 251
Makov, U.E., 225
Mandel, D.R., 80
Manski, C.F., 249
Manton, K.G., 220, 225, 253, 254
March, L., 139
Marec, Y., 122
Martin-Löf, P., 20, 21, 23
Martin, T., xiv
Masset, C., 176, 177, 182, 259
Masterman, M., 13, 192
Matalon, B., xiv, xv, 27, 82, 130
Maupin, M.G., 16, 38
Maxwell, J.C., 91
McCrimmon, K., 81
McCulloch, W.S., 226
McFadden, D.L., 249
McKinsey, J.C.C., 13
McLachlan, G.J., 224
Meadows, D.H., 256
Meadows, D.L., 256
Meier, P., 214
Mellenberg, G.J., 221
Menken, J., 214
Menzel, H., 220
Meusnier, N., xiii
Meyer, P.-A., 215
Mill, J.S., 205, 206, 251
Missiakoulis, S., xiv
Moheau, M., 160
Mongin, P., 4
Monjardet, B., 110

Morgenstern, O., 28, 53, 56, 60, 137, 260
Morrison, D., 81
Mortera, J., 128
Mosteller, F., 144
Mouchart, M., 218
Mueller, P., 226
Muliere, P., 59
Müller, H.-G., 176
Murrugarra, D., 257

N
Nadeau, R., 196
Nagel, E., 69, 103, 251
Narens, L., 63
Neuhaus, J.M., 206
Neveu, J., 215
Newton, I., 13, 112, 202, 238
Neyman, J., 24, 26, 102, 142
Ng, C.T., 111
Nicole, P., 44, 45, 47, 48, 56, 133, 134, 196
Novick, M.R., 67, 69

O
Oakes, D., 69
O'Donnel, R., 88
Oppenheim, P., 251
Orchard, T., 177

P
Palacios, A., 223
Paris, J.B., 104, 105, 131, 132
Parlebas, P., 252
Parmigiani, G., 59
Pascal, B., xiii, xiv, xvii, xviii, xx, xxi, xxv, xxvii, xxix, 3, 9, 28, 34, 43, 83, 85, 112, 133, 140–142, 145, 151, 156–159, 166, 195, 196, 245
Pasch, M., 20
Patilea, V., 226
Patil, G.P., 259
Pearl, J., 132, 247–251
Pearson, E.S., 26, 102, 142
Pearson, K., 25, 50, 52, 205–207, 223, 251
Peel, D., 224
Peirce, C.S., 137
Petty, W., xix, 28, 138, 151
Piaget, J., xiv, 82
Piantadosi, S., 206
Pitts, W., 226

Plott, C.R., 62
Poincaré, H., 15, 86
Poinsot, L., 127, 135, 151
Poisson, S.D., xxviii, 51, 116, 117, 122–128, 135, 136, 166
Polya, G., 92, 99, 134
Poole, C., 249
Popper, K., 12
Porter, T.M., xiv, 138
Post, W., 27
Poulain, M., 234
Prade, H., 55, 130
Pratt, D., 192, 252
Pratt, J.W., 59
Prentice, R.L., 69, 172, 215, 218
Pressat, R., 36, 168, 211
Preston, M.G., 63
Preston, S.H., 203
Prskawetz, A., 257
Pumain, D., vii, 222

Q
Quesnay, F., xix
Quetelet, A., xxviii, 122, 123, 203–205

R
Rabinovitch, N.L., xiii
Radon, J., 18
Railton, P., 251
Ramsey, F.P., 43, 52, 59, 60, 84, 88, 137
Randers, J., 256
Rao, C.R., 259
Rebollo, I., 223
Reeves, J., 257
Reichenbach, H., 222
Remenik, D., 257
Reungoat, S., 28, 138
Riandey, B., 234
Richardson, R.C., 251
Richardson, S., 226
Richards, T., 225
Ripley, B.D., 226, 227
Ripley, R.M., 226, 227
Robert, C.P., 71, 90, 103, 108, 130, 131, 173, 218, 258, 260
Roberts, H.V., 63
Robertson, B., 116, 128
Robinson, W.S., 173, 208
Rodriguez, G., 225
Roehner, B., 259, 260
Rohrbasser, J.-M., xiv, 32
Rolin, J.-M., 218

Rouanet, H., 69
Rousseau, J., 90, 103
Royall, R.M., 27
Rubin, D.B., 69, 236, 249
Russo, F., 235, 252
Ryder, N.B., 196, 210

S
Sadler, M.T., 204
Sahlin, N.-E., 54
Salmon, W.C., 131, 222, 251
Salome, D., 138, 139, 143
Sarkar, S., 97
Savage, L.J., xxvi, 11, 43, 49, 53, 56, 60–62, 82, 84, 137, 143, 144, 172
Schaafsma, W., 138, 139, 143
Scherl, R.B., 65
Schervish, M.J., 84
Schmid, J., 222
Schmitt, R.C., 169
Scott, D., 59
Séguy, I., 174, 187
Seidenberg, A., 59
Seidenfeld, T., 84, 108
Shafer, G., xxvi, 18, 23, 54, 55, 63–66, 82, 83, 87, 104, 106, 130, 137, 138, 145, 146, 152, 215, 245, 247–249
Shalizi, C.R., 216
Shannon, C.E., xxvii, 91, 109, 146
Shen, A., 215
Shore, J.E., 109
Silvey, R., 236
Simpson, E.H., 69
Singer, B., 152, 220, 225
Singh, N.K., 225
Sinha, D., 69, 172, 218, 219, 227, 235
Sjödin, G., 104, 130
Skilling, J., 109, 130
Slovic, B., 80
Slovic, P., 81
Slutsky, E., xxiv
Smets, P., xxvi, xxvii, 45, 55, 65, 66, 84, 130, 137
Smith, A.F.M., xix, 63, 137, 139, 207, 225
Smith, C.A.B., 54, 63
Smith, H.L., 236, 237
Snow, P., 130
Sobel, M.E., 249
Solomon, H., 128
Solomonoff, R.J., 246, 247
Spearman, C., 221
Stallard, E., 220
Starmer, C., 81

Author Index

Steerneman, A.G.M., 138, 139, 143
Steinhaus, H., xxiv, 22
Stephan, F.F., 176
Stern, H.S., 69
Stigler, S.M., xiv, 50, 122, 166, 206
Stigum, B.P., 63
Sugar, A., 13
Sugden, R., 62, 81
Suppe, F., 14
Suppes, P., xxvi, 12, 13, 22, 53, 54, 59, 62–64, 66, 97, 101, 137, 145, 238, 251
Susarla, V., 218
Susser, M., 152
Süssmilch, J.P., 39, 40
Swijtink, Z., xiv
Syme, T., 259, 260

T

Tabutin, D., 197
Thomas, D.H., 259
Thurstone, L.L., 221, 222
Tinbergen, J., 249
Titterington, D.M., 225
Tiwari, R.C., 225
Todhunter, I., xiv, 38
Torche, F., 236
Tornier, E., 20
Tretli, S., 255
Trussel, J., 214
Trussell, J., 225, 254
Tuma, N.B., 153
Turing, A.M., 245
Tversky, A., 62, 63, 80, 81

U

Ulam, S., xxiv

V

Van Fraassen, B., 14
van Herden, J., 221
Van Horn, K.S., 106, 130
Van Imhoff, E., 27
van Lambalgen, M., 20
van Rysin, J., 218
van Thillo, M., 223
Vaupel, J.W., 181, 220, 224
Vencovská, A., 131, 132
Venn, J., 9, 11, 51, 120, 136, 137, 167
Vernon, P.E., 223
Véron, J., xiv
Vetta, A., 152

Vidal, A., 191
Vignaux, G.A., 116, 128
Ville, J.A., 15, 21, 215
Vilquin, E., 156, 198
Volk, V., 18
Voltaire, 35
von Mises, R., xxiv, xxv, 4, 11, 19–21, 137, 144, 167
von Neumann, J., 28, 53, 56, 60, 137, 260
Von Wright, G.H., 251
Vovk, V., 18, 23, 130

W

Wachter, K., 257
Waismann, F., 20
Wald, A., 20, 21, 102, 143
Wallace, D.L., 144
Waller, L.A., 257
Walliser, B., 192
Wargentin, P.W., 34, 174, 234
Wasserman, L., 109
Wavre, R., 24
Weber, B., 82
Weber, M., 62
West, M., 226
Whelpton, P., 209, 210
Wieand, S., 206
Wilke, C.K., 257
Wilks, S.S., 102
Willems, J.C., 138, 139, 143
Willer, R., 257
Williamson, J., 131, 132, 235, 251, 252, 255
Wilson, M.C., 87, 88
Wolfe, J.H., 224
Woodbury, M.A., 177, 220, 225
Wright, S., 249
Wrinch, D., 88, 89
Wunsch, G., vii

Y

Yager, R.R., 54
Yashin, A.I., 224, 253, 254
Younes, H., 225
Yule, U., 152, 205–207

Z

Zadeh, L.A., 4, 55, 130
Zanotti, M., 53, 59
Zarkos, S.G., 178

Subject Index

A
Actuarial sciences, 259
Age, 10, 11, 26, 29, 31–33, 36, 46, 47, 57, 95, 123, 157, 158, 166, 168, 170, 174–190, 195, 199–205, 208, 209, 211, 212, 221, 222, 231, 236, 237, 254, 258–259
Agent-based model, 256
Algebra
  Boolean, 65, 91, 92, 100, 101, 104, 106, 110, 111, 147
  classical, 92
  propositions (of), 92, 100
Algorithmic probability, 246, 247
Analysis
  contextual, 228, 229
  counterfactual, 250–251
  cross-sectional, 40, 168, 169, 199, 208–210, 213, 214, 220, 222, 234, 239, 240
  event history, 40, 69, 170, 172, 173, 213, 214, 219, 221, 224, 225, 228, 229, 237, 239, 240, 253, 255
  hierarchical, 225
  logit, 70
  longitudinal, 37, 40, 157, 168–170, 189, 210–214, 234
  multilevel, 37, 40, 69, 73, 74, 173, 194, 237, 239, 240, 255, 256
  regression, 69, 163, 169, 205, 237
  semi-parametric, 172
Anthropology, 191, 259
A posteriori, 136, 159
Approach (of probability)
  direct, 48, 159
  indirect, 159
A priori, 10, 48, 88, 123, 156, 159

Argument
  mixed, 76, 78
  pure, 76, 78
  weight, 87
Arithmetical triangle, 34
Artificial intelligence, 243–245, 247, 249, 259
Astronomy, 114, 135, 136, 194, 203
Average man, 122, 204
Axiom
  additivity, 102
  commutativity, 111
  comparability, 102
  continuity, 22
  independence, 60, 79–81
  pure-rationality, 62
  structural, 62
  transitivity, 58, 102, 110
  ZFC theory (of), 17

B
Bayesian
  model, 71
  networks, 247, 249, 252
  probability, 64, 234, 235
  theorem, 25, 53, 56, 58, 111
Behavior
  aggregated, 194, 208
  individual, 173, 194, 197–200, 214, 220, 256
  rational, 60, 62–63, 79
Belief
  degree (of), 7, 12, 52, 54, 64, 65, 75, 83, 84, 86, 88, 89, 100, 137, 144, 145, 147, 225, 252

Belief (*cont.*)
  function, 45, 54, 55, 57, 65, 66, 77, 78, 130, 137
  individual, 86, 87
  level (of), 54, 65
  notion (of), 132
  rational, 86–88, 90, 102
  theory (of), 65, 130
Biological indicator, 174, 178, 189, 190
Biological sciences, 207, 256, 259
Biology, 13, 69, 192, 207, 244, 249, 251, 257
Birth, 7, 13, 26, 31, 32, 34, 35, 37, 39, 79, 114–116, 156, 158, 160–163, 167, 197–202, 204–206, 209–211, 216, 227, 228, 231, 236
Black-box, 227

C

Causality
  counterfactual, 235–238, 241, 250
  mechanistic, 251, 257
Census, 34, 36, 37, 40, 116, 122, 158, 160, 163, 168, 170, 189, 199, 207, 208, 211, 221, 234, 239
Central limit theorem, 165
Certainty
  degree (of), 10, 11, 44–46, 134
  moral, 15
Chance, 8, 14, 16, 18, 25–28, 34, 36, 38, 39, 43, 46, 47, 49, 51, 53, 61, 62, 73, 77, 82, 115, 125, 127, 128, 132, 140–142, 144, 145, 156–158, 163, 167, 168, 190, 195, 197, 198
Cheating, 38
Choice
  paradox, xxvii
  rational, 84
Circuit, 8
Classical probability, 27, 138, 142, 144, 145
Coherence, 52, 56, 58–61, 66, 79, 80, 84, 99, 100, 113, 137, 143
Cohort
  heterogeneous, 212
  homogeneous, 210, 212
Collective, 11, 19–24, 120, 173, 215, 217, 256
Communication theory, 259
Comprehension, 196
Concomitant variation, 37, 168, 169, 205, 206, 244

Confidence interval, 26, 47, 48, 142, 143, 155, 171, 182, 184, 228, 258
Consistency, 20, 53, 79, 82, 91, 99, 100, 102, 103, 111, 113, 114, 137, 143, 144
Contract, 44, 45
Correlation, 38, 176, 206–209, 212, 223
Covariance, 71, 165, 184
Credibility
  calculus, 45
  degrees (of), 45, 53
  intervals, 184
  rules (of), 53
Criminology, 259
Cumulativity, 111, 144–147, 231, 238–241, 260

D

Death, 7, 10, 11, 29–34, 36, 37, 45, 47, 52, 57, 95, 119, 134, 156–158, 160, 168, 177, 178, 181, 189, 195, 197, 198, 200, 202, 205, 206, 211, 216, 231, 253
Decision-making, 53, 54, 56, 60, 65, 66, 80, 102, 117, 137–144, 220, 256, 260
Deduction, 13, 19, 102, 112, 257
Demography, 7, 15, 28, 32, 37, 163, 166, 168, 169, 192, 193, 209, 212, 214, 221, 223, 224, 229, 234, 241, 252, 255–257, 259
Density
  posterior, 183
  prior, 130, 183
Distribution
  continuous, 96, 108
  Dirichlet, 185, 186, 218
  discrete, 94, 109
  multinomial, 183, 185
  non-informative, 90
  plausibility, 113
  posterior, 183–186, 218, 219
  prior, 89, 96, 97, 103, 108, 183–186, 216, 225–227, 235, 246
  uniform, 52, 86, 94, 130, 182, 185, 235
Dutch Book argument, 84

E

Economics, 28, 40, 49, 53, 56, 69, 135, 138, 191, 196, 200, 203, 208, 209, 217, 220, 224, 227, 234, 235, 237, 241, 244, 249, 252–254, 256, 257, 259
Educational sciences, 69, 191, 221, 228, 244, 256, 259

Subject Index 305

Entropy
  maximization, 96, 108, 109, 111, 131, 132
  Shannon's, 94, 100, 103, 107, 109, 147
  thermodynamical, 91
Epidemiology, 35, 69, 172, 214, 217, 224, 235, 236, 243, 244, 249, 256, 257, 259
Epistemic probability, 7, 16, 25, 38, 41, 43, 44, 46, 47, 51, 85, 86, 88, 137, 159–167, 221, 225, 235, 259
Epistemology, 39, 144
Equally
  likely, 50, 109, 124, 131, 229
  possible, 9, 14, 15, 19, 47, 51, 53, 165
  probable, 9, 15, 19, 86, 91, 101, 102, 114, 136, 160, 201
Error
  Type I, 25, 26
  Type II, 26
Event
  conditional, 58
  dependant, 212
  exchangeable, 53, 67
  independent, 14, 53, 55, 238, 240, 248
  objective, 44, 46, 159
  observed, 56, 85, 120, 124, 161, 212
  random, 22, 23, 50, 53
  repetitive, 225
  single, 170
  social, 254
  subjective, 44, 46, 167
  tree, 65
  uncertain, 56, 57
Exchangeability, 53, 59, 68, 69, 72–75, 113, 143, 245
Expected gain, 8, 141, 158
Extent, 61, 98, 131, 132, 136, 191, 196

F
Fairness, 8, 141
Fair wager, 8, 83–85
Fallacy
  atomistic, 173, 220, 228, 230
  ecological, 173, 208, 220, 228, 230
Fertility, 114, 160, 193, 198–203, 207, 212, 221, 222, 240, 252
Forecast, 112, 190, 241, 247, 249, 256–258
Frailty, 220, 224, 254
Frequency, 11, 12, 15, 16, 19, 22–24, 27, 44, 48, 49, 52, 56, 68, 90, 98, 132, 134, 139, 145, 165, 175, 178, 198, 207, 210, 223, 234, 258
Frequentist probability, 11, 167

G
Game
  cards, 9, 14
  chance (of), 8, 14, 39, 46, 53, 82, 128, 140–142, 145, 156, 157
  dices, 14
  fair, 9, 16, 23, 48, 49, 141, 159
  heads or tails, 9, 17, 21, 23, 53, 67, 68, 142, 215
  life (of), 156
  theory, 21, 28, 53, 142, 260
  whist (of), 162
Genetics, 205, 249
Geography, 191, 256, 257
Geometric probability, 10
Geometry of chance, 195

H
Historical demography, 32
History, 37, 38, 40, 57, 69, 91, 156, 163, 170–174, 181, 189–193, 195, 199, 201, 209–221, 224–230, 234, 235, 237–240, 248, 253–256, 259–260
Hypothesis
  dependence, 212
  independence, 211
  invariance, 176, 177, 181, 185
  probability (of a), 25, 26, 189
  testing, 25, 39, 142
  uniformity, 176

I
Implication, 60, 99, 101, 110, 140
Impossible
  mathematically, 16, 35
  physically, 16, 35
Individual
  observed, 174, 197, 218, 253, 260
  rational, 86
  statistical, 194, 197, 208, 211, 217, 218, 230, 253–254, 260
  theoretical, 79
Induction, 18, 102, 112, 135, 139, 195, 222, 246, 257
Industrial reliability, 172
Information, 25, 26, 45, 47, 51, 53–57, 65, 69, 70, 74, 75, 82–84, 86, 90, 93–97, 99, 100, 102, 103, 106, 107, 109–113, 128, 129, 131, 138–143, 146–155, 167, 173, 174, 184, 185, 187, 188, 190, 195, 214, 225, 234, 235, 246, 247, 250, 252, 258

J

Jurisprudence, 75, 116, 241
Jury, 117, 121–123
Justice, 8, 119, 122, 123

L

Lattice
   Boolean, 111
   distributive, 110, 111
   theory, 97, 110, 146
Law
   large numbers (of), 16, 127, 134, 135, 184
   succession (of), 136, 167
Legal science, 119, 259
Level
   credal, 65, 84
   decision, 65
   individual, 197, 224, 225, 227, 228, 237
   pignistic, 65, 137
   population, 224
Life
   expectancy, 46
   history, 234
   table, 28–32, 34, 158, 231, 234
Likelihood, 10, 62, 104, 109, 114, 115, 134, 165, 173, 178, 180, 181, 188, 207, 215, 224
Linguistics, 131, 259
Logic
   classical, 65, 88, 100
   deductive, 43, 94, 101, 102, 112, 130, 139, 140
   probability (of), 86–113, 116, 119, 123, 128, 129, 131, 132, 137, 138, 216, 218, 260
   propositions (of), 10, 146
Logical impossibility, 129
Logical probability
   axiom, 100–111
   paradigm, 98–100
Lower probability, 54, 64
Ludo aleae, xiii, xvi,

M

Market, 213
Martingale, 21, 214–218
Masculinity proportion, 115, 130, 160
Mathematical
   expectation, 40, 61, 62, 188
   limit, 21
   possibility, 16, 35

Mathematics, 17, 39, 112, 139, 258
Mean, 26, 29, 30, 35, 49, 53, 60, 85, 95–97, 141, 155, 164, 165, 178, 182, 183, 188, 199, 206, 216, 219, 222, 224, 253,
Measure, 7, 10, 12, 17, 18, 20–24, 40, 45, 46, 49, 52–54, 57–59, 61, 63–65, 82, 87, 91, 94, 96, 101, 105, 107, 109–111, 123, 128, 129, 138, 147, 155, 160, 164, 166, 168, 169, 171, 173, 193, 198, 200, 206, 209, 212, 215, 216, 218, 223, 228, 234, 237
Mechanism, 202, 247, 251, 252
Medical sciences, 244
Method
   ALK, 176, 178
   approximation, 178, 182, 259
   Bayesian, 69, 144, 173, 186, 188–190, 258, 259
   bootstrap, 71, 182
   concomitant variations (of), 168, 169, 205, 206, 244
   IALK, 177, 178, 181, 182, 188, 259
   IPFP, 176, 178
   least squares, 163–165, 180, 203–205, 207, 209
   maximum likelihood, 165, 178, 181
   MCMC, 71, 173
   microsimulation, 257
   Monte Carlo, 71, 173, 184
   multiplier, 160, 168
   probability-vectors, 176, 177, 187, 259
   regression, 189, 206, 211, 229, 245
Methodology, 53
Migration, 31, 73, 170, 198–201, 203, 208, 227–229, 231, 236, 237, 240, 252
Miracles, 7, 44
Mixture, 186, 225–227
Mobility, 237
Money, xvi, xvii, xxii
Moral matter, 120, 127
Mortality, 30, 32, 35, 40, 48, 114, 138, 158, 166, 168, 181, 182, 186, 188, 193, 195, 198–205, 214, 231, 240, 252, 258
Multiplier, 28, 29, 95, 107, 116, 157, 158, 160, 162, 163, 168, 180

N

Natural sciences, 204, 240
Number
   equal, 8, 29, 34
   normal, 21

Subject Index

O

Objective probability
   axiom, 12, 17–24, 52, 57, 90
   paradigm, 162–16, 43
Odds
   fair, 145
   game (of), 156
   personal, 145

P

Paleodemography, 174–189, 209, 235
Paradigm, 8, 12–24, 43, 44, 55–67, 79, 93, 98–111, 156, 191, 192, 197, 200, 208, 213, 214, 227, 230, 234, 235, 238–241, 252, 260
Paradox
   Bertrand's, 96
   Simpson's, 70
Petitio principi, 15, 86
Phenomenon, 9, 25–27, 39, 52, 54, 77, 89, 94, 99, 100, 112, 129, 136, 142, 156, 166, 167, 169–171, 199, 206, 209–212, 227, 235, 236, 244, 250, 252, 255, 257
Philosophy, 38, 119, 251
Physic, 12, 91, 97, 140
Physical sciences, 17, 203, 205, 207
Plausibility, 54, 55, 57, 65, 77, 78, 91, 93, 104, 106, 113, 128, 129, 137, 147, 236
Political arithmetics, 34, 138, 189, 199
Politics, 193
Population
   counts, 163
   finite, 10, 11
   genetics, 205
   heterogeneity, 173, 214, 217
   homogeneity, 213
   infinite, 11, 167
   observed, 27, 174–178, 181–183, 190, 203, 225
   reference, 174–178, 181–183, 187
   register, 34, 36, 37, 40, 158, 160, 174, 213, 234
   stationary, 31, 160, 181
   table, 176
Possible
   absolute, 159
   physically, 16
   posterior probability, 25
Price
   fair, 112
   personal, 112
Principle
   additivity, 53
   independence, 53
   indifference, 47, 88, 89, 100, 130
   insufficient reason (of), 47, 51, 56, 68, 86, 94, 100
   inverse probability (of), 50, 52, 87, 99, 102, 103, 164
   maximum expected utility, 80
   sufficient reason (of), 47, 130
   sure-thing, 61, 62, 79
Prior probability, 94, 103, 143, 182, 188
Probability function, 54, 84
Process, 12, 22, 28, 36, 46, 67, 80, 99, 110, 112, 144, 170, 172, 173, 191, 197, 203, 213–219, 224, 227, 251–255, 257
Psychology, 84, 99, 205, 214, 221, 223

R

Random contract, 44
Randomness, 20, 198
Rate, 32, 37, 69–71, 73, 74, 155, 171, 203, 206, 211, 214, 217, 222, 224, 226, 227, 229, 253–255
Reasoning
   deductive, 92, 134
   plausible, 92, 94, 99, 134
Register
   birth, 202
   civil, 166
   death, 34, 158, 160, 202
   parish, 166, 174
   population, 34, 36, 37, 40, 158, 160, 174, 213, 234
Regret theory, 81
Relation
   complete order, 80
   partial-order, 60
   semi-order, 64
   simple-order, 60
   weak-order, 57, 59
Risk, 26, 30, 34, 36, 45, 49, 53, 61, 63, 79, 80, 116, 121, 122, 157, 158, 168, 173, 182, 188, 198, 217, 218, 220, 226–229, 248, 253, 254

S

Saint Petersburg paradox, 48, 49, 141
Sampling, 160–162, 173, 182, 218, 234, 235, 258
School
   constructivist, 20
   formalist, 20
Semantic approach, 13

Series, 10, 11, 20, 23, 35, 64, 68, 71, 89, 102, 105, 117, 142, 161, 167, 204, 205, 209, 222, 230, 247
Set
   closed, 17
   countable, 17, 18, 21
   dense, 105
   empty, 18
   finite, 17, 59, 105
   fuzzy, 55, 130
   infinite, 59, 89, 98, 105
   measurable, 17
   null, 23
   partially ordered, 146
   theory, 17, 18, 22, 23, 57, 59, 135, 146, 245
Share, 8, 86, 111, 156, 169, 208
Signal processing, 75, 249
Social facts, 37, 200, 206, 208
Sociology, 37, 191, 214, 220, 224, 235, 241, 244, 252, 257
Standard deviation, 70, 89, 94, 96, 97, 103, 108, 155, 165, 207, 222, 229
State of nature, 60, 61, 63
Statistical inference
   logicist, 112–113
   objectivist, 24–28
   subjectivist, 53, 59, 67–69, 143, 144
Statistics, 28, 34, 36–38, 103, 122, 138–140, 145, 158, 165, 172, 206, 207, 211, 231, 234, 243–249, 258
Stochastic process, 12, 170, 213, 215, 216, 218, 219, 255
Subjective probability
   axiom, 53–67, 79, 82
   paradigm, 55–67, 79
Survey, 36, 86, 163, 166, 170, 190, 208, 213, 230, 234, 235, 238, 240, 256, 258
Syllogism, 92–94, 101, 139

T
Testimony, 44, 77, 114
Theological nature, 128

Theorem
   Bayesian, 25, 53, 56, 58, 111
   compound probability (of), 58
   total probability (of), 56, 58
Time, lived, 210, 239, 240
Total probability, 15, 56, 58
Transferable belief model, 55, 67
Transformation group, 96, 97, 107, 108, 132
Transitivity, 58, 80, 102, 110
Tribe, 17, 55
Two-aces puzzle, 82

U
Uncertainty, 7, 27, 32, 41, 43, 56, 57, 60, 79–81, 84, 94, 102, 109, 164, 203, 250, 255, 256, 258, 260
Upper probability, 54, 64
Urn, 48, 159, 167
Utility
   expectation, 56, 63, 80, 81
   function, 53, 60, 141
   notion, 48, 56, 59, 61, 79
   paradigm, 58, 59
   state-dependent, 84

V
Variable
   dependant, 247
   independent, 21
   random, 21, 71, 90, 130, 155, 170, 219
Variance, 36, 37, 40, 71, 72, 116, 155, 165, 168, 171, 178, 184–186, 188–190, 207, 208, 211, 212, 228–230, 254

W
Winnings, 8, 14, 45, 61, 62, 80, 81, 98, 156, 159

Z
Zero probability, 23, 129